readings in
statistics
for the
behavioral
scientist

readings in statistics

for the behavioral scientist

edited by
JOSEPH A. STEGER

State University of New York
at Albany

HOLT, RINEHART AND WINSTON, INC.

New York   Chicago   San Francisco   Atlanta
Dallas   Montreal   Toronto   London   Sydney

Preface

This book of readings is designed to supplement the basic courses in statistical methods and research design, or other undergraduate or first level graduate courses. It is for those who are not statisticians but who use statistics as tools in their field of study.

The readings have been selected for two reasons: first, to acquaint the student with original articles dealing with statistics, and second, to increase his sophistication in the use, misuse, and limitations of statistics. Although each article stands as a unique contribution, I have presented groups of related articles to give each chapter a contextual cohesiveness.

The readings have been edited to abridge the highly technical material where possible. The "where possible" should be stressed because in many instances the non-mathematical student must struggle to learn some of the required algebra.

It is hoped that this collection of articles, when used in conjunction with a statistical textbook, will throw more light onto the use of statistics and will enable the reader to appreciate the fact that understanding statistical techniques is not simply knowing how to compute statistics.

v

I am indebted to the authors of the various articles and to the following publishers for their permission to reprint or extract articles from their journals or books: the American Psychological Association, Inc. (*Psychological Bulletin, Psychological Review, American Psychologist, Journal of Educational Psychology*), *Technometrics, Biometrics, Biometrika, Science,* John Wiley & Sons, Inc., and Educational and Psychological Measurement, Jossey-Bass, Inc.

To the many others who helped by discussion or suggestions, I am also very grateful.

Special thanks are due my wife, Barbara, for her valuable assistance in the many phases of preparing this book for publication.

Albany, New York   JOSEPH A. STEGER
January 1971

Contents

PREFACE   v

INTRODUCTION   1

ON LEARNING STATISTICS   3

1 Measurement and Statistics

Introduction   6

  1.1 Scales of Measurement   *S. S. Stevens*   8

  1.2 On the Statistical Treatment of Football
     Numbers   *Frederic M. Lord*   19

  1.3 Scales and Statistics: Parametric and
     Nonparametric   *Norman H. Anderson*   23

  1.4 Weak Measurements vs Strong Statistics:
     An Empirical Critique of S. S. Stevens'
     Proscriptions on Statistics   *Bela O. Baker,*
     Curtis D. Hardyck, and Lewis F. Petrinovich   39

2 Chi-Square

Introduction   53

  2.1 The Use and Misuse of the Chi-Square Test
     Don Lewis and C. J. Burke   55

2.2   The Misuse of Chi-Square—A Reply to
       Lewis and Burke   *Charles C. Peters*   **102**
2.3   Some Comments on "The Use and Misuse of
       the Chi-Square Test"   *Nicholas Pastore*   **109**
2.4   On "The Use and Misuse of the
       Chi-Square Test"—The Case of the 2×2
       Contingency Table   *Allen L. Edwards*   **113**
2.5   Further Discussion of the Use and Misuse of
       the Chi-Square Test   *Don Lewis and*
       C. J. Burke   **120**
2.6   Letter to the Editor on Peters' Reply to
       Lewis and Burke   *C. J. Burke*   **130**
2.7   Some Methods for Strengthening the
       Common χ^2 Tests   *William G. Cochran*   **132**

3   **Parametric Techniques**

Introduction   **157**
3.1   The Lanarkshire Milk Experiment,
       "Student"   *W. S. Gosset*   **159**
3.2   The Use of Transformations   *M. S. Bartlett*   **169**
3.3   Analysis of Covariance: Its Nature and Uses
       William G. Cochran   **179**

4   **Assumptions and Statistical Inference**

Introduction   **195**
4.1   Two Kinds of Experiments Distinguished in
       Terms of Statistical Operations
       Melvin R. Marks   **199**
4.2   Tests of Hypotheses: One-Sided vs
       Two-Sided Alternatives   *Lyle V. Jones*   **207**
4.3   A Note on One-Tailed and Two-Tailed Tests
       W. E. Hick   **211**
4.4   One- and Two-Tailed Tests   *Melvin R. Marks*   **214**
4.5   A Brief Note on One-Tailed Tests   *C. J. Burke*   **216**
4.6   A Rejoinder on One-Tailed Tests
       Lyle V. Jones   **220**
4.7   Further Remarks on One-Tailed Tests
       C. J. Burke   **223**
4.8   The Fallacy of the Null-Hypothesis
       Significance Test   *William W. Rozeboom*   **228**
4.9   Testing the Null Hypothesis and the
       Strategy and Tactics of Investigating
       Theoretical Models   *David A. Grant*   **244**

4.10 Further Considerations on Testing the Null
      Hypothesis and the Strategy and
      Tactics of Investigating Theoretical Models
      Arnold Binder   **256**
4.11 Tactical Note on the Relation Between
      Scientific and Statistical Hypotheses
      Ward Edwards   **268**
4.12 Much Ado About the Null Hypothesis
      *Warner Wilson, Howard L. Miller, and
      Jerold S. Lower*   **273**
4.13 The Test of Significance in Psychological
      Research   *David Bakan*   **288**
4.14 The Effects of Violations of Assumptions
      Underlying the *t* Test   *C. Alan Boneau*   **311**
4.15 Some Consequences When the Assumptions
      for the Analysis of Variance Are Not Satisfied
      William G. Cochran   **330**

5 **Potpourri**

Introduction   **347**
5.1 Degrees of Freedom   *Helen M. Walker*   **349**
5.2 Conclusions vs Decisions   *John W. Tukey*   **364**
5.3 N = 1   *William F. Dukes*   **378**
5.4 R. A. Fisher (1890–1962): An Appreciation
     Jerzy Neyman   **387**

Index   **401**

Introduction

        The demand for courses in statistics has grown very rapidly in the last decade. Most college graduates will have encountered at least one course in statistics. Many people on the job are sent back to school or attend training sessions at their place of work to learn something about statistics. Why the demand? Why the interest in a topic that most students regard with great timidity? I think the answer lies in the increasing utility of the subject. More and more people are using statistics to support an argument, to substantiate a research conclusion, or to validate a polling result. Consequently, there is a growing need for people to be able to compute statistics and to understand what a given statistic means.

    Most students leave their first statistics course knowing many computational methods but having little understanding of when to apply the statistics. They know computational steps but they are unsure of what the statistics' limitations are, and what alternative statistics may be used in each situation.

    This collection of readings juxtaposes articles by authors who disagree either openly or implicitly. These disagreements are the catalyst for many developments in the area of science. The debates are professional controversies and are not motivated by injured personal feelings of the authors. The tolerance for argumentation is the keystone of research. When dogma or rigidity becomes established in a particular field of study, advances cease. The debates presented in this collection allow the reader to see the field of statistics as being neither rigid nor plagued by a sense of finality with regard to technique. It should be noted here that these disagreements, through reviewed journals, lead to new techniques, theories, and avenues of research. I hope that in reading these discussions, the reader will realize the limitations of specific statistics, know when specific statistics can be justifiably applied, and thus become better informed and more discriminating when statistics are encountered in the literature.

The plan of the book is simple. The readings have been divided into chapters which are designed to parallel those in any basic statistical textbook. The opening chapter is on measurement and its relationship to statistics. Although most statistical texts do not give this particular subject much attention, one is often faced with the question, "what effect has my scale of measurement had upon the statistical analysis I plan to conduct?"

The second chapter includes articles related to the chi-square statistic. Learning to compute the chi-square is easy for most students; however, its correct application is difficult to master. Chapter 2 should be helpful in this regard.

Since much course time is usually spent in learning the parametric methods such as Student's t statistic, Chapter 3 presents articles related to the standard parametric methods but covering topics not usually discussed in class. The use of transformations and the article by W. G. Cochran on analysis of covariance may be somewhat advanced for the first course in statistics but should be mandatory for the second course.

Perhaps the most valuable section for the beginning student is Chapter 4, entitled "Assumptions and Statistical Inference." This chapter deals with technical and conceptual problems in statistics and statistical application. Several current, and continuing, public debates are presented. For example, the articles by Grant, Binder, and Rozeboom present some of the arguments between the proponents of "the classical null" versus "Bayesian" approaches to hypothesis testing. This chapter also includes articles on the effects of violations of assumptions for the t and F tests.

The last chapter is a collection of articles that do not fit neatly into any of the other chapters; hence, the title "Potpourri." However, each article is related to those in one or more of the other chapters and may be read accordingly. There is one article in particular that should be singled out, namely H. Walker's article on degrees of freedom. This is a complex topic that is generally not given much coverage in standard statistical texts, and Walker does an excellent job of making it understandable.

On Learning Statistics

Probably the most difficult problem in learning statistics is the student's preconceptions. Most students are not aware of their everyday use of statistics and statistical concepts. Most people, for example, have computed averages (baseball, grades, and so forth) long before they have a formal course in statistics.

The second most common problem with learning statistics is what I call "symbol shock." For students who have had little or no mathematical background the strange array of symbols confronting them in their first statistics course is confusing. Unfortunately, some students decide at this point to withdraw. Granted that the symbols (or at least some of them) are at first strange, they will nevertheless be used over and over again, so learning them (after the initial dismay) is not so difficult. Ask the instructor to put the symbols and their verbal definitions on the board.

A small sample of the most common symbols is given below, with definitions. Most of the symbols used in statistics will have been encountered before in even the lowest-level algebra course.

X or Y Capital X or Y is usually used to denote a single score or numerical value for any given individual or plant or animal, and so forth.

Σ This sign (Sigma) denotes the sum, and is an instruction to add. The Σ is shown with a subsymbol below and above it, like this $\sum_{1}^{3}$. This means that one is to add all the numbers between 1 and 3. It is written in its general form as $\sum_{i}^{n}$ ("sum from i to n"). We specify the sum by substituting numbers for these letters. For example, $\sum_{i=1}^{n}$ means "Sum from i to n where i equals number 1 and n is all the numbers in our list of scores." Suppose we have these scores: 10, 20, 30, 40. If we list 10 as score 1, 20 as score 2, 30 as score 3 and 40 as score 4, $\sum_{1}^{4}$ means "Sum from score 1 to score 4," directing us to add 10, 20, 30, and 40.

n and N A small n usually denotes the number of scores in a single sample or single collection of data. A capital N is usually used to denote the total of all numerical values if there are several groups of numerical values or scores. N, then, is the sum of the n's. For example:

Number of Scores	Group I Value	Group II
1	10	10
2	20	20
3	30	30
$n = 3$	$n = 3$	$N = 6$

n_1 or n_2 (or n_a, n_b)

The use of subscripts is also a help in statistical notation. Thus, n_1 means "the n of a group 1"; n_a means "the n of a group a." Subscripts are also used with scores or individual numerical values so that X_1 could be the score of person number 1.

To further identify our scores or values we may combine subscripts for the individual and the group he is in. For example, if there were two groups with five persons each, we would label them as follows.

Group I Individual	Score	Group II Individual	Score
1	10	1	10
2	20	2	20
3	30	3	30
4	40	4	40
5	50	5	50

The notation X_{11} would specify the score of person number 1 in group I. The score of person number 2 in Group II would be X_{22}. What score would X_{12} specify?

$\overline{X}$

This symbol denotes the arithmetic average, commonly termed the mean. It should be noted that there are other types of means. This symbol simply stands for all the numerical values added together and divided by the number of these values.

md.

This denotes the median (some authors use med. or m.). The median is the point at which exactly half the scores are greater and half less in value.

$X > Y$ This sign denotes that X is greater in numerical value than Y.

$X < Y$ This denotes that X is less than Y.

$X \geq Y$ This denotes that X is equal to or greater than Y.

$X \leq Y$ This denotes that X is less than or equal to Y.

The third most common problem encountered in learning statistics is a result of cramming. Learning statistics is sequential; one must understand the first segments of the course before you can handle later segments. Those who count on cramming will have a difficult time. Unfortunately, I have no solution for this problem except to advise the student to keep up with the course. Learning statistics is not a passive situation, but calls for active involvement and problem-solving. There is no substitute for "doing" in learning statistics.

There is no doubt that for the researcher, businessman, politician, housewife, or anyone wishing to understand and critically examine information, a knowledge of statistics is imperative. This book is a collection of articles written by people working in various fields but all concerned with statistics. It is designed to augment the textbook used in your course and widen your readings in statistics, in order to enhance your sophistication in statistics and introduce you to original articles concerned with statistics. Hopefully, it will aid you in learning statistics and serve as a reference even after you have finished your formal course work.

1

Measurement and Statistics

INTRODUCTION

Assume, for a moment, that you have been timing yourself at some task, say, running the 100-yard dash. Your times from start to finish on three trials are 90, 100, and 110 seconds. Given these times, you can compute your average (arithmetic mean) time by adding the seconds (300) and then dividing this number by the number of times you ran the dash (3). You arrive at the conclusion that on the average you run the 100-yard dash in 100 seconds. Suppose, however, you were asked, What is the average fruit in a bowl of fruit containing 3 pears, 2 apples and 1 orange? Do you add the $3 + 2 + 1$ and divide by 3 to get an average of 2? Certainly one does not compute the "average" fruit in the bowl as he could compute average times. What is the difference between the two examples? Why in one case is the arithmetic average meaningful and in the second meaningless?

The fruit example represents a *nominal scale*. This scale is a category or classification of items and has no other properties. That does not mean that the items themselves do not have both qualitative and quantitative differences, of course, but only that when using a nominal we simply classify items.

The running scores, in contrast, represent some magnitude of a difference. We can say that a score of 110 is *greater* than a score of 100. Several types of scales of measurement reflect magnitudes. The most elementary is the *ordinal scale,* which requires that the categories or classes be ordered. Thus, for example, we can say that Harry is taller than John and John is taller than Steve. The rank, in terms of "tallness," is Harry first, John second, and Steve third. This scale tells us the order, but not the specific magnitude of the difference in height between each fellow. We can say that one is "more than" the other, but not how much more.

To be specific about the magnitude of the observations or items on our scale we must use at least an *interval measurement* or a *ratio scale.* The interval measurement incorporates the ordinal scale requirement of class order and requires that the distance between the classes be equal. On the equal interval scale, one unit of change (one interval) anywhere on the scale has the same value as one unit anywhere else on the scale. For example, a change of two degrees on a Fahrenheit scale of temperature is the same magnitude of change no matter what the temperature; that is, the interval from $32°$ to $34°$ is the same size *change* as the interval from $30°$ to $32°$. We can apply the rules and operations of addition and subtraction with the equal interval scale of measurement. We cannot, however, meaningfully multiply or divide scores. We cannot say, for example, that a person scoring 100 on an intelligence test is twice as intelligent as a person scoring 50, because there is no zero point on our scale of intelligence. The zero point is arbitrary. Arbitrary refers to the fact that the zero point on the scale does not correspond to a state where nothing of the attribute under study is present. For example, if $0°$ Centigrade is $32°$ Fahrenheit, then the zero on the Centrigrade measure must be arbitrary since we have some of the attribute present.

To be able to say that something is twice as much as something else (multiply) we must have our measures in terms of a ratio scale. This means we must have a meaningful zero point and equal intervals. Anyone dealing with budgets or money is familiar with a ratio scale. We arrive at a zero point very readily. We can also say car A costs twice as much as car B. Measures of length and weight are also examples of ratio scales.

S. S. Stevens, in the first article in this section, discusses in detail the types of scales, their properties, and the permissible operations with each. His "proscriptions" on the permissible statistical operations with each type of scale has become a topic of current interest. Not everyone agrees that Stevens' proscriptions need be followed.

Frederic M. Lord, in "On the statistical treatment of football numbers," presents in humorous fashion the argument that the particular scale of measurement may not be so clearly a consideration in the selection of the statistic to be used. That is not to say that the scale is *not* a consideration, but that the rules prescribed by Stevens are not absolute "laws."

In the third article, in a more formal and detailed manner than Lord's, Norman H. Anderson continues the argument that the use of statistical tests is not as specifically contingent upon the scale of measurement as Stevens suggests. Anderson states, for example, that ". . . the validity of a statistical inference cannot depend on the type of measuring scale used" (p. 28).

It should be made clear that Anderson is referring to the *validity of a statistical inference,* not the validity of the measuring device. The decision as to whether the difference between two means is real rather than due to chance is not a function of the scale. Anderson presents very convincing examples and arguments, but the most telling evidence contrary to Stevens' position is presented by Baker, Hardyck, and Petrinovich in the last article of this section.

In a very interesting and clever study they investigated the effects of distortions in the scale of measurement upon the validity of the statistical inference. They demonstrate, very clearly, that Stevens' (and others') "proscriptions" are very questionable. See V. C. Senders, *Measurement and Statistics.* New York: Oxford, 1958; and S. Siegel, *Nonparametric Statistics.* New York: McGraw-Hill, 1956.

1.1

Scales of measurement

S. S. Stevens

A rule for the assignment of numerals (numbers) to aspects of objects or events creates a *scale.* Scales are possible in the first place only because there exists an isomorphism between the properties of the numeral series and the empirical operations that we can perform with the aspects of objects. This isomorphism, of course, is only partial. Not *all* the properties of number and not *all* the properties of objects can be paired off in a systematic correspondence. But *some* properties of objects can be related by semantical rules to *some* properties of the numeral series. In particular, in dealing with the aspects of objects we can invoke empirical operations for determining equality (the basis for classifying things), for rank ordering, and for determining when differences and when ratios between the aspects of objects are equal. The conventional series of numerals—the series in which by definition each member has a successor—yields to analogous operations: We can identify the members

Reprinted from *Handbook of Experimental Psychology* (New York: John Wiley & Sons, Inc., 1951), Chap. I, pp. 23–30.

of the series and classify them. We know their order as given by convention. We can determine equal differences, as $7 - 5 = 4 - 2$, and equal ratios, as $10/5 = 6/3$. This isomorphism between the formal system and the empirical operations performable with material things justifies the use of the formal system as a *model* to stand for aspects of the empirical world.

The type of scale achieved when we deputize the numerals to serve as representatives for a state of affairs in nature depends upon the character of the basic empirical operations performed on nature. These operations are limited ordinarily by the peculiarities of the thing being scaled and by our choice of concrete procedures, but, once selected, the procedures determine that there will eventuate one or another of four types of scale: *nominal, ordinal, interval,* or *ratio.* Each of these four classes of scales is best characterized by its range of invariance—by the kinds of transformations that leave the "structure" of the scale undistorted. And the nature of the invariance sets limits to the kinds of statistical manipulations that can legitimately be applied to the scaled data. This question of the applicability of the various statistics is of great practical concern to several of the sciences.

The principal facts about scales are summarized in Table 1.* It will be noted that the column listing the basic operations needed to create each type of scale is cumulative: to an operation listed opposite a particular scale must be added all those operations preceding it. Thus, an interval scale can be erected only provided we have an operation for determining equality of intervals, for determining greater or less, and for determining equality (not greater and not less). To these operations must be added a method for ascertaining equality of ratios if a ratio scale is to be achieved.

In the column that records the group structure of each scale are listed the mathematical transformations that leave the scale form invariant (see Figure 1). Thus, any numeral x on a scale can be replaced by another numeral x', where x' is the function of x listed in this column. Each mathematical group in the column is contained in the group immediately above it.

The fourth column presents examples of the type of statistical operations appropriate to each scale. This column is cumulative in that *all* statistics listed are admissible for data scaled against a ratio scale. The

* A classification essentially equivalent to that in Table 1 was presented by the author before the International Congress for the Unity of Science, September 1941. (The present discussion follows Stevens, 1946.) The gist of this notion of relating scales of measurement to transformation groups is also contained in the recent book by von Neumann and Morgenstern on games and economic behavior (pp. 22–23). They omit mention of the group corresponding to the nominal scale, and they call attention specifically to the group in which no transformation would be tolerated, i.e. where, under the similarity group, a would be limited to unity.

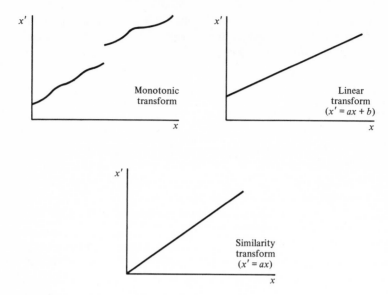

FIGURE 1.

Graphical examples of three groups of transformations. The increasing mono-tonic function, which may have discontinuities of the type shown, belongs to the isotonic group and is a transformation applicable to ordinal scales. The linear transformation, intersecting the ordinate at the value b, is applicable to interval scales. Ratio scales allow only the similarity transformation (multiplication by a constant).

criterion for the appropriateness of a statistic is *invariance* under the transformations in column 3. Thus the case that stands at the median (midpoint) of a distribution maintains its 'middleness' under all trans-formations that preserve order (isotonic group), but an item located at the mean remains at the mean (retains its 'meanness'!) only under trans-formations as restricted as those of the linear group. The ratio expressed by the coefficient of variation remains invariant only under the similarity transformation (multiplication by a constant).

We note that there are two kinds of invariance involved here. If the statistic in question is dimensionless* (e.g. product-moment correlation

---

* Dimensions and the lack thereof figure prominently in the formulas of physics. The dimensionless factors are generally those that are formed of ratios in which the dimensions of the numerator cancel those of the denominator, as in the case of decibels. Measures having dimensions are those like velocity (having the dimensions of length divided by time), area (length times length), etc. In his *Dimensional Analysis,* Bridgman (1931) created a kind of algebra of dimensions which proves helpful in the investigation of physical relations. The basic notion is that the two sides of an equation must be dimensionally equivalent and a discrepancy in dimensions may reveal an error that might be hard to detect by other means.

coefficient, coefficient of variation, etc.), the numerical value of the statistic remains fixed when the scales are subjected to their permissible transformations. But when the statistic has a dimension (e.g. mean, median, etc.) its numerical value changes in accordance with the transformations applied to the scale. Thus the height of the *average man* in centimeters is numerically different from his height in inches. On the other hand, the man (or men) whose height is average remains the same man (or men) whether the measurements are made in inches or in centimeters. The invariance required of those statistics that have dimensions, then, is the identity of the object or event to which the given statistic corresponds.

The numerical value that a statistic assumes depends, of course, upon the statistic and upon the transformation. For example, a linear transformation, $x' = ax + b$, operating on the original scale values, produces for the mean $m' = am + b$, for the standard deviation $\sigma' = a\sigma$, and for the variance $\sigma'^2 = a^2\sigma^2$.

The last column in Table 1 lists some typical examples for each type of scale. It is an interesting fact that the measurement of many physical quantities has progressed from scale to scale. When men knew temperature only by sensation, when things were only "warmer" or "colder" than other things, temperature belonged to the ordinal class of scales. It became an interval scale with the development of thermometry, and after thermodynamics had used the expansion *ratio* of gases to extrapolate to zero it became a ratio scale.

Let us now consider each scale in turn.

NOMINAL SCALE

The *nominal scale* represents the most unrestricted assignment of numerals. The numerals are used only as labels or type numbers, and words or letters would serve as well. Two types of nominal assignments are sometimes distinguished, as illustrated (type A) by the 'numbering' of football players for the identification of the individuals, and (type B) by the 'numbering' of types or classes, where each member of a class is assigned the same numeral. Actually, the first is a special case of the second, for when we label our football players we are dealing with unit classes of one member each. Since the purpose is just as well served when any two designating numerals are interchanged, the structure of this scale remains invariant under the general substitution or permutation group (sometimes called the symmetric group of transformations). The only statistic relevant to nominal scales of type A is the number of cases, e.g. the number of players assigned numerals. But once classes containing several individuals have been formed (type B) we can determine the

TABLE 1

Scales of Measurement

The basic operations needed to create a given scale are all those listed in the second column, down to and including the operation listed opposite the scale. The third column gives the mathematical transformations that leave the scale form invariant. Any numeral x on a scale can be replaced by another numeral x', where x' is the function of x listed in column 3. The fourth column lists, cumulatively downward, some of the statistics that show invariance under the transformations of column 3.

Scale	Basic Empirical Operations	Mathematical Group Structure	Permissible Statistics (*invariantive*)	Typical Examples
Nominal	Determination of equality	Permutation group $x' = f(x)$ [$f(x)$ means any one-to-one substitution]	Number of cases Mode Contingency correlation	"Numbering" of football players Assignment of type or model numbers to classes
Ordinal	Determination of greater or less	Isotonic group $x' = f(x)$ [$f(x)$ means any increasing monotonic function]	Median Percentiles Order correlation (type O)	Hardness of minerals Quality of leather, lumber, wool, etc. Pleasantness of odors
Interval	Determination of equality of intervals or differences	General linear group $x' = ax + b$	Mean Standard deviation Order correlation (type I) Product-moment correlation	Temperature (Fahrenheit and centigrade) Energy Calendar dates "Standard scores" on achievement tests (?)
Ratio	Determination of equality of ratios	Similarity group $x' = ax$	Geometric mean Coefficient of variation Decibel transformations	Length, weight, density, resistance, etc. Pitch scale (mels) Loudness scale (sones)

most numerous class (the mode), and under certain conditions we can test, by the contingency methods, hypotheses regarding the distribution of cases among the classes.

The nominal scale is a primitive form, and quite naturally there are many who will urge that it is absurd to attribute to this process of assigning numerals the dignity implied by the term measurement (cf. Campbell, 1938, p. 122). Certainly there can be no quarrel with this objection, for the naming of things is an arbitrary business. However we christen it, the use of numerals as names for classes is an example of the "assignment of numerals according to rule." The rule is: Do not assign the same numeral to different classes or different numerals to the same class. Beyond that, anything goes with the nominal scale.

The formation of *classes* of objects or events is based on the demonstration of *equality* in respect of some trait or other. As an empirical problem the forming of classes is by no means trivial, and endless arguments have ensued over taxonomic standards. The definition of any common noun, like "horse," raises problems of classification—what animals are included, what excluded? (Cf. Stevens, 1939a, p. 233.) At the formal level, the logical (mathematical) definition of classes and of equality is the subject of much writing and arguing. Curiously enough, the relation of equality, which was described above as reflexive, symmetrical, and transitive, is so "obvious" that not until the 1930's did it become customary in books on logic to include the postulates governing this basic relation (cf. Bell, 1945, p. 578). Mathematics has learned to pay attention to the obvious, and the empirical sciences need never apologize for following suit.

ORDINAL SCALE

The *ordinal scale* arises from the operation of rank ordering. Since any 'order-preserving' transformation will leave the scale form invariant, this scale has the structure of what may be called the isotonic or order-preserving group. This, of course, is a big group, for it includes transformations by all increasing monotonic functions, i.e., functions that never decrease and therefore do not have maxima. Thus the positive scale values on an ordinal scale may be replaced by their square or their logarithm or by a host of other functions, including in particular the "normalizing" transformation used in factor analysis (Thurstone, 1947, p. 368). All these transformations leave invariant the relation of "betweenness" for a given value with respect to its neighbors.

As a matter of fact, most of the scales used widely and effectively by psychologists are ordinal scales. In the strictest propriety the ordinary statistics involving means and standard deviations ought not to be used with these scales, for these statistics imply a knowledge of something more than the relative rank order of data. On the other hand, for this 'illegal' statisticizing there can be invoked a kind of pragmatic sanction: in numerous instances it leads to fruitful results. Although the outlawing of this procedure would probably serve no good purpose, it is proper to point out that means and standard deviations computed on an ordinal scale are in error to the extent that the successive intervals on the scale are unequal in size. When only the rank order of data is known, we should proceed cautiously with our statistics, and especially with the conclusions we draw from them.

Even in applying those statistics that are normally appropriate to ordinal scales, we sometimes find rigor compromised. Thus, although it is indicated in Table 1 that percentile measures may be applied to rank-ordered data, it should be pointed out that the customary procedure of assigning a value to a percentile by interpolating linearly within a class interval is, in all strictness, wholly out of bounds. Likewise, it is not strictly proper to determine the midpoint of a class interval by linear interpolation, because the linearity of an ordinal scale is precisely the property that is open to question.

Another matter also needs comment. In earlier discussions (e.g. Stevens, 1946) I expressed the opinion that rank-order correlation does not apply to ordinal scales because the derivation of the formula for this correlation involves the assumption that the differences between successive ranks are equal. My colleague, Frederick Mosteller, convinces me that this conservative view can be liberalized, provided the resulting coefficient (e.g. Spearman's ρ, or Kendall's τ) is *interpreted only* as a test function for a hypothesis about *order*. On the other hand, the interpretation of the coefficient as equivalent to r (the product-moment coefficient) would assume an underlying interval scale, and a bivariate normal distribution as well. In Table 1 I have allowed for these two interpretations by placing order correlation under both headings: type O for ordinal, type I for interval.

INTERVAL SCALE

With the interval scale we come to a form that is "quantitative" in the ordinary sense of the word. Almost all the usual statistical measures are applicable here, unless they are the kinds that imply a knowledge of a 'true' zero point. The zero point on an interval scale is a matter of convention or convenience, as is shown by the fact that the scale form remains invariant when a constant is added.

This point is illustrated by our two scales of temperature, centigrade and Fahrenheit. Equal intervals of temperature are scaled off by noting equal volumes of expansion; an arbitrary zero is agreed upon for each scale; and a numerical value on one of the scales is transformed into a value on the other by means of an equation of the form $x' = ax + b$. Similarly, energy is measured on an interval scale, for, as von Neumann and Morgenstern assert, "there is nothing in mechanics to fix a zero or a unit of energy" (p. 22). Our scales of *calendar time* offer another example.

Dates on one calendar are transformed to those on another by way of this same equation. On these scales, of course, it is meaningless to say that one value is twice or some other proportion greater than another.

Periods of time, however, can be measured on ratio scales, and one period may be correctly defined as double another. In like manner, *differences* in energy (which is what we mean by *work*) may be considered ratio magnitudes, measurable on ratio scales. *Differences* between values on an interval scale become ratio scale measures for the simple reason that the process of determining a difference (i.e. subtraction) gets rid of the additive constant b.

Most psychological measurement aspires to create interval scales, and it sometimes succeeds. The problem usually is to devise operations for equalizing the units of the scales—a problem not always easy of solution but one for which there are several possible modes of attack. The determination of what we call equal sense distances is one obvious procedure. Only occasionally in psychological scaling is there concern for the location of a 'true' zero point. Intelligence, for example, is usefully assessed on ordinal scales that try to approximate interval scales, and it is not necessary to define what zero intelligence would mean. (Both Thorndike and Thurstone have tried it, however.)

The variability of a psychological measure is itself sometimes used to equalize the units of a scale. This process smacks of a kind of magic—a rope trick for climbing the hierarchy of scales. The rope in this case is the *assumption* that in the sample of individuals tested the trait in question has a canonical distribution (e.g. "normal"). Then it is a simple matter to adjust the units of the scale so that the assumed distribution is recovered when the individuals are measured. But this procedure is obviously no better than the gratuitous postulate behind it, and we are reminded of what Russell said about the larcenous aspects of postulation. There are those who believe that the psychologists who make assumptions whose validity is beyond test are hoist with their own petard, but the fact remains that the assumption of normality has the advocacy of a certain pragmatic usefulness in the measurement of many human traits. Forced to live up to certain criteria of internal consistency, as in the method of paired comparisons (cf. Thurstone, 1948), such assumptions make it possible to lay hold on problems of preferences and the like, which seem recalcitrant to other treatments.

RATIO SCALE

Ratio scales are those most commonly encountered in physics, and they are possible only when there exist operations for determining all four relations: equality, rank order, equality of intervals, and equality of ratios. In the practical instance the determination of the last of these four relations—equal ratios—may take the form of the determination of successive equal intervals beginning at the zero value of the scale. This is *one* of the procedures by which we can assign numerals in such a way that equal ratios among them correspond to equal ratios of some attribute or other.

Once a ratio scale is erected, its numerical values can be transformed (as from inches to feet) only by multiplying each value by a constant. An absolute zero is always implied, even though the zero value on some scales (e.g. absolute temperature) may never be produced. All types of statistical measures are applicable to ratio scales, and only with these scales may we properly indulge in the kind of transformations involved in the use of decibels where we take the logarithm of the ratio of two amounts of power.

Foremost among the ratio scales is the scale of "number" itself—number in the empirical sense—the scale we use when we count such things as eggs, pennies, and apples. This scale of the numerosity of aggregates is so basic and so common that it is ordinarily not even mentioned in discussions of measurement. This neglect might surprise us had we not already learned, in connection with the notion of equality, that it is often the most common and obvious things that are longest overlooked!

On the scale of numerosity we ordinarily admit only the transformation that involves multiplication by unity—the identity element of the similarity group. In other words we ordinarily count by ones. But it is plain that we could equally well count by twos, threes, tens, etc. We could assign numerals to collections of objects by the rule that would lead us to say that the numerosity of our toes is two and a half, in which case we would be counting by fours.

It is conventional in physics to distinguish between two types of ratio scales: *fundamental* and *derived*. Fundamental scales are represented by length, weight, and electrical resistance (and we should add numerosity), whereas derived scales are represented by density, velocity, and elasticity.

These latter are *derived* magnitudes in the sense that they are mathematical functions of certain fundamental magnitudes. They are actually more numerous in physics than are the fundamental magnitudes, which are commonly held to be basic because they permit a physical operation of addition analogous to the mathematical operation of addition. Weights, lengths, and resistances can be added in the physical sense, but this important empirical fact is generally accorded more prominence in the theory of measurement than it deserves. The so-called fundamental scales are important instances of ratio scales, but they are only instances.

As a matter of fact, it can be demonstrated that the fundamental scales can be set up even if the physical operation or addition is ruled out as impossible of performance. As an example, consider the scale of weight, and assume that we live in a world of explosive stuff, such that if we ever separate two blobs of it and then put them together in the same scale pan they will blow up. (There are such materials, we now know.) We will require three balances as follows: (1) a balance like the standard variety of the laboratory, (2) one like the standard variety (fulcrum at the center of the horizontal arm) but on which one pan hangs below the other when the pans are empty, and (3) one like the standard variety except that the fulcrum is not at the center of the arm. The first balance will suffice to determine equality and order, the second will determine equal differences, and the third equal ratios. We can then measure out with the second balance any required number of samples separated by equal intervals. Call these samples a, b, c, d . . . . With the first balance we can find the order of this series from least to greatest. Then with the third balance we can get samples (A, B, C, D . . .) related by equal, but unknown, ratios. At this point we apply the first balance to find which members of the one series are equal to which members of the other. Suppose it is

$$C = d, \quad D = j, \quad R = h, \quad S = p$$

Then C/D = R/S, and we can replace the capital letters by the letter d, plus the number of equal intervals between d and the other lower case letters corresponding to the appropriate capital letters. Then C/D = R/S becomes

$$\frac{d}{d + 6} = \frac{d + 4}{d + 12}$$

Solving, we find $d = 12$. Then, since the intervals $a–b–c–d–e$ . . . are equal, it follows that $c = 11$, $b = 10$, $a = 9$, and $e = 13$, . . . $h = 16$, . . . , $j = 18$, . . . , $p = 24$, etc. These values form a true ratio scale, and the weights they attach to could serve as "standard" weights for measuring other things.

By this procedure we could achieve a set of weights whose properties would be completely isomorphic with the properties of a set determined by what the physicist calls fundamental measurement—built on the process of equating weights and of "adding" them.

This highly condensed description of a possible procedure is given here simply to show that *physical* addition is not necessarily the basis of all measurement. Valid measuring goes on where resort can never be had to the process of laying things end to end or of piling them in a heap.

*   *   *   *

We conclude, then, that the most liberal and useful definition of measurement is the assignment of numerals to things so as to represent facts and conventions about them. The problem of what is and what is not measurement then reduces to the simple question: What are the rules, if any, under which numerals are assigned? If we can point to a consistent set of rules, we are obviously concerned with measurement of some sort, and we can then proceed to the more interesting question: what kind of measurement is it? In most cases a formulation of the rules of assignment discloses directly the kind of measurement and hence the kind of scale involved. If there remains any ambiguity, we may seek the final and definitive answer in the mathematical group structure of the scale form: in what ways can we transform its values and still have it serve all the functions previously fulfilled? We know that the numerical values on all scales can be multiplied by a constant, which changes the numerical size of the unit (unless the multiplier is itself unity). If, in addition, a constant can be added (or a new zero point chosen), it is proof positive that we are not concerned with a ratio scale. Then, if the purpose of the scale is still served when its values are squared or cubed, it is not even an interval scale. And finally, if any two values may be interchanged at will, the ordinal scale is ruled out and the nominal scale is the sole remaining possibility.

This proposed solution to the problem of classifying scales is not meant to imply that all scales belonging to the same mathematical group are equally precise or accurate or useful or "fundamental," or even that in the practical instance it can always be decided into which category a given scale falls. Measurement is never better than the empirical operations by which it is carried out, and operations range from bad to good. Any particular scale, psychological or physical, may be objected to on the grounds of bias, low precision, restricted generality, and other factors, but the objector should remember that these are relative and practical matters and that no scale used by mortals is perfectly free of their taint.

1.2

On the statistical treatment of football numbers

Frederic M. Lord

Professor X sold "football numbers." The television audience had to have some way to tell which player it was who caught the forward pass. So each player had to wear a number on his football uniform. It didn't matter what number, just so long as it wasn't more than a two-digit number.

Professor X loved numbers. Before retiring from teaching, Professor X had been chairman of the Department of Psychometrics. He would administer tests to all his students at every possible opportunity. He could hardly wait until the tests were scored. He would quickly stuff the scores in his pockets and hurry back to his office where he would lock the door, take the scores out again, add them up, and then calculate means and standard deviations for hours on end.

Professor X locked his door so that none of his students would catch him in his folly. He taught his students very carefully: "Test scores are ordinal numbers, not cardinal numbers. Ordinal numbers cannot be added. *A fortiori*, test scores cannot be multiplied or squared." The professor required his students to read the most up-to-date references on the theory of measurement (e.g., Coombs, 1951; Stevens, 1951; Weitzenhoffer, 1951). Even the poorest student would quickly explain that it was wrong to compute means or standard deviations of test scores.

When the continual reproaches of conscience finally brought about a nervous breakdown, Professor X retired. In appreciation of his careful teaching, the university gave him the "football numbers" concession, together with a large supply of cloth numbers and a vending machine to sell them.

Frederic M. Lord, "On the Statistical Treatment of Football Numbers," *American Psychologist,* **8,** 1953, 750–751. Copyright 1953 by the American Psychological Association, and reproduced by permission.

The first thing the professor did was to make a list of all the numbers given to him. The University had been generous and he found that he had exactly 100,000,000,000,000,000 two-digit cloth numbers to start out with. When he had listed them all on sheets of tabulating paper, he shuffled the pieces of cloth for two whole weeks. Then he put them in the vending machine.

If the numbers had been ordinal numbers, the Professor would have been sorely tempted to add them up, to square them, and to compute means and standard deviations. But these were not even serial numbers; they were only "football numbers"—they might as well have been letters of the alphabet. For instance, there were 2,681,793,401,686,191 pieces of cloth bearing the number "69," but there were only six pieces of cloth bearing the number "68," etc., etc. The numbers were for designation purposes only; there was no sense to them.

The first week, while the sophomore team bought its numbers, everything went fine. The second week the freshman team bought its numbers. By the end of the week there was trouble. Information secretly reached the professor that the numbers in the machine had been tampered with in some unspecified fashion.

The professor barely had time to decide to investigate when the freshman team appeared in a body to complain. They said they had bought 1,600 numbers from the machine, and they complained that the numbers were too low. The sophomore team was laughing at them because they had such low numbers. The freshmen were all for routing the sophomores out of their beds one by one and throwing them in the river.

Alarmed at this possibility, the professor temporized and persuaded the freshmen to wait while he consulted the statistician who lived across the street. Perhaps, after all, the freshmen had gotten low numbers just by chance. Hastily he put on his bowler hat, took his tabulating sheets, and knocked on the door of the statistician.

Now the statistician knew the story of the poor professor's resignation from his teaching. So, when the problem had been explained to him, the statistician chose not to use the elegant nonparametric methods of modern statistical analysis. Instead he took the professor's list of the 100 quadrillion "football numbers" that had been put into the machine. He added them all together and divided by 100 quadrillion.

"The population mean," he said, "is 54.3."

"But these numbers are not cardinal numbers," the professor expostulated. "You can't add them."

"Oh, can't I?" said the statistician. "I just did. Furthermore, after squaring each number, adding the squares, and proceeding in the usual fashion, I find the population standard deviation to be exactly 16.0."

"But you can't multiply 'football numbers,'" the professor wailed. "Why, they aren't even ordinal numbers, like test scores."

"The numbers don't know that," said the statistician. "Since the numbers don't remember where they came from, they always behave just the same way, regardless."

The professor gasped.

"Now the 1,600 'football numbers' the freshmen bought have a mean of 50.3," the statistician continued. "When I divide the difference between population and sample means by the population standard deviation. . . ."

"Divide!" moaned the professor.

". . . And then multiply by $\sqrt{1,600}$, I find a critical ratio of 10," the statistician went on, ignoring the interruption. "Now, if your population of 'football numbers' had happened to have a normal frequency distribution, I would be able rigorously to assure you that the sample of 1,600 obtained by the freshmen could have arisen from random sampling only once in 65,618,050,000,000,000,000,000 times; for in this case these numbers obviously would obey all the rules that apply to sampling from any normal population."

"You cannot . . ." began the professor.

"Since the population is obviously not normal, it will in this case suffice to use Tchebycheff's inequality,"[1] the statistician continued calmly. "The probability of obtaining a value of 10 for such a critical ratio in random sampling from any population whatsoever is always less than .01. It is therefore highly implausible that the numbers obtained by the freshmen were actually a random sample of all numbers put into the machine."

"You cannot add and multiply any numbers except cardinal numbers," said the professor.

"If you doubt my conclusions," the statistician said coldly as he showed the professor to the door, "I suggest you try and see how often you can get a sample of 1,600 numbers from your machine with a mean below 50.3 or above 58.3. Good night."

To date, after reshuffling the numbers, the professor has drawn (with replacement) a little over 1,000,000,000 samples of 1,600 from his machine. Of these, only two samples have had means below 50.3 or above 58.3. He is continuing his sampling, since he enjoys the computations. But he has put a lock on his machine so that the sophomores cannot tamper with the numbers again. He is happy because, when he has added together a sample of 1,600 "football numbers," he finds that the resulting sum obey the same laws of sampling as they would if they were real honest-to-God cardinal numbers.

Next year, he thinks, he will arrange things so that the population

---

[1] Tchebycheff's inequality, in a convenient variant, states that in random sampling the probability that a critical ratio of the type calculated here will exceed any chosen constant, c, is always less than $1/c^2$, irrespective of the shape of the population distribution. It is impossible to devise a set of numbers for which this inequality will not hold.

distribution of his "football numbers" is approximately normal. Then the means and standard deviations that he calculates from these numbers will obey the usual mathematical relations that have been proven to be applicable to random samples from any normal population.

The following year, recovering from his nervous breakdown, Professor X will give up the "football numbers" concession and resume his teaching. He will no longer lock his door when he computes the means and standard deviations of test scores.

References

Coombs, C. H. Mathematical models in psychological scaling. *J. Amer. Stat. Ass.*, 1951, **46**, 480–489.

Stevens, S. S. Mathematics, measurement, and psychophysics. In S. S. Stevens (Ed.), *Handbook of experimental psychology*. New York: Wiley, 1951. Pp. 1–49.

Weitzenhoffer, A. M. Mathematical structures and psychological measurements. *Psychometrika*, 1951, **16**, 387–406.

1.3

Scales and statistics: parametric and nonparametric[1]

Norman H. Anderson

The recent rise of interest in the use of nonparametric tests stems from two main sources. One is the concern about the use of parametric tests when the underlying assumptions are not met. The other is the problem of whether or not the measurement scale is suitable for application of parametric procedures. On both counts parametric tests are generally more in danger than nonparametric tests. Because of this, and because of a natural enthusiasm for a new technique, there has been a sometimes uncritical acceptance of nonparametric procedures. By now a certain degree of agreement concerning the more practical aspects involved in the choice of tests appears to have been reached. However, the measurement theoretical issue has been less clearly resolved. The principal purpose of this article is to discuss this latter issue further. For the sake of completeness, a brief overview of practical statistical considerations will also be included.

A few preliminary comments are needed in order to circumscribe the subsequent discussion. In the first place, it is assumed throughout that the data at hand arise from some sort of measuring scale which gives numerical results. This restriction is implicit in the proposal to compare parametric and nonparametric tests since the former do not apply to strictly categorical data (but see Cochran, 1954). Second, parametric tests will mean tests of significance which assume equinormality, i.e., normality and some form of homogeneity of variance. For convenience,

Norman H. Anderson, "Scales and Statistics: Parametric and Nonparametric," *Psychological Bulletin,* **58**, 1961, 305–316. Copyright (1961) by the American Psychological Association, and reproduced by permission.

[1] An earlier version of this paper was presented at the April 1959 meetings of the Western Psychological Association. The author's thanks are due F. N. Jones and J. B. Sidowski for their helpful comments.

parametric test, F test, and analysis of variance will be used synonymously. Although this usage is not strictly correct, it should be noted that the t test and regression analysis may be considered as special applications of F. Nonparametric tests will refer to significance tests which make considerably weaker distributional assumptions as exemplified by rank order tests such as the Wilcoxon T, the Kruskal-Wallis H, and by the various median-type tests. Third, the main focus of the article is on tests of significance with a lesser emphasis on descriptive statistics. Problems of estimation are touched on only slightly although such problems are becoming increasingly important.

Finally, a word of caution is in order. It will be concluded that parametric procedures constitute the everyday tools of psychological statistics, but it should be realized that any area of investigation has its own statistical peculiarities and that general statements must always be adapted to the prevailing practical situation. In many cases, as in pilot work, for instance, or in situations in which data are cheap and plentiful, nonparametric tests, shortcut parametric tests (Tate & Clelland, 1957), or tests by visual inspection may well be the most efficient.

PRACTICAL STATISTICAL CONSIDERATIONS

The three main points of comparison between parametric and nonparametric tests are significance level, power, and versatility. Most of the relevant considerations have been treated adequately by others and only a brief summary will be given here. For more detailed discussion, the articles of Cochran (1947), Savage (1957), Sawrey (1958), Gaito (1959), and Boneau (1960) are especially recommended.

Significance Level

The effects of lack of equinormality on the significance level of parametric tests have received considerable study. The two handiest sources for the psychologist are Lindquist's (1953) citation of Norton's work, and the recent article of Boneau (1960) which summarizes much of the earlier work. The main conclusion of the various investigators is that lack of equinormality has remarkably little effect although two exceptions are noted: one-tailed tests and tests with considerably disparate cell n's may be rather severely affected by unequal variances.[2]

A somewhat different source of perturbation of significance level

[2] The split-plot designs (e.g., Lindquist, 1953) commonly used for the analysis of repeated or correlated observations have been subject to some criticism (Cotton, 1959; Greenhouse & Geisser, 1959) because of the additional assumption of equal correlation which is made. However, tests are available which do not require this assumption (Cotton, 1959; Greenhouse & Geisser, 1959; Rao, 1952).

should also be mentioned. An overall test of several conditions may show that something is significant but will not localize the effects. As is well known, the common practice of t testing pairs of means tends to inflate the significance level even when the over-all F is significant. An analogous inflation occurs with nonparametric tests. There are parametric multiple comparison procedures which are rigorously applicable in many such situations (Duncan, 1955; Federer, 1955) but analogous nonparametric techniques have as yet been developed in only a few cases.

Power

As Dixon and Massey (1957) note, rank order tests are nearly as powerful as parametric tests under equinormality. Consequently, there would seem to be no pressing reason in most investigations to use parametric techniques for reasons of power *if* an appropriate rank order test is available (but see Snedecor, 1956, p. 120). Of course, the loss of power involved in dichotomizing the data for a median-type test is considerable.

Although it might thus be argued that rank order tests should be generally used where applicable, it is to be suspected that such a practice would produce negative transfer to the use of the more incisive experimental designs which need parametric analyses. The logic and computing rules for the analysis of variance, however, follow a uniform pattern in all situations and thus provide maximal positive transfer from the simple to the more complex experiments.

There is also another aspect of power which needs mention. Not infrequently, it is possible to use existing data to get a rough idea of the chances of success in a further related experiment, or to estimate the N required for a given desired probability of success (Dixon & Massey, 1957, Ch. 14). Routine methods are available for these purposes when parametric statistics are employed but similar procedures are available only for certain nonparametric tests such as chi square.

Versatility

One of the most remarkable features of the analysis of variance is the breadth of its applicability, a point which has been emphasized by Gaito (1959). For present purposes, the ordinary factorial design will serve to exemplify the issue. Although factorial designs are widely employed, their uses in the investigation and control of minor variables have not been fully exploited. Thus, Feldt (1958) has noted the general superiority of the factorial design in matching or equating groups, an important problem which is but poorly handled in current research (Anderson, 1959). Similarly, the use of replications as a factor in the design makes it possible to test and partially control for drift or shift in apparatus, procedure, or subject population during the course of an experiment. In the same way, taking experimenters or stimulus materials as a factor allows tests which bear on the adequacy of standardization of the experimental procedures and on the generalizability of the results.

An analogous argument could be given for latin squares, largely rehabilitated by the work of Wilk and Kempthorne (1955), which are useful when subjects are given successive treatments; for orthogonal polynomials and trend tests for correlated scores (Grant, 1956) which give the most sensitive tests when the independent variable is scaled; as well as for the multivariate analysis of variance (Rao, 1952) which is applicable to correlated dependent variables measured on incommensurable scales.

The point to these examples and to the more extensive treatment by Gaito is straightforward. Their analysis is more or less routine when parametric procedures are used. However, they are handled inadequately or not at all by current nonparametric methods.

It thus seems fair to conclude that parametric tests constitute the standard tools of psychological statistics. In respect of significance level and power, one might claim a fairly even match. However, the versatility of parametric procedures is quite unmatched and this is decisive. Unless and until nonparametric tests are developed to the point where they meet the routine needs of the researcher as exemplified by the above designs, they cannot realistically be considered as competitors to parametric tests. Until that day, nonparametric tests may best be considered as useful minor techniques in the analysis of numerical data.

Too promiscuous a use of F is, of course, not to be condoned since there will be many situations in which the data are distributed quite wildly. Although there is no easy rule with which to draw the line, a frame of reference can be developed by studying the results of Norton (Lindquist, 1953) and of Boneau (1960). It is also quite instructive to compare p values for parametric and nonparametric tests of the same data.

It may be worth noting that one of the reasons for the popularity of nonparametric tests is probably the current obsession with questions of statistical significance to the neglect of the often more important questions of design and power. Certainly some minimal degree of reliability is generally a necessary justification for asking others to spend time in assessing the importance of one's data. However, the question of statistical significance is only a first step, and a relatively minor one at that, in the over-all process of evaluating a set of results. To say that a result is statistically significant simply gives reasonable ground for believing that some nonchance effect was obtained. The meaning of a nonchance effect rests on an assessment of the design of the investigation. Even with judicious design, however, phenomena are seldom pinned down in a single study so that the question of replicability in further work often arises also. The statistical aspects of these two questions are not without importance but tend to be neglected when too heavy an emphasis is placed on p values. As has been noted, it is the parametric procedures which are the more useful in both respects.

MEASUREMENT SCALE CONSIDERATIONS

The second and principal part of the article is concerned with the relations between types of measurement scales and statistical tests. For convenience, therefore, it will be assumed that lack of equinormality presents no serious problem. Since the F ratio remains constant with changes in unit or zero point of the measuring scale, we may ignore ratio scales and consider only ordinal and interval scales. These scales are defined following Stevens (1951). Briefly, an ordinal scale is one in which the events measured are, in some empirical sense, ordered in the same way as the arithmetic order of the numbers assigned to them. An interval scale has, in addition, an equality of unit over different parts of the scale. Stevens goes on to characterize scale types in terms of permissible transformations. For an ordinal scale, the permissible transformations are monotone since they leave rank order unchanged. For an interval scale, only the linear transformations are permissible since only these leave relative distance unchanged. Some workers (e.g., Coombs, 1952) have considered various scales which lie between the ordinal and interval scales. However, it will not be necessary to take this further refinement of the scale typology into account here.

As before, we suppose that we have a measuring scale which assigns numbers to events of a certain class. It is assumed that this measuring scale is an ordinal scale but not necessarily an interval scale. In order to fix ideas, consider the following example. Suppose that we are interested in studying attitude toward the church. Subjects are randomly assigned to two groups, one of which reads Communication A, while the other reads Communication B. The subjects' attitudes towards the church are then measured by asking them to check a seven category pro-con rating scale. Our problem is whether the data give adequate reason to conclude that the two communications had different effects.

To ascertain whether the communications had different effects, some statistical test must be applied. In some cases, to be sure, the effects may be so strong that the test can be made by inspection. In most cases, however, some more objective method is necessary. An obvious procedure would be to assign the numbers 1 to 7, say, to the rating scale categories and apply the F test, at least if the data presented some semblance of equinormality. However, some writers on statistics (e.g., Siegel, 1956; Senders, 1958) would object to this on the ground that the rating scale is only an ordinal scale, the data are therefore not "truly numerical," and hence that the operations of addition and multiplication which are used in computing F cannot meaningfully be applied to the scores. There are three different questions involved in this objection, and much of the controversy over scales and statistics has arisen from a failure to keep them separate. Accordingly, these three questions will be taken up in turn.

Question 1. Can the F *test be applied to data from an ordinal scale?* It is convenient to consider two cases of this question according as the assumption of equinormality is satisfied or not. Suppose first that equinormality obtains. The caveat against parametric statistics has been stated most explicitly by Siegel (1956) who says:

> The conditions which must be satisfied . . . before any confidence can be placed in any probability statement obtained by the use of the t test are at least these: . . . 4. The variables involved must have been measured in *at least* an interval scale . . . (p. 19).

This statement of Siegel's is completely incorrect. This particular question admits of no doubt whatsoever. The F (or t) test may be applied without qualm. It will then answer the question which it was designed to answer: can we reasonably conclude that the difference between the means of the two groups is real rather than due to chance? The justification for using F is purely statistical and quite straightforward; there is no need to waste space on it here. The reader who has doubts on the matter should postpone them to the discussion of the two subsequent questions, or read the elegant and entertaining article by Lord (1953). As Lord points out, the statistical test can hardly be cognizant of the empirical meaning of the numbers with which it deals. Consequently, the validity of a statistical inference cannot depend on the type of measuring scale used.

The case in which equinormality does not hold remains to be considered. We may still use F, of course, and as has been seen in the first part, we would still have about the same significance level in most cases. The F test might have less power than a rank order test so that the latter might be preferable in this simple two group experiment. However, insofar as we wish to inquire into the reliability of the difference between the measured behavior of the two groups in our particular experiment, the choice of statistical test would be governed by purely statistical considerations and have nothing to do with scale type.

Question 2. Will statistical results be invariant under change of scale? The problem of invariance of result stems from the work of Stevens (1951) who observes that a statistic computed on data from a given scale will be invariant when the scale is changed according to any given permissible transformation. It is important to be precise about this usage of invariance. It means that if a statistic is computed from a set of scale values and this statistic is then transformed, the identical result will be obtained as when the separate scale values are transformed and the statistic is computed from these transformed scale values.

Now our scale of attitude toward the church is admittedly only an ordinal scale. Consequently, we would expect it to change in the direction of an interval scale in future work. Any such scale change would correspond to a monotone transformation of our original scale since only such transformations are permissible with an ordinal scale. Suppose then that a monotone transformation of the scale has been made subsequent to the experiment on attitude change. We would then have two sets of data: the responses as measured on the original scale used in the experiment, and the transformed values of these responses as measured on the new, transformed scale. (Presumably, these transformed scale values would be the same as the subjects would have made had the new scale been used in the original experiment, although this will no doubt depend on the experimental basis of the new scale.) The question at issue then becomes whether the same significance results will be obtained from the two sets of data. If rank order tests are used, the same significance results will be found in either case because any permissible transformation leaves rank order unchanged. However, if parametric tests are employed, then different significance statements may be obtained from the two sets of data. It is possible to get a significant F from the original data and not from the transformed data, and vice versa. Worse yet, it is even logically possible that the means of the two groups will lie in reverse order on the two scales.

The state of affairs just described is clearly undesirable. If taken uncritically, it would constitute a strong argument for using only rank order tests on ordinal scale data and restricting the use of F to data obtained from interval scales. It is the purpose of this section to show that this conclusion is unwarranted. The basis of the argument is that the naming of the scales has begged the psychological question.

Consider interval scales first, and imagine that two students, P and Q, in an elementary lab course are assigned to investigate some process. This process might be a ball rolling on a plane, a rat running an alley, or a child doing sums. The students cooperate in the experimental work, making the same observations, except that they use different measuring scales. P decides to measure time intervals. He reasons that it makes sense to speak of one time interval as being twice another, that time intervals therefore form a ratio scale, and hence a fortiori an interval scale. Q decides to measure the speed of the process (feet per second, problems per minute). By the same reasoning as used by P, Q concludes that he has an interval scale also. Both P and Q are aware of current strictures about scales and statistics. However, since each believes (and

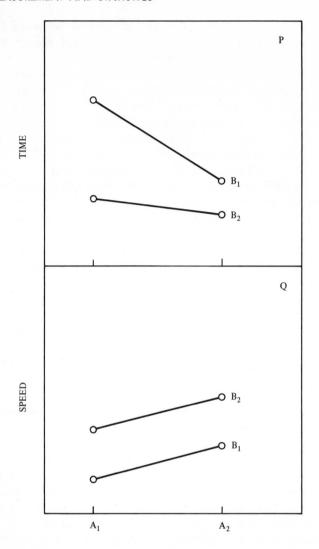

FIGURE 1.

Temporal aspects of some process obtained from a 2 × 2 design. (The data are plotted as a function of Variable A with Variable B as a parameter. Subscripts denote the two levels of each variable. Note that Panel P shows an interaction, but that Panel Q does not.)

rightly so) that he has an interval scale, each uses means and applies parametric tests in writing his lab report. Nevertheless, when they compare their reports they find considerable difference in their descriptive statistics and graphs (Figure 1), and in their F ratios as well. Consultation with a statistician shows that these differences are direct consequences of the difference in the measuring scales. Evidently, then, possession of an interval scale does not guarantee invariance of interval scale statistics.

For ordinal scales, we would expect to obtain invariance of result by using ordinal scale statistics such as the median (Stevens, 1951). Let us suppose that some future investigator finds that attitude toward the church is multidimensional in nature and has, in fact, obtained interval scales for each of the dimensions. In some of his work he chanced to use our original ordinal scale so that he was able to find the relation between this ordinal scale and the multidimensional representation of the attitude. His results are shown in Figure 2. Our ordinal scale is represented by the curved line in the plane of the two dimensions. Thus, a greater distance from the origin as measured along the line stands for a higher value on our ordinal scale. Points A and B on the curve represent the medians of Groups A and B in our experiment, and it is seen that Group A is more pro-church than Broup B on our ordinal scale. The median scores for these two groups on the two dimensions are obtained simply by projecting Points A and B onto the two dimensions. All is well on Dimension 2 since there Group A is greater than Group B. On Dimension 1, however, a reversal is found: Group A is less than Group B, contrary to our ordinal scale results. Evidently, then, possession of an ordinal scale does not guarantee invariance of ordinal scale statistics.

A rather more drastic loss of invariance would occur if the ordinal scale were measuring the resultant effect of two or more underlying processes. This could happen, for instance, in the study of approach-avoidance conflict, or ambivalent behavior, as might be the case with attitude toward the church. In such situations, two people could give identical responses on the one-dimensional scale and yet be quite different as regards the two underlying processes. For instance, the same resultant could occur with two equal opposing tendencies of any given strength. Representing such data in the space formed by the underlying dimensions would yield a smear of points over an entire region rather than a simple curve as in Figure 2.

Although it may be reasonable to think that simple sensory phenomena are one-dimensional, it would seem that a considerable number of psychological variables must be conceived of as multidimensional in nature as, for instance, with "IQ" and other personality variables. Accordingly, as the two cited examples show, there is no logical guarantee that the use of ordinal scale statistics will yield invariant results under scale changes.

It is simple to construct analogous examples for nominal scales. However, their only relevance would be to show that a reduction of all results to categorical data does not avoid the difficulty with invariance.

It will be objected, of course, that the argument of the examples has

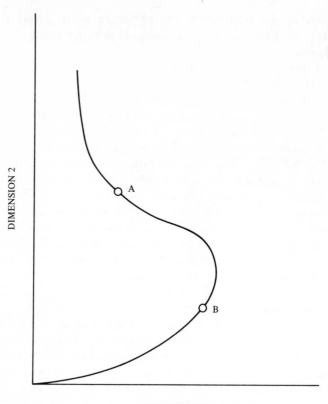

DIMENSION 1

FIGURE 2.
The curved line represents the ordinal scale of attitude toward the church plotted in the two-dimensional space underlying the attitude. (Points A and B denote the medians of two experimental groups. The graph is hypothetical, of course.)

violated the initial assumption that only "permissible" transformations would be used in changing the measuring scales. Thus, speed and time are not linearly related, but rather the one is a reciprocal transformation of the other. Similarly, Dimension 1 of Figure 2 is no monotone transformation of the original scale. This objection is correct, to be sure, but it simply shows that the problem of invariance of result with which one is acually faced in science has no particular connection with the invariance of "permissible" statistics. The examples which have been cited show that knowing the scale type, as determined by the commonly accepted criteria, does not imply that future scales measuring the same phenomena will be "permissible" transformations of the original scale. Hence the use of "permissible" statistics, although guaranteeing invariance of result over the class of "permissible" transformations, says little about invariance of result over the class of scale changes which must actually be considered by the investigator in his work.

This point is no doubt pretty obvious, and it should not be thought that those who have taken up the scale-type ideas are unaware of the problem. Stevens, at least seems to appreciate the difficulty when, in the concluding section of his 1951 article, he distinguishes between psychological dimensions and indicants. The former may be considered as intervening variables whereas the latter are effects or correlates of these variables. However, it is evident that an indicant may be an interval scale in the customary sense and yet bear a complicated relation to the underlying psychological dimensions. In such cases, no procedure of descriptive or inferential statistics can guarantee invariance over the class of scale changes which may become necessary.

It should also be realized that only a partial list of practical problems of invariance has been considered. Effects on invariance of improvements in experimental technique would also have to be taken into account since such improvements would be expected to purify or change the dependent variable as well as decrease variability. There is, in addition, a problem of invariance over subject population. Most researches are based on some handy sample of subjects and leave more or less doubt about the generality of the results. Although this becomes in large part an extrastatistical problem (Wilk & Kempthorne, 1955), it is one which assumes added importance in view of Cronbach's (1957) emphasis on the interaction of experimental and subject variables. In the face of these assorted difficulties, it is not easy to see what utility the scale typology has for the practical problems of the investigator.

The preceding remarks have been intended to put into broader perspective that sort of invariance which is involved in the use of permissible statistics. They do not, however, solve the immediate problem of whether to use rank order tests or F in case only permissible transformations need be considered. Although invariance under permissible scale transformations may be of relatively minor importance, there is no point in taking unnecessary risks without the possibility of compensation.

On this basis, one would perhaps expect to find the greatest use of rank order tests in the initial stages of inquiry since it is then that measuring scales will be poorest. However, it is in these initial stages that the possibly relevant variables are not well-known so that the stronger experimental designs, and hence parametric procedures, are most needed. Thus, it may well be most efficient to use parametric tests, balancing any risk due to possible permissible scale changes against the greater power and versatility of such tests. In the later stages of investigation, we would be generally more sure of the scales and the use of rank order procedures would waste information which the scales by then embody.

At the same time, it should be realized that even with a relatively crude scale such as the rating scale of attitude toward the church, the possible permissible transformations which are relevant to the present discussion are somewhat restricted. Since the F ratio is invariant under change of zero and unit, it is no restriction to assume that any transformed scale also runs from 1 to 7. This imposes a considerable limitation on the permissible scale transformations which must be considered. In addition, whatever psychological worth the original rating scale possesses will limit still further the transformations which will occur in practice.

Although rank order tests do possess some logical advantage over parametric tests when only permissible transformations are considered, this advantage is, in the writer's opinion, very slight in practice and does not begin to balance the greater versatility of parametric procedures. The problem is, however, an empirical one and it would seem that some historical analysis is needed to provide an objective frame of reference. To quote an after-lunch remark of K. MacCorquodale, "Measurement theory should be descriptive, not proscriptive, nor prescriptive." Such an inquiry could not fail to be fascinating because of the light it would throw on the actual progress of measurement in psychology. One investigation of this sort would probably be more useful than all the speculation which has been written on the topic of measurement.

Question 3. Will the use of parametric as opposed to nonparametric statistics affect inferences about underlying psychological processes? In a narrow sense, Question 3 is irrelevant to this article since the inferences in question are substantive, relating to psychological meaning, rather than formal, relating to data reliability. Nevertheless, it is appropriate to discuss the matter briefly in order to make explicit some of the considerations involved because they are often confused with problems arising under the two previous questions. With no pretense of covering all aspects of this question, the following two examples will at least touch some of the problems.

The first example concerns the two students, P and Q, mentioned above, who had used time and speed as dependent variables. We suppose that their experiment was based on a 2×2 design and yielded means as plotted in Figure 1. This graph portrays main effects of both variables which are seen to be similar in nature in both panels. However, our principal concern is with the interaction which may be visualized as measuring the degree of nonparallelism of the two lines in either panel. Panel P shows an interaction. The reciprocals of these same data, plotted in Panel Q, show no interaction. It is thus evident in the example, and true in general, that interaction effects will depend strongly on the measuring scales used.

Assessing an interaction does not always cause trouble, of course.

Had the lines in Panel P, say, crossed each other, it would not be likely that any change of scale would yield uncrossed lines. In many cases also, the scale used is sufficient for the purposes at hand and future scale changes need not be considered. Nevertheless, it is clear that a measure of caution will often be needed in making inferences from interaction to psychological process. If the investigator envisages the possibility of future changes in the scale, he should also realize that a present inference based on significant interaction may lose credibility in the light of the rescaled data.

It is certainly true that the interpretation of interactions has sometimes led to error. It may also be noted that the usual factorial design analysis is sometimes incongruent with the phenomena. In a 2×2 design it might happen, for example, that three of the four cell means are equal. The usual analysis is not optimally sensitive to this one real difference since it is distributed over three degrees of freedom. In such cases, there will often be other parametric tests involving specific comparisons (Snedecor, 1956) or multiple comparisons (Duncan, 1955) which are more appropriate. Occasionally also, an analysis of variance based on a multiplicative model (Williams, 1952) will be useful (Jones & Marcus, 1961). A judicious choice of test may be of great help in dissecting the results. However, the test only answers set questions concerning the reliability of the results; only the research worker can say which questions are appropriate and meaningful.

Inferences based on nonparametric tests of interaction would presumably be less sensitive to certain types of scale changes. However, caution would still be needed in the interpretation as has been seen in Question 2. The problem is largely academic, however, since few nonparametric tests of interaction exist.[3] It might be suggested that the question of interaction cannot arise when only the ordinal properties of the data are considered since the interaction involves a comparison of differences and such a comparison is illegitimate with ordinal data. To the extent that this suggestion is correct, a parametric test can be used to the same purposes equally well if not better; to the extent that it is not correct, nonparametric tests will waste information.

One final comment on the first example deserves emphasis. Since both time and speed are interval scales, it cannot be argued that the difficulty in interpretation arises because we had only ordinal scales.

The second example, suggested by J. Kaswan, is shown in Figure 3. The graph, which is hypothetical, plots amount of aggressiveness as a

---

[3] There is a nomenclatural difficulty here. Strictly speaking, nonparametric tests should be called more-or-less distribution free tests. For example, the Mood-Brown generalized median test (Mood, 1950) is distribution free, but is based on a parametric model of the same sort as in the analysis of variance. As noted in the introduction, the usual terminology is used in this article.

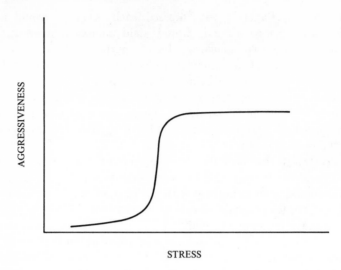

FIGURE 3.

Aggressiveness plotted as a function of stress. (The curve is hypothetical. Note the hypothetical threshold effect.)

function of amount of stress. A glance at the graph leads immediately to the inference that some sort of threshold effect is present. Under increasing stress, the organism remains quiescent until the stress passes a certain threshold value, whereupon the organism leaps into full scale aggressive behavior.

Confidence in this interpretation is shaken when we stop to consider that the scales for stress and aggression may not be very good. Perhaps, when future work has given us improved scales, these same data would yield a quite different function such as a straight line.

One extreme position regarding the threshold effect would be to say that the scales give rank order information and no more. The threshold inference, or any inference based on characteristics of the curve shape other than the uniform upward trend, would then be completely disallowed. At the other extreme, there would be complete faith in the scales and all inferences based on curve shape, including the threshold effect, would be made without fear that they would be undermined by future changes in the scales. In practice, one would probably adopt a position between these two extremes, believing, with Mosteller (1958), that our scales generally have some degree of numerical information worked into them, and realizing that to consider only the rank order character of the data would be to ignore the information that gives the strongest hold on the behavior.

From this ill-defined middleground, inferences such as the threshold effect would be entertained as guides to future work. Such inferences, however, are made at the judgment of the investigator. Statistical techniques may be helpful in evaluating the reliability of various features of the data, but only the investigator can endow them with psychological meaning.

SUMMARY

This article has compared parametric and nonparametric statistics under two general headings practical statistical problems, and measurement theoretical considerations. The scope of the article is restricted to situations in which the dependent variable is numerical, thus excluding strictly categorical data.

Regarding practical problems, it was noted that the difference between parametric and rank order tests was not great insofar as significance level and power were concerned. However, only the versatility of parametric statistics meets the everyday needs of psychological research. It was concluded that parametric procedures are the standard tools of psychological statistics although nonparametric procedures are useful minor techniques.

Under the heading of measurement theoretical considerations, three questions were distinguished. The well-known fact that an interval scale is not prerequisite to making a statistical inference based on a parametric test was first pointed out. The second question took up the important problem of invariance. It was noted that the practical problems of invariance or generality of result far transcend measurement scale typology. In addition, the cited example of time and speed showed that interval scales of a given phenomenon are not unique. The discussion of the third question noted that the problem of psychological meaning is not basically a statistical matter. It was thus concluded that the type of measuring scale used had little relevance to the question of whether to use parametric or nonparametric tests.

References

Anderson, N. H. Education for research in psychology. *American Psychologist,* 1959, **14**, 695–696.

Boneau, C. A. The effects of violations of assumptions underlying the *t* test. *Psychological Bulletin,* 1960, **57**, 49–64.

Cochran, W. G. Some consequences when the assumptions for the analysis of variance are not satisfied. *Biometrics,* 1947, **3**, 22–38.

Cochran, W. G. Some methods for strengthening the common χ^2 tests. *Biometrics,* 1954, **10**, 417–451.

Coombs, C. H. A theory of psychological scaling. *Bulletin of the Engineering Research Institute of the University of Michigan,* 1952, No. 34.

Cotton, J. W. A re-examination of the repeated measurements problem. Paper read at American Statistical Association, Chicago, December 1959.

Cronbach, L. J. The two disciplines of scientific psychology. *American Psychologist,* 1957, **11,** 671–684.

Dixon, W. J., & Massey, F. J., Jr. *Introduction to statistical analysis.* (2nd ed.) New York: McGraw-Hill, 1957.

Duncan, D. B. Multiple range and multiple *F* tests. *Biometrics,* 1955, **11,** 1–41.

Federer, W. T. *Experimental design.* New York: Macmillan, 1955.

Feldt, L. S. A comparison of the precision of three experimental designs employing a concomitant variable. *Psychometrika,* 1958, **23,** 335–354.

Gaito, J. Nonparametric methods in psychological research. *Psychological Reports,* 1959, **5,** 115–125.

Grant, D. A. Analysis-of-variance tests in the analysis and comparison of curves. *Psychological Bulletin,* 1956, **53,** 141–154.

Greenhouse, S. W., & Geisser, S. On methods in the analysis of profile data. *Psychometrika,* 1959, **24,** 95–112.

Jones, F. N., & Marcus, M. J. The subject effect in judgments of subjective magnitude. *Journal of Experimental Psychology,* 1961, **61,** 40–44.

Lindquist, E. F. *Design and analysis of experiments.* Boston: Houghton Mifflin, 1953.

Lord, F. M. On the statistical treatment of football numbers. *American Psychologist,* 1953, **8,** 750–751.

Mood, A. M. *Introduction to the theory of statistics.* New York: McGraw-Hill, 1950.

Mosteller, F. The mystery of the missing corpus. *Psychometrika,* 1958, **23,** 279–290.

Rao, C. R. *Advanced statistical methods in biometric research.* New York: Wiley, 1952.

Savage, I. R. Nonparametric statistics. *Journal of the American Statistical Association,* 1957, **52,** 331–344.

Sawrey, W. L. A distinction between exact and approximate nonparametric methods. *Psychometrika,* 1958, **23,** 171–178.

Senders, V. L. *Measurement and statistics.* New York: Oxford, 1958.

Siegel, S. *Nonparametric statistics.* New York: McGraw-Hill, 1956.

Snedecor, G. W. *Statistical methods.* (5th ed.) Ames, Iowa: Iowa State College Press, 1956.

Stevens, S. S. Mathematics, measurement, and psychophysics. In S. S. Stevens (Ed.), *Handbook of experimental psychology.* New York: Wiley, 1951.

Tate, M. W., & Clelland, R. C. *Nonparametric and shortcut statistics.* Danville, Ill.: Interstate, 1957.

Wilk, M. B., & Kempthorne, O. Fixed, mixed, and random models. *Journal of the American Statistics Association,* 1955, **50,** 1144–1167.

Williams, E. J. The interpretation of interactions in factorial experiments. *Biometrika,* 1952, **39,** 65–81.

1.4

Weak measurements
vs. strong statistics:
an empirical critique of
S. S. Stevens' proscriptions
on statistics[1,2]

Bela O. Baker

The disagreement between those who belong to what Lubin (1962) called the "school of 'weak measurement' theorists" and those who belong to what might be called the school of "strong statistics" has persisted for a number of years with little apparent change of attitude on either side. Stevens, as the leading spokesman for the weak measurement school, has asserted (1951) and reasserted (1959, 1960) the view that measurement scales are models of objects relationships and, for the most part, rather poor models which can lead one far astray from the truth if the scores they yield are added when they should only be counted. At least two current statistics texts intended for psychologists (Senders, 1958; Siegel, 1956) present this view as gospel.

Opposing this view, an assortment of statistically minded psychologists—e.g., Lord (1953), Burke (1953), Anderson (1961), McNemar (1962), and Hays (1963) have argued that statistics apply to num-

Bela O. Baker, Curtis D. Hardyck, and Lewis F. Petrinovich, "Weak Measurements vs. Strong Statistics: An Empirical Critique of S. S. Stevens' Proscriptions on Statistics," *Educational and Psychological Measurement,* **26,** 1966, 291–309.

[1] This research was supporter by research grants from the National Institutes of Health, U. S. Public Health Service (MH 07310) and the Research Committee, University of California Medical Center. Preliminary work was accomplished by a grant of free computer time by the Computer Center, University of California, Berkeley.

[2] We are grateful to Professor Jack Block, Professor Quinn McNemar, and Miss Mary Epling for their many helpful suggestions throughout this study. We are also indebted to Mrs. Eleanor Krasnow who developed and tested the computer programs used in this study.

bers rather than to things and that the formal properties of measurement scales, as such, should have no influence on the choice of statistics. Savage (1957), a statistician, has supported this point of view, stating: "I know of no reason to limit statistical procedures to those involved arithmetic operations consistent with the scale properties of the observed quantities." In other words, a statistical test answers the question it is designed to answer whether measurement is weak or strong.

In his widely cited discussion of measurement, Stevens (1951) distinguished four classes of scales: Nominal, ordinal, interval, and ratio, and specified the arithmetic operations (and hence the statistics) which are permissible for each scale. Nominal scales consist simply of class names and can be treated only by counting operations and frequency statistics. Ordinal scales are developed by demonstrating that some objects have more of a particular quality than do other objects and representing numerically this order among objects. Lacking units, the numbers of an ordinal scale cannot be added, subtracted, multiplied, or divided, but they can be treated by order statistics such as the median or the rank-order correlation. Interval scales represent equal increments in the magnitudes of an object property by equal numerical increments. An increase of one unit in any region of an interval scale represents the same increment in the object property as does an increase of one unit in any other region of the scale. Scores from interval scales can be added and substracted and hence such statistics as the mean, the standard deviation, and the product-moment correlation can be used. Ratio scales add a true zero point to equal intervals and can be multiplied, divided, and treated by subtle statistics which are of little concern to most psychologists.

Although Stevens develops his rationale for relating measurements and statistics almost exclusively in terms of descriptive statistics, he introduces the issue of hypothesis testing somewhat obliquely in his discussions of invariance of results under scale transformations (1951, 1959). He says, "The basic principle is this: Having measured a set of items by making numerical assignments in accordance with a set of rules, we are free to change the assignments by whatever group of transformations will preserve the empirical information in the scale. These transformations, depending on which group they belong to, will upset some statistical measures and leave others unaffected (1959, p. 30)."

If parametric significance tests, such as t or F are used, the permissible transformations are linear. Only then will invariant results be found in comparing groups. An implication of this point of view, which is not made explicit by Stevens, is that if a scale is viewed as a model of object relationships, then any scale transformation is a transformation of those relationships. Hence the problem of invariance of results under scale transformations raises the following question: Can we make correct decisions about the nature of reality if we disregard the nature of the measurement scale when we apply statistical tests?

This aspect of Stevens' position has apparently been ignored by many of his critics. Anderson (1961) dismisses out of hand any restriction on the uses of t arising from the nature of the measurements to which it is applied but discusses the question of invariance of results under scale transformations seriously and at length before concluding that: "The practical problems of invariance or generality of results far transcend measurement scale typology" (p. 316).

The aspects of the problem as related to descriptive and inferential statistics are as follows: The problem for descriptive statistics as presented by Stevens (1951, 1959, 1960) concerns the relationship of the value of a particular statistic computed on obtained measurements to the value of the same statistic computed under conditions of perfect measurement. The argument is that the farther the measurement model departs from the underlying properties of the objects being measured, the less accurate the statistics. In other words, this aspect is concerned with precision of measurement.

In making statistical inferences, however, the issue is whether one will arrive at the same probability estimates from different types of measurement scales. Given the condition that a measurement scale may be a very poor model indeed of the properties of the objects under study, the question of the effect of the scale on the sampling distribution of a statistic remains unanswered. Where hypothesis testing is the issue, the appropriate question is: Do statistics computed on measures which are inaccurate descriptions of reality distribute differently than the same statistics computed under conditions of perfect measurement? If not, then a research worker who has nothing better than an ordinal scale to work with may have to face the problem of more precise measurement for descriptive purposes, but at least the probability decisions he may make from his ordinal measurements will not be inappropriate for parametric statistical models.

In view of the importance of the issue raised by Stevens for users of statistics, it is surprising, as Lubin (1962) notes, that so few detailed discussions of the problem are available. If Stevens is correct, then psychologists should be disturbed about the state of their research literature. Since it can be safely asserted that most measurements in psychology yield scales which are somewhere between ordinal and interval scales, many psychologists may have been propagating fiction when they have made statistical inferences based on significance tests inappropriate to anything less than interval measurements. If Stevens' position is correct, it should be emphasized more intensively; if it is incorrect, something should be done to alleviate the lingering feelings of guilt that plague research workers who deliberately use statistics such as t on weak measurements.

A test of the issue would seem to require a comparison of the sampling distribution of a statistic computed under conditions of "perfect" measurement with the sampling distribution of the same statistic based on imperfect measurements. Since it is not possible to obtain such "perfect" measurements, this comparison is manifestly impossible. As noted above, however, Stevens has suggested that the main issue is that of invariance of results when measurement scales are transformed. Cast in these terms, the problem can be examined empirically. All that is required is that the sampling distribution of a statistic based on one set of scores be compared with the sampling distribution of the same statistic based on scores which are not "permissible" transformations of the first set. If Stevens is right, these sampling distributions should differ in some important way. If they do not, then the nature of the measurement scale is, within potentially determinable limits, an irrelevant consideration when one chooses an hypothesis testing statistic.

METHOD

The statistic selected for study was Student's t. Not only is this one of the most commonly used statistics in psychology but it also has the advantage of having been demonstrated empirically to be relatively robust in the face of violations of the assumptions of normality and equality of variances (Norton, 1953; Boneau, 1960).

The first set of scores used—which will be referred to as criterion (unit-interval) scores—comprised the cardinal numbers from 1 to 30. Three populations of 1000 scores each were constructed by assigned frequencies to the unit-interval scores to approximate as closely as possible the expected frequencies for (1) a normal distribution, (2) a rectangular distribution, and (3) an exponential distribution ($f = 1 + 275^{(e-.25x)}$).

According to Stevens (1951), when one uses the mean or standard deviation the permissible score transformations are linear. And so, to evaluate this proscription, 35 non-linear transformations of the unit-interval scores were constructed. These fell into three subsets, each designed to simulate a common measurement problem in psychology. The rationale for this approach is that an investigator can almost always develop a measuring device that looks as if it yields scores with equal intervals. However, relations among the objects represented by these numbers may not be equal—this is, of course, Stevens' main concern—and consequently by producing non-linear transformations of cardinal numbers, a class of situations is produced directly analogous to the situation that obtains when a measurement scale correctly represents the order among objects but incorrectly represents the magnitude of differences between objects. A statistic such as t, it is argued, cannot be used with such a measuring device, since the operations of addition, subtraction, multiplication, and division are inappropriate—a condition which rather limits the investigator seeking statistical support for his conclusions.

A total of 36 t values were computed for each pair of samples drawn: One value for the unit-interval or criterion scores and one for each of the 35 transformations of the criterion. This is analogous to sampling from a pool of 1000 subjects with scores on 36 variables, the first variable representing a set of measurements with equal intervals and the remaining 35 variables representing measurement scales standing in varying non-linear relationships to the first set.

The following notation will be used throughout the paper:

N,R,E,	the type of distribution used: Normal, Rectangular, or Exponential
5,5; 15,15; 5,15	size of first and second samples
C	the criterion set of values
T_n	the nth transformation
T_{j-k}	transformations j through k.

The sequence of the notation is as follows: $N,15,15,T_{6-10}$; a normal distribution with sample sizes of 15 and 15 for transformations 6 through 10.

The computations were summarized in three forms: (1) Contingency tables showing the relationship of the criterion t value to each transformation t value for the .01 and .05 significance levels were tabulated. These tables allow the determination of the difference in the percentage of t's reaching given significance levels for a particular transformation and for the criterion scores. (2) Frequency distributions of all sets of t values were tabulated. Figure [1] shows the frequency distributions of t for N, 5, 5, C; N, 5, 5, T_5; N, 5, 5, T_{20}; and N, 5, 5, T_{35}. (3) Pearson correlation coefficients between the criterion t values and each of the transformation t values were calculated over the total number of sample pairs drawn for a given condition. The standard error of estimate was also computed. These statistics provide estimates of the degree to which t's based on transformed scores varied from t's based on unit-interval scores for the same pairs of samples.

Parts (1) and (2) of the computation summaries are directly relevant to the question of the effect of the measurement model on the sampling distribution of t. Part (3) is an attempt to represent the deviation of the various types of "inappropriate" measurement models from a true value (represented by the criterion).

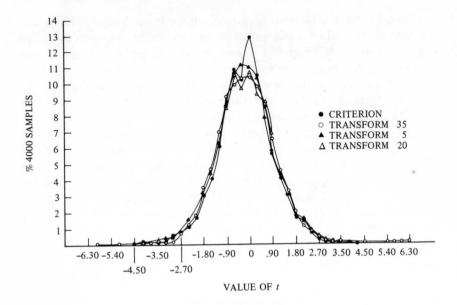

FIGURE 1.

Empirical sampling distributions of t for 4000 pairs of samples with NA = NB = 5, from a normal distribution for the Criterion and Transformations 5, 20, and 35.

As has been mentioned earlier, three types of distributions—normal, rectangular, and exponential—were used. Three variations in sample size were studied—NA = NB = 5; NA = NB = 15; and NA = 5, NB = 15. These combinations are identical with those used by Boneau (1960) and permit the present results to be compared directly with his.

RESULTS

As a first step, the distributions of t on the criterion measurements for all conditions were compared with the theoretical distributions for the appropriate degrees of freedom. Table 1 contains the percentage of t values falling in the 5 per cent and 1 per cent regions of the distribution for the criterion scores.

The deviations of the empirical distributions from the expected theoretical values for the normal curve are quite small. The results are very similar to those reported by Boneau (1960), including the underestimates for the exponential distribution. Somewhat surprisingly, the deviations are reduced only slightly from those reported by Boneau, despite the fact that the present results are based on 4000 t's as compared to Boneau's 1000.

TABLE 1

Per Cent of t's Based on Criterion Unit Interval Scores Falling in the 5% and 1% Regions of Rejection for 4000 Random Sampling Runs

| | | | Sample Sizes | | | |
| | $NA = NB = 5$ | | $NA = NB = 15$ | | $NA = 5, NB = 15$ | |
Population Distribution	5% Level	1% Level	5% Level	1% Level	5% Level	1% Level
Normal	4.8	.8	5.4	1.0	5.3	1.1
Rectangular	5.4	1.6	4.6	.9	5.1	.8
Exponential	3.9	.9	5.1	1.0	4.2	.9

The results for the first set of transformations, which were constructed to simulate a situation where intervals vary randomly throughout the range of the measuring instrument, are given in Table 2. Since the tabulation of results for individual transformations within and across sets T_{1-5}, T_{6-10}, and T_{11-15} revealed little variation, only mean values for all transformations are given in Table 2.[3] The mean value tabled for each transformation is based on 60,000 t's.

Examination of Table 2 indicates that random variations tend to have little effect on the number of t's falling in the 5 per cent and 1 per cent regions of rejection. Columns (1) and (3) contain the total percentages for the 5 per cent and 1 per cent levels of the t distribution for all distributions and sample sizes. For the first group of transformations, the total percentages in the critical regions are very close to the theoretical expectation for a normal distribution. The largest discrepancy present is for E, 5, 5,—a discrepancy of only 1.1 per cent. Columns (2) and (4), which contain the percentage in the larger tail, show similar minimal variations. If one allows for the effects of sampling and takes the deviations of each transformation from the obtained percentages of the C distribution, the discrepancies become almost nonexistent. The largest deviation is .3 per cent for N, 5, 15. It is evident that random variations in interval sizes, regardless of the magnitude of those variations, have virtually no effect on the percentage of t's reaching conventional significance levels.

One condition reported in Table 2 (E, 5, 15) did result in an asymmetrical t distribution. However, as can be seen by examining the tabled values for E, 5, 15, the C distribution is equally asymmetrical. For E, 5, 15, the majority of the t values reaching the 5 per cent level and all of the t's at the 1 per cent level are in one tail of the distribution. The direction of the skewing is negative, indicating that where large differ-

[3] Complete tables are available on request from the authors.

TABLE 2

Per Cent of t's Falling in the 5% and 1% Regions of Rejection When Interval Sizes Vary Randomly (4000 Samples Per Condition)

		5% Level		1% Level	
		Total %	% in Larger Tail	Total %	% in Larger Tail
N, 5, 5:	C	4.8	2.4	.8	.4
	T_{1-15}	4.8	2.4	.8	.4
N, 15, 15:	C	5.4	3.0	1.0	.5
	T_{1-15}	5.3	2.9	.9	.5
N, 5, 15:	C	5.3	2.7	1.1	.6
	T_{1-15}	5.0	2.6	1.1	.6
R, 5, 5:	C	5.4	3.0	1.6	.9
	T_{1-15}	5.3	3.0	1.6	.9
R, 15, 15:	C	4.6	2.6	.9	.5
	T_{1-15}	4.6	2.5	.9	.5
R, 5, 15:	C	5.1	2.7	.8	.7
	T_{1-15}	5.0	2.6	.9	.7
E, 5, 5:	C	3.9	2.0	.9	.5
	T_{1-15}	3.9	2.0	.8	.4
E, 15, 15:	C	5.1	2.6	1.0	.6
	T_{1-15}	5.2	2.6	1.0	.6
E, 5, 15:	C	4.2	3.8	.9	.9
	T_{1-15}	4.2	3.8	1.0	1.0

ences between sample means occurred, the higher mean tended to be the mean of the smaller sample. On the basis of this finding, an experimenter would be ill-advised to use a one-tailed test when he is using samples of unequal sizes (at least if the sizes are of the magnitudes used in this study). However, it makes little difference whether there is an interval scale of measurement or not.

The results for the more irregular and extreme transformations (16 through 35) are presented in Table 3. Again, since there was little variation within sets only mean values for transformations 16–25 and transformations 26–35 are presented.

An inspection of Table 3 also permits the conclusion that the magnitude of variations in interval sizes has little effect on the t distribution. At the same time, it is apparent that t is affected more by these types of transformations than was the situation for simple random variation. However, the discrepancies are still far from extreme. In columns (1) and (3) the largest obtained discrepancy is 2.3 per cent for E, 15,15,T_{26-35} at the 5 per cent level. In columns (2) and (4), the largest discrepancy is again at the 5 per cent level for E, 15,15,T_{26-35}, a value of 1.1 per cent.

When compared to the 5.1 per cent of t's falling in the 5 per cent region for the E, 15, 15, C distribution, this discrepancy of 2.3 per cent seems rather large. However, it seems slight compared to the discrepancies obtained when more serious violations of the assumptions for the use of t are made. For example, Boneau (1960) reported 16 per cent of obtained t's at the 5 per cent level for samples of 5 and 15 drawn from normally distributed populations with unequal variances.

When Table 3 is examined for asymmetry, it is evident that the transformations in which the intervals in one half of a scale stand for substantially smaller variations in the objects being measured than do intervals in the other half of the scale—T_{26-35}—yield seriously skewed distributions of t for all conditions where unequal sample sizes are used. For E distributions, skewing is present for most of the transformations. These transformations provide the only situation where the nature of the scale transformation affected the sampling distribution of t to a more serious degree than could be attributable to the use of unequal sample sizes drawn from an exponential distribution. Even for this condition the effect is quite small. For any real-life situation in which the possibility of such a measurement scale exists, an experimenter should be chary of using t to make a one-tailed test between means based on unequal sample N's. Fortunately, this problem occurs only rarely and when it does occur the use of equal sample sizes will minimize the distortion.

In reviewing the results presented so far, the following generalizations seem warranted:

1. The percentage of t's reaching the theoretical 5 per cent and 1 per cent levels of significance is not seriously affected by the use of non-equal interval measurements.[4]

2. To the extent that there is any influence of the scale transformation on the percentage of t's reaching theoretical significance levels, the influence is more marked when intervals in one broad region of a scale are larger than intervals in another region of the scale than it is when interval sizes vary randomly.

3. If an investigator has a measuring instrument which produces either an interval scale or an ordinal scale with randomly varied interval

---

[4] It is possible that the effects of the scale transformations used in this study are actually due to changes in the shape of the distributions which the different transformations produced. However, if this is the case, the arguments presented regarding the insignificant effects of the nature of measurement scales on probability statements are strengthened even more.

TABLE 3
Per Cent of t's Falling in the 5% and 1% Regions of Rejection When Interval Sizes Vary More in Some Regions of the Scale Than in Others (4000 Samples Per Condition)

		5% Level		1% Level	
		Total %	% in Larger Tail	Total %	% in Larger Tail
N, 5, 5:	C	4.8	2.4	.8	.4
	T_{16-25}	3.4	1.8	.4	.3
	T_{26-35}	4.2	2.2	.8	.4
N, 15, 15:	C	5.4	3.0	1.0	.5
	T_{16-25}	4.6	2.6	.6	.4
	T_{26-35}	5.1	2.8	.9	.5
N, 5, 15:	C	5.3	2.7	1.1	.6
	T_{16-25}	5.0	2.7	1.0	.6
	T_{26-35}	4.7	3.6	1.0	.9
R, 5, 5:	C	5.4	3.0	1.6	.9
	T_{16-25}	4.9	2.7	1.0	.6
	T_{26-35}	4.4	2.6	.8	.5
R, 15, 15	C	4.6	2.6	.9	.5
	T_{16-25}	4.6	2.4	.8	.4
	T_{26-35}	4.9	2.8	.8	.4
R, 5, 15:	C	5.1	2.7	.8	.7
	T_{16-25}	4.9	2.6	.8	.6
	T_{26-35}	4.3	3.1	1.1	1.1
E, 5, 5:	C	3.9	2.0	.9	.5
	T_{16-25}	5.2	2.8	1.4	.7
	T_{26-35}	3.3	1.6	.7	.4
E, 15, 15:	C	5.1	2.6	1.0	.6
	T_{16-25}	5.4	2.7	1.2	.6
	T_{26-35}	2.8	1.4	.4	.3
E, 5, 15:	C	4.2	3.8	.9	.9
	T_{16-25}	4.6	3.4	.8	.7
	T_{26-35}	3.8	3.6	.6	.6

sizes, he can safely use t for statistical decisions under all circumstances examined in this study. The single exception is that t should not be used to do a one tailed test when samples of unequal size have been drawn from a badly skewed population.

4. If a measurement scale deviates from reality in such a fashion that the magnitude of trait differences represented by intervals at the extremes of the scale may be greater than those represented by equal-appearing intervals in the middle of the scale (T_{16-25}), it seems reasonably safe to use t. Unequal sample sizes can even be used if the population is symmetrical, but the proscriptions against using one-tailed tests for unequal sample sizes from exponential populations still apply.

5. If the scale is of the kind represented by the relationship between C and T_{26-35} (in which inequality of units is present in one-half of the distribution only), it is still safe to use t, with a somewhat stricter limitation on the use of one-tailed tests. This arises from the finding that for all population distributions these transformations yielded skewed distributions of t when unequal sample sizes were used.

6. As a maximally conservative empirical set of rules for using t, the following restrictions would seem to be sufficient to compensate for almost any violation of assumptions investigated up to this time:

a. Have equal sample sizes.

b. Use a two-tailed test.

7. Returning to the question as originally formulated: Do statistics computed on a measurement scale which is at best a poor fit to reality distribute differently than the same statistics computed under conditions of perfect measurement? The answer is a firm "no," provided that the conditions of equal sample sizes and two-tailed tests are met. The research worker who has nothing better than an ordinal scale to work with may have an extremely poor fit to reality, but at least he will not be led into making incorrect probability estimates if he observes a few simple precautions.

As a final step, a different sort of analysis will be cited. The previous results and discussion related to one aspect of the measurement problem as posed by Stevens (1951); a second aspect remains. This concerns the accuracy of the descriptive statistics when the measurement model is a poor fit. Stevens has presented his point of view almost exclusviely in terms of descriptive statistics and has tended to use illustrations from descriptive statistics to support his arguments. In the last analysis, this would seem to raise an epistemological question, since it is concerned with the relationship of measurement to a true value which cannot be known. However, evidence as to the correctness or incorrectness of the point of view can be examined from the data of the present study, even though the results are of no help in solving the problems faced by an experimenter who is wondering how to evaluate the validity and the precision of his measuring instrument.

The question of the accuracy of representation can be evaluated by defining the unit interval criterion t values as true measures and the values calculated on the various transformations as those obtained on a measurement model which misrepresents reality. Then the degree of relationship between the values of t calculated on specific samples for C and the values calculated on T_{1-35} can be obtained. This is a correlational question and the results are reported in Table 4.

Columns (1), (3), and (5) contain for each of the distributions the correlations between values of t for each set of transformations and the corresponding values of t for the criterion. The correlations are impressively high. However, because of the broad range of values in the t distribution, the standard errors of estimate in columns (2), (4), and (6) are more informative statistics.

Several points can be noted in connection with Table [2]: There is

TABLE 4

Mean[a] Correlation Coefficients and Standard Errors of Estimate for the Prediction of t's Based on Transformed Scores from t's Based on Criterion Unit-interval Scores

Population Distribution	$NA = NB = 5$		$NA = NB = 15$		$NA = 5, NB = 15$	
	Mean r (1)	Mean $s_{x \cdot y}$ (2)	Mean r (3)	Mean $s_{x \cdot y}$ (4)	Mean r (5)	Mean $s_{x \cdot y}$ (6)
N: T_{1-5}	.997	.089	.997	.082	.997	.084
T_{6-10}	.996	.111	.995	.100	.995	.104
T_{11-15}	.992	.146	.991	.138	.991	.142
T_{16-20}	.975	.244	.966	.265	.970	.260
T_{21-25}	.968	.271	.964	.274	.966	.278
T_{26-30}	.935	.401	.933	.380	.933	.386
T_{31-35}	.914	.462	.911	.434	.912	.439
R: T_{1-5}	.999	.056	.988	.048	.999	.033
T_{6-10}	.996	.094	.996	.081	.996	.084
T_{11-15}	.994	.117	.994	.088	.988	.104
T_{16-20}	.973	.256	.973	.231	.973	.233
T_{21-25}	.973	.258	.975	.227	.976	.225
T_{26-30}	.948	.368	.943	.339	.944	.348
T_{31-35}	.927	.430	.922	.394	.924	.404
E: T_{1-5}	.994	.121	.993	.113	.992	.117
T_{6-10}	.992	.138	.992	.126	.992	.129
T_{11-15}	.984	.199	.985	.181	.983	.189
T_{16-20}	.970	.283	.946	.342	.951	.324
T_{21-25}	.963	.313	.953	.314	.954	.309
T_{26-30}	.981	.218	.930	.382	.940	.325
T_{31-35}	.966	.288	.885	.483	.922	.405

[a] Median values do not differ until the third decimal place for the majority of transformations.

a regular progression in the size of the standard errors of estimate across the sets of transformations used, such that they are smallest for T_{1-15}, and largest for T_{26-35}. These standard errors also become larger as the magnitude of variations in interval size increases, but this is less striking than the differences among types of transformations. Variations in sample sizes and in the shape of the population distribution do not seem to have much influence on the standard errors of estimate; consequently these results seem to show a specific influence of scale transformations on the values of t. The correspondence between values of t based on the criterion unit interval scores and values of t based on transformations decreases regularly and dramatically—from standard errors of estimate on the order of .08 to standard errors of estimate on the order of .45— as the departure from linear transformations becomes more extreme. Here, then, is a finding consistent with Stevens' expectations: The value of t determined for a comparison of samples of non-interval scores does tend to be different from the value of t based on interval scores for the same samples and the discrepancy tends to become greater as the departure from equal intervals is more marked.

In conclusion, the views presented by Stevens (1951, 1959, 1960) and by advocates of his position such as Senders (1958), Siegel (1956), and Stake (1960) state that, when one uses t, the measurement model should have equal intervals representing linear transformations of the magnitudes of the characteristics being measured, or the statistic will be "upset." This view may be correct if one considers single specific determinations of a statistic in a descriptive sense—this seems to be the significance of the standard errors of estimate reported in Table [2]— but it is incorrect when applied to the problem of statistical inference.

The present findings indicate that strong statistics such as the t test are more than adequate to cope with weak measurements—and, with some minor reservations, probabilities estimated from the t distribution are little affected by the kind of measurement scale used.

References

Anderson, N. H. Scales and statistics: Parametric and nonparametric. *Psychological Bulletin*, 1961, **58**, 305–316.

Boneau, C. A. The effects of violations of assumptions underlying the t test *Psychological Bulletin*, 1960, **57**, 49–64.

Burke, C. J. Additive scales and statistics. *Psychological Review*, 1953, **60**, 73–75.

Hays, W. L. *Statistics for psychologists.* New York: Holt, Rinehart and Winston, 1963.

Lord, F. M. On the statistical treatment of football numbers. *American Psychologist*, 1953, **8**, 750–751.

Lubin, A. Statistics. In *Annual Review of Psychology*. Palo Alto, Calif.: Stanford University Press, 1962.

McNemar, Q. *Psychological statistics* (3rd ed.) New York: Wiley, 1962.

Norton, D. W. An empirical investigation of some effects of non-normality and heterogeneity on the F-distribution. Unpublished Doctoral Dissertation, State University of Iowa, 1952. In E. F. Lindquist, *Design and analysis of experiments in psychology and education*. Boston: Houghton-Mifflin, 1953.

RAND Corporation. *A million random digits*. New York: The Free Press, 1955.

Savage, I. R. Non-parametric statistics. *Journal of the American Statistical Association*, 1957, **52**, 331–344.

Senders, V. L. *Measurement and statistics*. London: Oxford University Press, 1958.

Siegel, S. *Nonparametric statistics*. New York: McGraw-Hill, 1956.

Stake, R. E. Review of *Elementary Statistics by P. G. Hoel. Educational and Psychological Measurement*, 1960, **20**, 871–873.

Stevens, S. S. Mathematics, measurement and psychophysics. In S. S. Stevens (Ed.), *Handbook of experimental psychology*. New York: Wiley, 1951.

Stevens, S. S. Measurement, psychophysics and utility. In Churchman, G. W., and Ratoosh, P. (Eds.), *Measurement: Definitions and theories*. New York: Wiley, 1959.

Stevens, S. S. Review of *Statistical Theory* by Lancelot Hogben. *Contemporary Psychology*, 1960, **5**, 273–276.

2

Chi-Square

INTRODUCTION

Probably one of the most frequently used statistics is the chi-square. Unfortunately it is also one of the most frequently misused. It is easy to learn to compute, but its correct application is not so easily learned. This is attested to by the first article in this chapter, by D. Lewis and C. J. Burke. When it appeared in 1949, this article caused much embarrassment among psychologists since it pointed out the incorrect use of the chi-square statistic in much of their published research.

The most common mistake in the application of the chi-square statistic and yet the most critical for its correct application is the violation of independence between measures or events. This assumption of independence is not to be confused with the chi-square as a *test* of independence. The assumption of independence refers to the individual observations or frequencies and means that the occurrence of one event has no effect upon the occurrence of any other event. Another way of stating this meaning of independence is that the probability of each event's occurrence is independent of the probability of occurrence all other events. In statistical terms we say that the joint probability of two random events is equal to the product of the probabilities of these events.

The chi-square as a test of independence refers to the statistical test of the possibility of a relationship between two variables. This is often called a test of association; the question tested is whether the frequencies observed of one category are contingent upon another category. For example, is the number of "yes" answers to some question contingent upon the age of the respondent?

A lesson can be drawn from the findings of the Lewis and Burke article. You cannot simply use a statistic because you know how to calculate it; you must understand the rationale behind its development and the limitations on its application imposed by the assumptions underlying it. Certainly, articles like that of Lewis and Burke helped convince the skeptics that statistical training was needed in the social science curriculum.

Aside from the article by W. H. Cochran on techniques to strengthen the chi-square, the other articles in Chapter 2 are a sequence of replies and counter-replies following from the Lewis and Burke article. Charles C. Peters' article is largely a reply to the criticisms that Lewis and Burke leveled at Peters' textbook and its presentation of chi-square. Lewis and Burke charge that Peters (and Von Voorhis) in their statistical textbook give six incorrect examples out of seven for the use of chi-square. Peters' answer replies to all the criticisms and concludes that none of the six is justified. Nicholas Pastore's comments on Lewis and Burke's article includes reference to their confusing use of independence and six specific criticisms of their article.

A. L. Edwards adds more comment to the growing public discussion with his article on the application of chi-square to the 2×2 contingency table. He argues that one need not have as large a theoretical cell frequency as Lewis and Burke suggest and yet one will still be correctly applying the chi-square statistic.

Naturally, Lewis and Burke made a reply to Peters, Pastore, and Edwards. This article should be read for a complete picture of the debate and for conclusions regarding the "correctness" of each of the participants' comments. A brief letter to the editor by C. J. Burke is also included in this chapter because it clarifies an important point that was lost due to an editorial mistake.

The final article in Chapter 2, by W. H. Cochran, is essential reading for the graduate student or researcher using chi-square in his own field of study. In this article Cochran provides some techniques and examples that aid in the use of chi-square tests by strengthening the application of the test in various situations.

2.1

The use and misuse
of the chi-square test

Don Lewis
C. J. Burke

It has become increasingly apparent over a period of several years that psychologists, taken in the aggregate, employ the chi-square test incorrectly. The number of applications of the test does not seem to be increasing, but the number of misapplications does. This paper has been prepared in hopes of counteracting the trend. Its specific aims are to show the weaknesses in various applications that have been made and to set forth clearly the circumstances under which χ^2 can be legitimately applied in testing different hypotheses.

To confirm a general impression that the number of misuses of χ^2 has become surprisingly large, a careful survey was made of all papers published in the *Journal of Experimental Psychology*[1] during the three years 1944, 1945, and 1946. Fourteen papers were found which contained one or more applications of the chi-square test. The applications in only three of these papers (Anastasi & Foley, 1944; Holt, 1946; Kuenne, 1946) were judged to be acceptable. In one other paper (Lewis & Franklin, 1944), the several applications could be called "correct in principle" but they involved extremely small theoretical frequencies. In nine of the fourteen papers (Arnold, 1944; Chen & Irwin, 1946; Great & Norris, 1946; Irwin & Chen, 1946; King, Landis, & Zubin, 1944; Lewis, 1944; Seward, Dill, & Holland, 1944; Stram & Spooner, 1945; Smith, 1946) the

Don Lewis and C. J. Burke, "The Use and Misuse of the Chi-Square Test," *Psychological Bulletin,* **46,** 1949, 433–489. Copyright (1949) by the American Psychological Association, and reproduced by permission.

[1] The choice of this particular journal resulted from a belief that the psychologists who publish in it are probably better versed, on the average, in statistical methodologies than are those publishing in other journals. No criticism of the journal nor of individual authors nor of experimental findings is intended. The sole purpose is to illustrate correct and incorrect applications of the chi-square test.

applications were clearly unwarranted. In the remaining case,[2] it was not possible to determine what had been done; and the author, when questioned twice by letter, did not choose to reply.

The principal sources of error (or accuracy) in the fourteen papers just referred to, as well as in papers published in other journals, are as follows:

1. Lack of independence among the single events or measures[3]
2. Small theoretical frequencies
3. Neglect of frequencies of non-occurrence
4. Failure to equalize ΣF_o (the sum of the observed frequencies) and ΣF_t (the sum of the theoretical frequencies)
5. Indeterminate theoretical frequencies
6. Incorrect or questionable categorizing
7. Use of non-frequency data
8. Incorrect determination of the number of degrees of freedom
9. Incorrect computations (including a failure to weight by N when proportions instead of frequencies are used in the calculations)

These errors will be explained in detail and illustrated with examples taken for the most part from books and published papers.

It is not surprising that errors of the types listed are frequently made; several of the standard texts to which psychologists turn for statistical guidance contain faulty illustrations. For example, Peters and Van Voorhis (1940) make four applications of χ^2, only one of which is without flaws; and in one of the applications made by Guilford (1936, p. 91), there is a failure to equalize ΣF_o and ΣF_t and to calculate the number of degrees of freedom correctly.

A single application made by Peters and Van Voorhis contains the first four errors in the above list. Table 1 is based on their Table XXXV (1940, p. 411). Twelve dice were thrown fourteen times, and a record was kept of the number of aces appearing at each throw. The observed frequencies F_o are entered in the second column of the table. A value of χ^2, given in the last column, was calculated for each of the fourteen throws. A composite value of χ^2 was obtained by summing the separate values. The first of the four errors in this application is that the observed frequencies lack independence. They lack independence because the same

---

[2] This paper, by Pronko (1945), fails to provide the reader with any basis whatever for forming an independent judgment relative to the correctness or incorrectness of the two applications of χ^2 which were made. In this respect it is a good example of the current trend in papers published in psychological journals to reduce explanations of methods of analyzing data to a point where they are quite unintelligible.

[3] The term independence, as here used, has reference to individual or single events. In contrast, the hypothesis of independence that is tested by means of χ^2 specifies a lack of relationship (that is, an absence of interaction) between the variates represented in a contingency table. The events that occur to yield the frequencies of a contingency table must be mutually independent even though the variates are related.

twelve dice were thrown each time. This means that, when the frequencies are grouped, it is impossible to take into account the effects of individual differences in the dice and possible compensating effects from one die to another. As a consequence, no statements can be made about the behavior of an individual die, nor is it possible to generalize the findings to any population of dice from which the twelve can be considered a sample. Therefore, only hypotheses which relate specifically to the twelve dice as a group can be tested. More will be said later about this kind of error.

The second flaw in the application comes from using theoretical frequencies of 2. These values are too low to yield a quantity whose

TABLE 1
Application of the Chi-Square Test by Peters and Van Voorhis (1940)

Throw	F_o	F_t	$(F_o - F_t)^2$	$\dfrac{(F_o - F_t)^2}{F_t}$
1	1	2	1	0.5
2	3	2	1	0.5
3	2	2	0	0.0
4	3	2	1	0.5
5	1	2	1	0.5
6	4	2	4	2.0
7	2	2	0	0.0
8	4	2	4	2.0
9	1	2	1	0.5
10	0	2	4	2.0
11	3	2	1	0.5
12	2	2	0	0.0
13	3	2	1	0.5
14	1	2	1	0.5
Σ	30	28		10

distribution approximates the χ^2 distribution. The third mistake is the failure to equalize ΣF_o and ΣF_t, which are shown in the table as 30 and 28 respectively. This mistake is related to a fourth one—a failure to take account of the frequencies of non-occurrence of aces. Any of the four errors is sufficient to invalidate this use of the χ^2 test.[4]

The two most basic requirements in any application of the chi-square

[4] If the observed and theoretical frequencies of non-occurrence of aces had been used in the calculations, the composite value of χ^2 would have been 12 instead of 10. This difference happens not to be large. But in another illustration used by Peters and Van Voorhis (1940, Table XXXVI, p. 414), the difference is large. The value of 14.52 is given in the text. When χ^2 is correctly computed by taking account of the frequencies of non-occurrence, the resulting value is 29.72. The number of degrees of freedom remains the same, but the calculated value of χ^2 is more than doubled.

test are (a) independence among the separate measures and (b) theoretical frequencies of reasonable size. These requirements can be shown in an elementary way by examining a two-category distribution of measures. But first, an unequivocal definition of χ^2 is needed.

If z is a normal deviate in standard form defined in relation to population parameters m and σ, then

$$z = \frac{X - m}{\sigma}$$

and

$$\chi^2 = z^2 = \frac{(X - m)^2}{\sigma^2}, \text{ with } df = 1. \tag{1}[5]$$

Chi-square with 1 degree of freedom is thus defined as the square of a deviation from the population mean divided by the population variance.

If there are r *independent* measures of the variate X, there will be r independent values of z, and the resulting formula for χ^2 is

$$\chi^2 = \sum_{i=1}^{r} z_i^2 = \sum \frac{(X_i - m)^2}{\sigma^2}, \text{ with } df = r. \tag{2}$$

Values of χ^2 may range from 0 to ∞, and they have frequency distributions which depend upon the value of r. The distribution function of χ^2, in general form, will be given later. It will then be made clear that the chi-square tests of independence and goodness of fit can be applied unequivocally only to frequency data (or to proportions derived from frequency data). For the present, the plausibility of the two basic requirements stated above will be shown through an examination of the two-category case.

THE TWO-CATEGORY CASE

Consider a population of N *independent* events (things; measures), each of which may fall either into category A or into category B. It is assumed that these categories are clearly defined before samples are drawn and that the category in which a given event falls can be unequivocally determined. A sample is drawn from the population and the sample data are to be employed in determining whether or not a certain hypothesis regarding the proportion of cases in each category is tenable. If p is the expected (theoretical) proportion for category A and q the expected proportion for category B, it follows that

$$p + q = 1.$$

---

[5] Equation (1) verifies the statement that the square root of χ^2 with $df = 1$ is distributed as z (or Student's t) with $df = \infty$.

The probability $P(n)$ that n of the N events will fall into category A is given by the binomial distribution function

$$P(n) = \frac{N!}{n!(N-n)!} p^n q^{N-n}, \tag{3}$$

which is the general expression for obtaining the successive terms arising from the expansion of the binomial $(p+q)^N$. The limiting form of equation (3), as N becomes indefinitely large, is a normal distribution function having a mean of Np and a variance of Npq. In symbols,

$$\lim_{N \to \infty} P(n) = \frac{1}{\sqrt{2\pi} \sqrt{Npq}} e^{-(n-Np)^2/2Npq}. \tag{4}[6]$$

Np is the *population* mean of category A. It is the expected value of n, that is, the theoretical frequency to be associated with category A. Nq is the corresponding theoretical frequency to be associated with category B. Npq is the *population* variance. As stated in equation (1), the square of a deviation from the population mean divided by the population variance is distributed as χ^2 with 1 *df*. Thus, from equation (4), it is seen that the quantity

$$\chi^2 = \frac{(n-Np)^2}{Npq} \tag{5}$$

would be distributed exactly as χ^2 with 1 *df*, provided that N is indefinitely large. In the two-category case, equation (5) may be employed to calculate an approximate value of χ^2. It should be noted that both p and q appear in the denominator of the right-hand term. No restriction is placed during the calculation; the equation gives an approximate solution for any value of N.

The formula that is commonly used in the two-category case to obtain a value of χ^2 is

$$\chi'^2 = \frac{(n-Np)^2}{Np} + \frac{(N-n-Nq)^2}{Nq}, \tag{6}$$

where Np and Nq are theoretical frequencies and n and $(N-n)$ are the corresponding observed frequencies. The prime symbol is placed on χ^2 in (6) to distinguish between χ^2 as *defined* by equation (5) and χ^2 as ordinarily *calculated* with formula (6). When (6) is used, the number of *df* is 1 less than the number of categories because one restriction $(\Sigma F_t = N)$ is imposed on the theoretical frequencies. Therefore, the number of *df* for (6) is 1, just as it is for equation (5).

It can readily be shown that χ^2 and χ'^2 are identical quantities. It

---

[6] If the investigator is concerned with the probability $P(N-n)$ that $(N-n)$ events will fall into category B, the limiting form of equation (3) would be written

$$\lim_{N \to \infty} P(N-n) = \frac{1}{\sqrt{2\pi} \sqrt{Npq}} e^{-(N-n-Nq)^2/2Npq}. \tag{4a}$$

is for this reason, and this reason alone, that formula (6) may be used in obtaining an estimate of χ^2 in the two-category case.[7]

The foregoing discussion reveals the two limitations that hold in any application of the chi-square test. The first limitation is that χ^2 is correcly used only if the N events or measures are independent. Equations (3) and (4) are valid statements only when independence exists. The second basic limitation relates to the size of the theoretical frequencies. If Np (or Nq) remains small as N becomes large, the limiting form of the binomial distribution function is not a normal distribution, as was assumed in writing equation (4). If, for any reason, either Np or Nq is small, the limiting form of equation (3) is the Poisson distribution function. Under such circumstances, the quantity on the right of equation (5) would not be the square of a normal deviate divided by the population variance and, consequently, would not be distributed as χ^2 with 1 df.

It should be emphasized that the categories are assumed to be designated in the population before the individual sample is drawn. It should also be emphasized that the equating of the sums of observed and theoretical frequencies *and* the use of the frequency of non-occurrence (in this case, the frequency with which measures fall in category B) are necessary to establish the identity between the quantities defined in equation (5) and equation (6).

THE MORE GENERAL CASE

It is common to have frequency data that fall into several categories instead of just two. This fact requires an extension of the ideas discussed in the two-category case to encompass any number of categories. The

---

[7] If observed and theoretical proportions are used in calculating values of χ^2, equations (5) and (6) become

$$\chi^2 = N \frac{(p_o - p_t)^2}{p_t q_t} \tag{5a}$$

or

$$\chi^2 = N \frac{(q_o - q_t)^2}{p_t q_t}; \tag{5b}$$

and

$$\chi^2 = N \frac{(p_o - p_t)^2}{p_t} + N \frac{(q_o - q_t)^2}{q_t}. \tag{6a}$$

In these equations, N is the total number of cases while p_o and q_o are the observed proportions and p_t and q_t are the theoretical proportions for categories A and B, respectively. The equations may be derived from (5) and (6), and they reveal that if proportions are used instead of frequencies, the values calculated from the proportions must be multiplied by N.

basic features of this extension will now be presented. Actually, no new ideas enter into the development. The proof is mathematically more complex, but the underlying ideas are the same.

Consider a population of N *independent* events, with k possible outcomes, $v_1, v_2, v_3 \ldots v_k$. Assume that, in the population,

v_1 occurs with a probability of p_1
v_2 occurs with a probability of p_2
v_3 occurs with a probability of p_3
. . . .
v_k occurs with a probability of p_k

The *joint* probability $P(n_i)$ that out of N events exactly n_1 will fall in category v_1, n_2 will fall in category v_2, n_3 in v_3, *and* . . . n_k in v_k, is given by the multinomial distribution function

$$P(n_i) = \frac{N!}{n_1!n_2!n_3! \cdots n_k!} p_1{}^{n_1} p_2{}^{n_2} p_3{}^{n_3} \cdots p_k{}^{n_k}, \tag{7}$$

where

$$\sum_{i=1}^{k} n_i = N.$$

This is the fundamental expression from which the distribution function of χ^2 is derived. It confirms the statement that the measures (frequencies) in the various cells (categories, classes, etc.) of a multidimensional table must be mutually independent to enable a legitimate application of χ^2 in testing any hypothesis concerning the table. And because equation (7) is written in terms of the frequencies $n_1, n_2, n_3 \ldots n_k$, the chi-square tests of independence and goodness of fit, based as they are on a distribution function derived from (7), may be used unequivocally only in relation to frequency data.[8]

The distribution function of χ^2, here symbolized by $g_r(\chi^2)$, may be written

$$g_r(\chi^2) = C_r(\chi^2)^{(r+2)/2} e^{-(\chi^2/2)}, \tag{8}$$

where C_r, a coefficient which changes with r, is given by

$$C_r = \frac{1}{2^{(r/2)} \Gamma(r/2)}. \tag{9}[9]$$

[8] This statement limiting the use of χ^2 to frequency data is not meant to exclude certain special applications such as finding confidence limits for a population variance from a known sample variance or testing several sample variances for homogeneity. These special applications are mentioned again toward the end of the paper.

[9] This expression for C_r contains the gamma function $\Gamma(r/2)$. A gamma function is a function which reduces to a factorial whenever the argument is an integer. In the general case, $(n-1)! = \Gamma(n)$ Equation (9) may be written

It should be emphasized that equation (8) is an exact distribution function for the quantity defined in equations (2) and (5) but only an approximation of the distribution of the quantity defined by equation (7). Its use in relation to equation (7) requires three separate approximations, each of which assumes a theoretical frequency of reasonable size. The three approximations are:

1. Replacing each of the factorials in equation (7) by its Stirling approximation.
2. Taking a step similar to the one whereby $(1 + [X/n])^n$ is replaced by e^x when n is large.
3. Substituting a continuous integral for a summation of discrete quantities.

All of these approximations are quite acceptable and lead to inconsequential errors so long as Np is reasonably large. This reaffirms the fundamental requirement that *the theoretical frequencies must not be small,* if any calculated value of χ^2 is to be distributed as χ^2.[10]

GENERAL COMPUTATIONAL FORMULA

The formula that is commonly employed in calculating values of χ^2 is

$$\chi^2 = \sum \frac{(F_o - F_t)^2}{F_t}, \tag{10}$$

where F_o and F_t, as usual, are observed and theoretical frequencies and the summation extends over all cells (categories) of the table. The

---

$$C_r = \frac{1}{2^{(r/2)}\left(\frac{r-2}{2}\right)!}. \tag{9a}$$

Whenever r is an even number, the factorial in (9a) is an integral number, and its value along with the value of Cr, can be determined in a straightforward manner. On the other hand, if r is an odd number, the "factorial" is fractional, and its value must be determined either by referring to a table of the gamma function or by using the equation

$$\left(\frac{r-2}{2}\right)! = \left(\frac{r-2}{2}\right)\left(\frac{r-4}{2}\right)\left(\frac{r-6}{2}\right)\left(\frac{r-[r-1]}{2}\right)(\sqrt{\pi}).$$

A reader who is interested in plots of the χ^2 distribution function for various values of r is referred to Lewis (19).

[10] A derivation of equation (8) from equation (7) is presented in considerable detail by Greenhood (11) and can be followed by persons familiar with advanced calculus. Greenhood's development, which is more complete but similar to one given by Fry (8), indicates clearly the limitations which hold in applications of the chi-square test to frequency data. It is not necessary, of course, to derive equation (8) from the multinomial distribution function; it may be derived directly from the joint normal distribution function in n variables.

equation holds for any number of categories and reduces to the form of (6) when the number of categories is 2. It serves as a constant reminder that the chi-square tests of independence and goodness of fit can be applied unequivocally only to frequency data.

APPLICATIONS: I. THE GOODNESS OF FIT OF DISTRIBUTION FUNCTIONS

One of the commonest applications of the chi-square test is in evaluating the hypothesis that a set of frequency data can be satisfactorily represented by some specified distribution function. It makes little difference what the function is, so long as its fundamental properties are known. The goodness of fit of binomial, Poisson, and normal distribution functions is often tested.

Two-Category Case

The correct use of χ^2 in connection with a symmetrical binomial distribution function (where $p = q$) can be illustrated with data obtained in a coin-guessing "experiment." A coin was tossed, and 96 students of elementary psychology each guessed whether the coin came up "heads" or "tails." The hypothesis to be tested is that the guess of each student, like the fall of the coin itself, was a purely chance occurrence and that each student was as likely to say heads as to say tails. The results are

TABLE 2
Data from a Coin-Guessing Experiment

H_o (= F_o for heads)	68
H_t (= F_t for heads)	48
T_o (= F_o for tails)	28
T_t (= F_t for tails)	48

shown in Table 2. As seen in the first and third rows, 68 students guessed heads and 28 guessed tails. The theoretical frequencies are 48 and 48. The value of χ^2 is obtained as follows:

$$\chi^2 = \frac{(H_o - H_t)^2}{H_t} + \frac{(T_o - T_t)^2}{T_t} = \frac{(68 - 48)^2}{48} + \frac{(28 - 48)^2}{48} = 16.67.[11]$$

[11] Two alternative formulas that may be used in the two-category case when $p = q$ are as follows:

$$\chi^2 = (H_o - H_i)^2(2/H_i);$$
$$\chi^2 = \frac{(H_o - T_o)^2}{H_o + T_o}.$$

These formulas are the exact equivalents of the one used above.

With 1 df, this value is significant at better than the 0.1% level of confidence; so it must be concluded that the guesses of the 96 students were somehow biased in favor of heads.

Additive Property of χ^2, Illustrated with Two-Category Data

A fundamental property of χ^2 is indicated by the rule which states that the sum of any number of separate and independent values of χ^2 is distributed as χ^2, the number of *df* being the sum of the separate *df*'s. The rule will now be applied in relation to coin-guessing data. Two applications will be made, the first of which is incorrect. *This incorrect application is purposely included* as a means of showing how it differs from a correct application and also of revealing the source of many of the errors made by investigators when employing the chi-square test.

Ninety-six students of elementary psychology were given five successive "trials" in coin guessing. A single coin was tossed five times. After each toss, the 96 students each guessed whether the coin came up "heads" or "tails." Each student wrote his five guesses in order on a sheet of paper. The turn of the coin was never revealed. The results are summarized in Table 3. The frequencies for toss 1 are the same as those given in Table 2.

With the ten pairs of observed and theoretical frequencies in Table 3 (two pairs for each toss), it is possible to test five separate hypotheses—that the guesses of the 96 students on *any one* of the five tosses were chance occurrences and were as apt to be heads as tails. Five values of χ^2 are given in the bottom row of the table. Each of the values was computed with the formula employed with the data of Table 2. The number of degrees of freedom is 1 in each case. For 1 *df*, the value of χ^2 at the 5% level of confidence is 3.841. Four of the five calculated values are less than this and provide no satisfactory basis for rejecting the hypothesis of chance occurrence.

TABLE 3
Data from a Second Coin-Guessing Experiment

	Tosses					
	1	*2*	*3*	*4*	*5*	
H_o	68	49	39	54	54	
H_t	48	48	48	48	48	
T_o	28	47	57	42	42	
T_t	48	48	48	48	48	
χ^2	16.67	.04	3.37	1.50	1.50	$\Sigma = 23.08$

To secure an estimate of χ^2 which makes the probability of rejecting a false hypothesis large, it is desirable, when conditions warrant, to summate separate estimates of χ^2 and obtain a composite estimate. The number of df for the composite estimate is always the sum of the separate df's. The sum of the five values of Table 3 is 23.08. The number of df is 5. The hypothesis under test is that the guesses of students *on five successive tosses* of a coin are purely chance occurrences, with the probability of a guess of heads (by any student on any toss) equal to the probability of a guess of tails. The composite value of 23.08 is significant at better than the 0.1% level of confidence, and would warrant a rejection of the hypothesis if the application of the test were correct.

But it is not correct to summate the five values of χ^2 in Table 3. The reason is that the responses of the 96 students from one toss to the next cannot be assumed to have been independent. In other words, the tabulated values of H_o and T_o from toss to toss are interdependent. Consequently, there is no way of obtaining an unbiased estimate of the theoretical frequency for any toss beyond the first (unless previous guesses are completely ignored). It is unreasonable to assume that a student's knowledge of his guess on one toss did not influence his guess on succeeding tosses. (It will be shown from the data that such an assumption is unsound.) The χ^2 test should never be based on an assumption that is already known to be false. On a single toss, the guess of each student was independent of the guess of the other students; but between tosses, the five guesses of each student were undoubtedly interrelated.

A lack of independence between separate events (measures) is the commonest flaw in the applications of χ^2 that are made by psychologists. Six of the fourteen papers (2, 10, 17, 28, 29, 30) referred to in the opening paragraphs contain applications having this shortcoming.

There are two ways in which the responses of subjects may be interdependent. They may be related from trial to trial, as they were in the coin-guessing illustration, or they may be internally linked within a single trial. Whenever individual subjects each make more than one response per trial, linkages among the measures within the trial must result unless there are no individual differences. Many investigators ignore this restriction and apply the χ^2 test even though the same subjects are used from trial to trial and make several responses on each trial.

The Correct Use of the Additive Principle with Two-Category Data

The conditions under which separate values of χ^2 may be legitimately summated can be illustrated with two-category data. Five non-overlapping groups of subjects made a single guess on each of five successive tosses of a coin. The number of subjects per group was 86. Each subject wrote down his guesses in order on a sheet of paper, the sequence of his guesses remaining on the sheet before him. The results for the five groups are summarized in Tables 4 (A to E, inclusive). Twenty-five separate values of χ^2 appear in the bottom rows of the five sections of the table. The question to be answered is: *Which of these 25 values may be legitimately summated and which may not be?* The five values in any one of the sections cannot be meaningfully added for reasons already given in

TABLE 4
Data from a Third Coin-Guessing Experiment

	Tosses				
	1	*2*	*3*	*4*	*5*
		A. Group 1			
H_o	63	56	36	45	41
H_t	43	43	43	43	43
T_o	23	30	50	41	45
T_t	43	43	43	43	43
χ^2	18.60	7.86	2.28	0.19	0.19
		B. Group 2			
H_o	63	56	42	48	45
H_t	43	43	43	43	43
T_o	23	30	44	38	41
T_t	43	43	43	43	43
χ^2	18.60	7.86	0.05	1.16	0.19
		C. Group 3			
H_o	65	55	40	48	46
H_t	43	43	43	43	43
T_o	21	31	46	38	40
T_t	43	43	43	43	43
χ^2	22.51	6.70	0.42	1.16	0.42
		D. Group 4			
H_o	68	54	38	52	41
H_t	43	43	43	43	43
T_o	18	32	48	34	45
T_t	43	43	43	43	43
χ^2	29.07	5.62	1.16	3.77	0.19
		E. Group 5			
H_o	72	57	30	62	38
H_t	43	43	43	43	43
T_o	14	29	56	24	48
T_t	43	43	43	43	43
χ^2	39.12	9.12	7.86	16.79	1.16

the discussion of the data in Table 3. However, the five values in the five sections for *any single toss* may be summated to yield a single composite value having 5 *df*. For example, a composite value may be obtained for the third toss and may be used to test the hypothesis that, on the third toss of the coin, the guesses of the members of the five groups were random occurrences with the probability of a heads guess equalling that of a tails guess. Neither the presence nor the absence of individual biases would nullify a meaningful test of this hypothesis. If the hypothesis could not be retained, it would be correct to conclude that the guesses were not chance occurrences and were perhaps influenced by what had gone before. The difference between this case and the one illustrated in Table 3 is clear-cut. The data in Table 3 are for five successive guesses by one group, and the successive theoretical probabilities cannot be established without making specific assumptions concerning prior events. In contrast, when the five χ^2 values for a single toss are taken from the five sections of Table 4 and summated, the only assumption that is made is that the guesses on that particular toss were in accordance with a theoretical probability for heads of .50.

Composite values of χ^2 for the five tosses are given in Part A, Table 5. The number of *df* in each case is 5. Except for toss 5, the composite values are all significant at far better than the 1% level of confidence. The value for toss 5 falls near the 80% level and lends support for the belief that on a fifth successive toss of a coin, the probability of a heads guess is .50. The values for tosses 1 and 2 show that, on these tosses, the guesses were strongly biased toward heads. Conclusions regarding tosses 3 and 4 should

TABLE 5
Values of Chi-Square Based upon Data from Table IV

Toss	Part A. Composite Values of χ^2 χ^2	df
1	127.90	5
2	37.16	5
3	11.77	5
4	23.07	5
5	2.15	5
Toss	Part B. Values of χ^2 for Combined Frequencies χ^2	df
1	125.17	1
2	36.92	1
3	7.82	1
4	14.88	1
5	.149	1

be made in the light of the large contributions made to the composite values for these tosses by the guesses of group 5 alone. As seen in Table 4-E, the χ^2 values for tosses 3 and 4 are 7.86 and 16.79, respectively. With 1 df in each case, both values are significant at better than the 1% level. The corresponding values in the other four parts of the table all fall below the 5% level. However, the deviations on tosses 3 and 4 are in the same direction for all five groups, and this fact indicates a definite departure from chance expectations.

It is legitimate to combine for each toss separately the empirical and theoretical frequencies listed in Table 4-A-B-C-D-E, and use the resulting sums to compute values of χ^2, each with 1 df. Values of χ^2 obtained in this way are given in Part B of Table 5. As in the case of the composite values, all of the values except the one for toss 5 are significant at better than the 1% level of confidence. Nevertheless, the procedure of combining frequencies is not recommended except where the theoretical frequencies for each of several duplicated experiments are too small to yield satisfactory individual estimates of χ^2. Other things equal, the greater the number of degrees of freedom is, the more stable is a value of χ^2 and the greater is the probability of rejecting a false hypothesis.

Five values of χ^2 may be selected from Table 4-A-B-C-D-E and used compositely to test the hypothesis that the guesses of persons on *five successive tosses* of a coin are chance occurrences, with the probability of a heads guess equalling that of a tails guess. The hypothesis is inclusive enough to cover the entire population from which the groups of subjects were randomly selected. To provide for independence between tosses, it is necessary to choose the five values of χ^2 so that there is one value for each toss and so that no two values are based on the guesses of a single group. To this end, numbers from 1 to 5 were assigned to the five sections of the table. The χ^2 value for toss 1 was taken from the section whose number first appeared in a table of random numbers; the χ^2 value for toss 2 from the section whose number next appeared, and so on. The values thus chosen are shown in Table 6. The composite value is 33.66, with 5 df. It is highly significant and leaves no grounds for believing that the hypothesis is true.

The discussion in the preceding pages on the necessity for independence between measures can perhaps be further clarified through a consideration of coin-tossing. Three different situations will be described to reveal unmistakable differences in hypotheses to be tested and methods of handling data. Suppose, first, that a single penny is selected at random from a large collection of pennies. This penny is tossed successively, say 100 times, and a record is kept of the way it turns. The probability of a head (or a tail) is .50. The χ^2 test may be applied to determine whether or not the empirical results conform to this theoretical probability. If they do conform it may justifiably be concluded that the penny is "unbiased." The test is an unequivocal one. The extent to which the investigator

TABLE 6
Composite Value of Chi-Square, Based on Data in Table IV

	Tosses				
	1	*2*	*3*	*4*	*5*
H_o	63	54	40	48	46
H_t	43	43	43	43	43
T_o	23	32	56	38	40
T_t	43	43	43	43	43
χ^2	18.60	5.62	7.86	1.16	0.42

Composite $\chi^2 = 33.66$

should generalize to the collection of pennies from which the single penny was chosen is a matter for personal judgment. The χ^2 test, as made, reveals nothing concerning the probability that the selected penny either represents or misrepresents the collection of pennies.

Suppose next that *two* pennies are randomly selected from a collection of pennies and that each penny is tossed 50 times to give a total of 100 tosses. The frequencies of occurrence of heads and of tails are combined (pooled) for the two coins. It may be assumed that the probability of a head is .50, but the χ^2 test cannot be *meaningfully* applied to test this theoretical probability.[12] The reason is that each penny makes its own unique contribution to the results. If one of them is biased while the other is unbiased, the obtained value of χ^2 could easily be significant and lead to a rejection of the hypothetical probability, even though it is correct for one of the pennies. Furthermore, one penny could be strongly biased for heads and the other equally strongly biased for tails, and the obtained value of χ^2 would turn out to be insignificant. The possible presence of individual idiosyncrasies precludes an unequivocal application of χ^2. The same thing would be true if five pennies were randomly selected, each one tossed, say 20 times, and the results pooled; or if 10 pennies were selected and each tossed 10 times, or 20 selected and each tossed five times.

Suppose, finally, that 100 pennies are randomly selected from a collection of pennies, that each penny is tossed a single time, and that the number of heads is recorded. It may be assumed that the probability of a turn of heads, in the population from which the pennies are selected, is .50. The fall of each coin is clearly independent of the fall of every other coin. The χ^2 test may be legitimately applied to determine whether

[12] The pooling of two or more sets of frequencies to obtain a single value of χ^2 is warranted if the aim is to study the "heterogeneity" or "interaction" aspects of the data. In this connection, see Snedecor's discussion (1946, pp. 191–192) of "pooled" and "total" chi-squares.

or not the observed frequency of heads conforms to the hypothetical frequency. The results of the test can be generalized to the entire collection of pennies. This would hold even though less than 100 pennies were selected, so long as a sufficient number was chosen to provide theoretical frequencies of the occurrence of heads and the occurrence of tails of sufficient magnitude to warrant an application of the χ^2 test. No statements can be made, of course, regarding the tendencies of any individual penny.

The crucial point is that frequencies obtained from individuals, whether pennies or subjects in psychological experiments, should not be pooled if the χ^2 test is to be used, except when it can be shown that there is an absence of biases or idosyncrasies among them[13] or when "interaction" effects are specifically under scrutiny. Results on individuals may be combined, but the combining should be done *after* the χ^2 test has been applied to the data on individuals separately. For example, if two pennies are each tossed 50 times, the χ^2 test may be applied to the results for each penny separately, and then the two values of χ^2 may be added to provide a composite value. Similarly, separate values of χ^2 may be obtained from the guesses made by two individuals. The separate values may then be combined to furnish a single composite value. As in the well-known analysis of variance techniques where each source of variability contributes to the total variability, each source of variability should be allowed to make its contribution to the value of χ^2. Unfortunately, χ^2 procedures provide no way, as analysis of variance techniques do, of introducing *statistical* controls over individual subjects as a source of variability. Therefore, in the use of χ^2, the control over individuals must be introduced as an intrinsic part of the sampling process.

Multi-Category Case (Single Dimension)

Frequency data sometimes fall into several categories along a single dimension. If the frequencies from category to category are independent and if some hypothesis regarding their distribution can be meaningfully set up, then the chi-square test may be used to evaluate the hypothesis, provided that the theoretical frequencies for the various categories are of reasonable magnitude. The correct application of the test in such a situation will be illustrated with data obtained in die throwing. A single die was thrown 120 times. There was no reason for believing that any throw was influenced by any other. The results are given in Table 7, where the first column lists the six faces of the die and the second column

---

[13] If an investigator firmly intends to restrict all generalizations to the group of persons studied—the group considered *in toto*, as a sort of amorphous mass—then the pooling of individual frequencies may be logically defended. In such a situation, the group is analogous to a single individual and must be treated as such.

TABLE 7
Data Obtained in Die Throwing

Face	F_o	F_t	$\dfrac{(F_o - F_t)^2}{F_t}$
1	23	20	.45
2	20	20	.00
3	22	20	.20
4	15	20	1.25
5	18	20	.20
6	22	20	.20
Σ	120	120	2.30

gives the number of times that each face turned up. The theoretical probability, on each throw, that any specified face of the die would turn up was 1/6. Consequently, the theoretical frequency of occurrence of each of the six faces was 20, as shown in the third column of the table. A single restriction, the sum of the observed frequencies, was placed in calculating the theoretical frequencies. Consequently, there are 5 df. With 5 df, the obtained value of χ^2 falls near the 80% level of confidence, and there is no basis for rejecting the hypothesis that the fall of the die was, on each throw, a strictly chance occurrence.[14]

Another illustration of a multi-category frequency distribution (along a single dimension) comes from results on the coin-guessing experiment. A total of 439 subjects each guessed heads or tails on each of five successive tosses of a coin. All but 9 of the subjects were the ones whose guesses were tabulated in Table 4. The χ^2 test will be applied to evaluate the hypothesis that chance factors operated in determining the

[14] It would be possible to use the data in Table 7 to test six separate hypotheses—that the appearance of *each* of the six faces was a chance occurrence, with a probability of 1/6. For example, to test the hypothesis that the appearance of the ace (one-spot) was a chance occurrence, a value of χ^2 would be computed as follows:

$$\chi^2 = \frac{(23 - 20)^2}{20} + \frac{(97 - 100)^2}{100} = .45 + .09 = .54.$$

The number of df is 1. Observe that in the calculation, the frequency of non-occurrence of the ace was taken into account. (The probability of occurrences of the ace was 1/6 [$=p$] while the probability of its non-occurrence was 5/6 [$=q$]. This is an example of an asymmetrical binomial.) It is clear that six separate values of χ^2 could be computed in the way just indicated. But these values could not then be legitimately summated to yield a composite value of χ^2 with 6 df. They could not be combined because they would lack independence; the frequency of non-occurrence in each calculation would include the frequency of occurrence of the other five faces.

frequencies of occurrence of the various possible "patterns" of guesses. With five successive tosses and five successive guesses, there were 32 possible patterns, as shown in the first column of Table 8. If there was no biasing of the guesses (that is, if the guess of every subject on every trial was as likely to be heads as to be tails), then each of the 32 patterns was as probable as any other; and with $N = 439$, the theoretical frequency for each pattern was $1/32 \times 439 = 13.7$, rounded to the first

TABLE 8
Analysis of Coin-Guessing Responses of 439 Subjects*

Patterns	F_o	F_t	$\dfrac{(F_o - F_t)^2}{F_t}$
H H H H H	25	(13.8)	9.089
H H H H T	12	13.7	0.211
H H H T H	22	13.7	5.028
H H H T T	18	13.7	1.350
H H T H H	29	(13.8)	16.742
H H T H T	96	(13.8)	489.626
H H T T H	22	(13.8)	4.872
H H T T T	14	13.7	0.007
H T H H H	5	13.7	5.525
H T H H T	15	13.7	0.123
H T H T H	12	13.7	0.211
H T H T T	4	13.7	6.868
H T T H H	33	(13.8)	26.713
H T T H T	17	13.7	0.795
H T T T H	12	13.7	0.211
H T T T T	3	13.7	8.357
T H H H H	3	13.7	8.357
T H H H T	7	13.7	3.277
T H H T H	10	13.7	0.999
T H H T T	14	13.7	0.007
T H T H H	5	13.7	5.525
T H T H T	1	13.7	11.773
T H T T H	6	13.7	4.328
T H T T T	0	13.7	13.700
T T H H H	4	13.7	6.868
T T H H T	6	13.7	4.328
T T H T H	25	(13.8)	9.089
T T H T T	7	13.7	3.277
T T T H H	2	13.7	9.992
T T T H T	5	13.7	5.525
T T T T H	2	13.7	9.992
T T T T T	3	13.7	8.357
Σ	439	439.0	$\chi^2 = 681.122$

* All but nine of the subjects are the same as those whose guesses are analyzed in Table 4.

decimal place. This is the value shown in the third column of the table, except where parentheses appear. Note that 32×13.7 does not equal **439** exactly, but equals **438.4**. One of the requirements in the application of χ^2 is that $\Sigma F_o = \Sigma F_t$. Therefore, six of the theoretical frequencies are given as 13.8. These are the theoretical frequencies corresponding to the six largest observed frequencies. This insures that any slight error that may result from equalizing ΣF_o and ΣF_t operates to make the test more conservative.

In the calculation of χ^2 for Table 8, a value of $(F_o - F_t)^2/F_t$ was secured for each row. These values are given in the fourth column, and their sum ($=681.122$) is the desired estimate. The number of df is $(32 - 1) = 31$. The only restriction placed in figuring the theoretical frequencies was ΣF_o, and this meant the loss of a single degree of freedom. Even with 31 df, the value of χ^2 is so large as to leave no basis whatever for retaining the hypothesis that the guesses were chance occurrences.

Normal Distribution Function

The chi-square test is often used in evaluating the fit of a normal curve to a set of frequency data. Applications of this type are usually correct except for an occasional failure to equalize ΣF_o and ΣF_t, a tendency to use some theoretical frequencies that are too small, and, most importantly, an incorrect specification of the number of degrees of freedom. The correct procedure will now be illustrated.

The distribution of the midterm scores of 486 students in a course in elementary psychology is shown in Table 9. The mid-points of class intervals of ten score units are given in the X-column, frequencies in the F_o-column. The mean M of the distribution of scores is 104.0, while the standard deviation is 16.1. There are two methods that can be used in fitting a normal curve to the data (that is, in calculating the theoretical frequencies that correspond to the observed frequencies). One method involves the estimation of areas under segments of the normal curve through the process of multiplying ordinate values by the class interval.[15] This is an approximation procedure. A more exact method (and the one used here) is to obtain the values for the areas from proportions taken from a table of the probability integral.

The column in Table 9 labeled X' gives the upper limits of the various score categories. Deviate scores and z scores based on the values of X' are shown in the fourth and fifth columns. Proportions of the total area under the normal curve from $-\infty$ to z are given in the P column. Proportions of the area in the segments corresponding to the various score intervals are shown in column P' and were obtained by taking the

[15] For an example of this method, see Guilford (1936, p. 91).

TABLE 9

Application of the Chi-Square Test in Evaluating the Fit of a Normal Curve to a Set of Frequency Data

X	F_o	X'	x $(=X'-M)$	z $(=x/\sigma)$	P	P'	F_t $(=P'N)$	$\dfrac{(F_o-F_t)^2}{F_t}$
				$(+\infty)$	(1.0000)			
144.5	1 ⎫					.0141	6.9 ⎫	
		139.5	35.5	2.195	.9859			.926
134.5	22 ⎭					.0437	21.2 ⎭	
		129.5	25.5	1.574	.9422			
124.5	56					.1125	54.7	.031
		119.5	15.5	0.953	.8297			
114.5	112					.1960	95.3	2.926
		109.5	5.5	0.342	.6337			
104.5	111					.2440	118.5	.475
		99.5	−4.5	−0.280	.3897			
94.5	94					.2032	98.8	.233
		89.5	−14.5	−0.891	.1865			
84.5	54					.1211	58.8	.392
		79.5	−24.5	−1.512	.0654			
74.5	27 ⎫					.0489	23.8 ⎫	
		69.5	−34.5	−2.133	.0165			
64.5	7 ⎬					.0135	6.6 ⎬	.555
		59.5	−44.5	−2.754	.0030			
54.5	2 ⎭					.0030	1.4 ⎭	
				$(-\infty)$	$(.0000)$			
Σ	486					1.0000	486.0	5.538

differences between the successive values of P. The theoretical frequencies came from multiplying the values of P' by N which is 486 in this case.

Because the first and the last two values of F_t are less than 10, they were combined with the adjacent values, as were the corresponding values of F_o. The sum of the last column in the table (=5.538) is the value of χ^2. Seven differences between F_o and F_t entered into the calculations. The number of degrees of freedom is $7-3=4$. Three degrees were lost because three restrictions were placed in determining the theoretical frequencies.[16] The restrictions were the computed values of ΣF_o, M, and σ.

[16] Some statistics tests (Edwards, 1946; Peatman, 1947; Peters & Van Voorhis, 1940) perpetuate the view, erroneously attributed to Pearson (1932), that the number of restrictions imposed in fitting a normal curve is *1* or *3*, depending upon the hypothesis that the investigator wishes to test. There is only one hypothesis open to test—that the frequency data arose from a normal population. If the mean and standard deviation of the fitted function are estimated from the data, three restrictions are imposed and 3 *df* are lost. The same tests give a similar misinterpretation of the number of restrictions imposed when χ^2 is applied in testing independence.

The hypothesis being tested is that the frequency data arose from a normal population. With 4 df, the probability of obtaining, by chance, a value of χ^2 greater than 5.538 is around .25; so the hypothesis is tenable.

The Poisson Distribution Function

If the probability of the occurrence of an event is quite small, so that Np remains small even though N is relatively large, the distribution of observed frequencies in samples of size N may be of the Poisson type. The equation for the Poisson distribution may be written

$$P(n) \doteq \frac{m^n}{n!} e^{-m} \qquad (11)^{[17]}$$

where $m = Np$ and e has its conventional meaning. As in equations (3) and (4), the symbol $P(n)$ represents the probability of n occurrences out of N possible occurrences. The symbol $\doteq$ is used in place of the equal sign to indicate that (11) is an approximation formula. The errors introduced by the approximation are negligible, provided that N is quite large and provided also that N is very much larger than the largest value of n that may reasonably be expected in random sampling.

The χ^2 test may be applied in relation to the data of Table 10. The column labeled F_o gives the number of samples, in a total of 120 short samples of American speech, that contained n occurrences of the consonant "th" (as in thin). Each of the 120 samples was 400 sounds in length. As seen from the table, 31 of the samples did not contain any "th"

TABLE 10

Analysis of Frequencies of Occurrence of the Consonant "th" in Samples of American Speech

n	F_o	$P(n)$	F_t	$(F_o - F_t)^2$	$\dfrac{(F_o - F_t)^2}{F_t}$
0	31	.1868	22.42	73.61	3.283
1	31	.3138	37.66	44.35	1.178
2	30	.2636	31.63	2.66	.084
3	11	.1476	17.71	45.02	2.542
4	11 ⎫	.0620	7.44 ⎫		
6	3 ⎬	.0209	2.51 ⎬	41.22	3.896
6	3 ⎭	.0050	.60 ⎬		
>6		.0003	.03 ⎭		
Σ	120	1.0000	120.00		$\chi^2 = 10.983$

[17] A derivation of this formula is given by Lewis (1948, pp. 168–169).

sounds; 31 samples contained one "th" sound each; and so on. As a first step, the data in Table 10 will be compared with some results obtained by Voelker (1934). In a study of over 600,000 sounds occurring in almost 6,000 announcements over the radio, Voelker found the proportion of "th" sounds to be .0065. Each of the 120 samples represented in Table 10 contained 400 sounds. This made a total of 48,000 sounds. The use of the proportion obtained by Voelker leads to 260 as the predicted, or theoretical, number of "th" sounds among the 48,000. The observed number was 201.[18] The χ^2 test may be applied in evaluating the hypothesis that the sounds in the present over-all sample, were drawn from a general population of American speech sounds which is assumed to be characterized exactly by the value of m obtained by Voelker. The value of χ^2 is computed as follows:

$$\chi^2 = \frac{(201 - 260)^2}{260} + \frac{(47{,}799 - 47{,}740)^2}{47{,}740} = 13.388 + 0.073 = 13.461.$$

Note that the observed theoretical frequencies of non-occurrence of the "th" enter into the calculation. Wih 1 df, the obtained value of χ^2 is significant at better than the 0.1% level of confidence: so there is a firm basis for the rejecting the hypothesis.

The observed proportion of .0042 ($=201/48{,}000$) may be used in testing the hypothesis that the 120 samples were all drawn from the same Poisson distribution. If $p_o = .0042$ and $N = 400$, then $m = Np = 1.68$, and the equation for the hypothetical distribution function may be written

$$P(n) = \frac{(1.68)^n}{n!} e^{-1.68}.$$

Values of $P(n)$ computed with this formula are given in the third column of Table 10, and the corresponding values of F_t are given in the fourth column. As shown in the last column of the table, the value of χ^2 was computed in accordance with equation (10). The last 4 values of F_t and the last 3 values of F_o were combined to avoid the use of theoretical frequencies of less than 10. The computed value of χ^2 is 10.982. Five differences between F_o and F_t were used in the computations. Two restrictions (N and m)[19] were imposed in calculating values of F_t. This leaves 3 df. With this number of df, the obtained value χ^2 falls at about the 3% level of confidence. The hypothesis may, therefore, be tentatively retained or may be rejected, depending upon the level of confidence that has been prescribed.

---

[18] $\Sigma n F_o = 0(31) + 1(31) + 2(30) + 3(11) + 4(11) + 5(3) + 6(3) = 201.$

[19] If the value of p had not been estimated from the empirical data, but had been taken from the Voelker study or from some other completely independent source, the only restriction that would have been placed would have been N, and there would have been 4 df instead of 3 df.

APPLICATIONS: II. THE CHI-SQUARE TEST OF INDEPENDENCE

A common application of the χ^2 test enables an examination of the frequencies of a contingency table to determine whether or not the two variables or attributes represented in the table are independent. The number of cells in the table may range from four (as in a 2×2 table) to an indefinitely large value. The χ^2 test is perhaps most commonly applied by psychologists in relation to 2×2 tables. The chief weaknesses in such applications are (a) a strong tendency to use excessively small theoretical frequencies and (b) an occasional failure to categorize adequately. The same two weaknesses are apt to occur when the number of categories in either, or both, of the "dimensions" of the table is greater than two.

Illustrations of the Chi-Square Test of Independence

Comparison of coin- and die-coin guessing. To obtain data for illustrating the χ^2 test of independence, 384 students of psychology were each asked to guess heads or tails on five successive tosses of a coin, where the tosses were interspersed among five throws of a die. The die was thrown; a guess was made as to the face that turned up. This guess was written on one edge of a sheet of paper. The edge of the paper was then folded under, to hide the guess. The coin was then tossed, the guess being written down. Again the paper was folded under, to hide the guess. The die was thrown a second time, the guess made, the paper folded under. The coin was then thrown a second time, the guess made, the paper folded under. Each of the five guesses on the coin was preceded by a throw of the die and a guess on its fall. The paper was folded under after each guess on the die and each guess on the coin. Thus, the guesses on the coin were not only separated by guesses on the die, but the sequence of guesses was hidden from view. The subjects were never informed as to how the coin or the die actually fell.

The frequencies of occurrence of the 32 possible patterns of guesses on the five successive tosses of the coin (with guesses on the die ignored) are given in the third column of Table 11. The frequencies in the second column of this table were copied directly from Table 8 and are based on guesses on five successive tosses of a coin alone.[20] The subjects for the two conditions of guessing were completely different.

[20] As will be seen, the 32 patterns in Table 11 have been divided into groups of four patterns each. The basis for the division will be discussed later. Short horizontal lines divide the corresponding observed frequencies. A numeral is placed to the right of each of these lines. Each numeral is the sum of the observed frequencies for the corresponding group of four patterns. Each is a sub-sum and will be used in Table 13.

TABLE 11
Results from Coin- and Die-Coin Guessing Experiments

	F_o			F_t	
Patterns	Coin Guessing	Die-Coin Guessing	Totals	Coin Guessing	Die-Coin Guessing
H H H H H	25	29	54	28.8	25.2
H H H H T	12	9	21	11.2	9.8
H H H T H	22	9	31	16.5	14.5
H H H T T	18	5	23	12.3	10.7
	—	—			
	77	52			
H H T H H	29	38	67	35.7	31.3
H H T H T	96	59	155	82.7	72.3
H H T T H	22	24	46	24.5	21.5
H H T T T	14	8	22	11.7	10.3
	—	—			
	161	129			
H T H H H	5	7	12	6.4	5.6
H T H H T	15	9	24	12.8	11.2
H T H T H	12	2	14	7.5	6.5
H T H T T	4	9	13	6.9	6.1
	—	—			
	36	27			
H T T H H	33	12	45	24.0	21.0
H T T H T	17	20	37	19.7	17.3
H T T T H	12	3	15	8.0	7.0
H T T T T	3	4	7	3.7	3.3
	—	—			
	65	39			
Sub-totals	(339)	(247)			
T H H H H	3	3	6	3.2	2.8
T H H H T	7	5	12	6.4	5.6
T H H T H	10	9	19	10.1	8.9
T H H T T	14	3	17	9.1	7.9
	—	—			
	34	20			
T H T H H	5	4	9	4.8	4.2
T H T H T	1	8	9	4.8	4.2
T H T T H	6	4	10	5.3	4.7
T H T T T	0	1	1	0.5	0.5
	—	—			
	12	17			

TABLE II (Continued)

Patterns	F_o Coin Guessing	Die-Coin Guessing	Totals	F_t Coin Guessing	Die-Coin Guessing
T T H H H	4	13	17	9.1	7.9
T T H H T	6	19	25	13.3	11.7
T T H T H	25	37	62	33.1	28.9
T T H T T	7	13	20	10.7	9.3
	—	—			
	42	82			
T T T H H	2	4	6	3.2	2.8
T T T H T	5	5	10	5.3	4.7
T T T T H	2	1	3	1.6	1.4
T T T T T	3	8	11	5.9	5.1
	—	—			
	12	18			
Sub-totals	(100)	(137)			
Totals	439	384	823	438.8	384.2

The first 16 patterns listed in Table 11 begin with a guess of heads, the last 16 with a guess of tails. A 2×2 contingency table, shown in Table 12, was set up, the division along one "dimension" being between first-guess-heads and first-guess-tails and along the other "dimension" between the two conditions of guessing. The four sub-totals in Table 11 consitute the observed frequencies appearing in the four cells of the 2×2 table.

TABLE 12

2×2 Contingency Table Based upon the Coin- and Die-Coin Guessing Experiments

	First Guess H	T	
Coin Guessing	339 (312.6)	100 (126.4)	ⁿ 439
Die-Coin Guessing	247 (273.4)	137 (110.6)	384
	586	237	823

The hypothesis to be tested in this case is that the occurrence of a first guess of heads was independent of the condition under which the guessing was done. If the observed frequencies in Table 12 were independent of the conditions of guessing, the probability of a guess of heads was 586/823. The probability of a subject's being in the coin guessing group was 439/823. The joint probability that a subject would be in the coin guessing group and would also guess heads was 439/823 × 586/823. The theoretical frequency for the upper left-hand cell of the table was obtained by multiplying this joint probability by 823. The other three theoretical frequencies were automatically determined by this single calculated frequency and by the restrictions of the border sums. The four values of F_t are shown in parentheses in the table. The value of χ^2 was computed as follows.

$$\chi^2 = \frac{(339 - 312.6)^2}{312.6} + \frac{(100 - 126.4)^2}{126.4} + \frac{(247 - 273.4)^2}{273.4} + \frac{(137 - 110.6)^2}{110.6}$$

$$= (26.4)^2 \left(\frac{1}{312.6} + \frac{1}{126.4} + \frac{1}{273.4} + \frac{1}{110.6} \right)$$

$$= 696.96(.02381) = 165.946.$$

The number of df is 1. This follows because there are four cells and because three restrictions were placed in determining the theoretical frequencies. The three restrictions were the total number of subjects ($=823$) and two border sums, one for a row and one for a column.[21] The calculated value of χ^2 is highly significant and leads immediately to a rejection of the hypothesis that the conditions of guessing had no influence on the tendency to guess heads on the first guess.

Table 11 may be regarded as a 32×2 contingency table, 32 "patterns" by 2 "conditions of guessing." In order to use the χ^2 test in evaluating the hypothesis that the patterns were independent of the conditions of guessing, the border sums would be used in calculating theoretical frequencies. For example, the two theoretical frequencies for the pattern $H\,H\,H\,H\,H$ are given by the relations: $F_t = (54 \times 439)/823 = 28.8$; and $F_t = 54 - 28.8 = 25.2 = (54 \times 384)/823$. These two values of F_t are shown in the top row of the last two columns of Table 11. The theoretical frequencies for the other 31 patterns were obtained in a similar way and are listed in the table. These frequencies are included to emphasize the fact that the χ^2 test cannot be legitimately applied to this table as it stands. The reason is that 37 of the 64 theoretical frequencies are less than 10 (some of them very much less than 10) and cannot be depended upon to yield quantities distributed as χ^2. To make

---

[21] The sums 439 and 237 (or 384 and 237, or 384 and 586) could have been used instead of 586 and 439. The values of F_t for all but one cell may be obtained by subtraction.

a legitimate application of χ^2 in this particular case, it would be necessary to increase the number of subjects to a point where the smallest of the theoretical frequencies was close to 10.

It must now be decided whether or not the observed frequencies in Table 11 can be combined so as to permit the use of the χ^2 test in evaluating the hypothesis of non-relationship between the patterning of the guesses and the conditions of guessing. The frequencies have already been combined in a gross way to yield the observed frequencies in Table 12. This division was on the basis of the first guess. Divisions might be made on the basis of the first two guesses or the first three guesses or the first four guesses. It is not possible to use the patterning on all five guesses because the theoretical frequencies become too small, as already seen. It turns out that a division on the basis of the first four guesses also leads to several theoretical frequencies that are less than 10. Consequently, a division based on the patterning of the first three guesses will be illustrated. In the division that was made, the following rule held: There would be a decreasing number of heads in the pattern and, contrariwise, there would be an increasing number of tails, from the first guess on. The resulting division of the patterns is shown in Table 11 and also in Table 13, the one to be used in applying the χ^2 test.

As seen, there are eight different patterns listed in Table 13. The

TABLE 13

Combination of the Frequencies Shown in Table 11, Based on the First Three Guesses

Pattern on 1st Three Guesses	Coin Guessing	Die-Coin Guessing	Totals
H H H	77 (68.8)	52 (60.2)	129
H H T	161 (154.7)	129 (135.3)	290
H T H	36 (33.6)	27 (29.4)	63
H T T	65 (55.5)	39 (48.5)	104
T H H	34 (28.8)	20 (25.2)	54
T H T	12 (15.5)	17 (13.5)	29
T T H	42 (66.1)	82 (57.9)	124
T T T	12 (16.0)	18 (14.0)	30
Totals	439	384	823

observed frequencies for the two conditions of guessing are shown (together with parenthesized theoretical frequencies) in the second and third columns. The hypothesis to be tested is that the patterns of guessing on the first three of five consecutive guesses were independent of the conditions of guessing. The theoretical frequencies were secured in the usual way by employing border sums and the value of $N(=823)$. The value of χ^2 was computed as follows:

$$\chi^2 = \frac{(77 - 68.8)^2}{68.8} + \frac{(161 - 154.7)^2}{154.7} + \cdots$$
$$+ \frac{(82 - 57.9)^2}{57.9} + \frac{(18 - 14.0)^2}{14.0} = 31.162.$$

The number of df is 7. The computed value of χ^2 is significant at far better than the 1% level of confidence and leads to a rejection of the hypothesis. It may be confidently concluded that the patterning of the guesses through the first three guesses was somehow influenced by the conditions under which the guessing was done.

The number of df for a value of X^2 obtained from a contingency table is always the number of cells in the table minus the number of restrictions imposed during the calculation of the theoretical frequencies. In Table 13, for example, there are 16 cells. Nine restrictions must be imposed in obtaining values of F_t. These restrictions are: 7 of the row sums, 1 of the column sums, and the total number of cases. Thus, $df = 16 - 9 = 7$.

A convenient formula for determining the number of df for a contingency table when the χ^2 test is applied is

$$df = (n_c - 1)(n_r - 1), \tag{12}$$

where n_c and n_r are the number of columns and the number of rows, respectively. There is only one hypothesis to be tested—that the variables are independent in the population from which the samples arise; so the number of df is always given by (12). (See footnote 16.)

Contingency table with more than two categories in each direction. For the sake of completeness, a contingency table having five categories in one direction and four in the other will be included. A total of 2,274 eighth-grade pupils, enrolled in 91 different schools, took an English Correctness Test. A summary of the scores obtained by these pupils has been taken from a report by Lindquist (1934). The scores were divided into five categories and symbolized by numbers from 1 to 5, as seen in Table 14. The schools were divided into four enrollment groups, labeled A to D in the table. The observed frequencies in the 20 cells of the table range from 3 to 342. These frequencies would obviously have been different if the enrollment groups had been differently established, and if the scores had been divided into different categories. The enrollment

TABLE 14

5 × 4 Contingency Table Based upon Lindquist's Data (23)

Scores	Enrollment Groups*				Totals
	A	B	C	D	
1	36	40	20	3	99
	(23.5)	(38.4)	(28.2)	(8.9)	
2	76	108	59	11	254
	(60.2)	(96.8)	(72.3)	(22.9)	
3	150	181	111	28	470
	(111.4)	(182.5)	(133.7)	(42.4)	
4	211	342	285	88	926
	(219.5)	(359.6)	(263.5)	(83.5)	
5	66	212	172	75	525
	(124.4)	(203.8)	(149.4)	(47.3)	
Totals	539	883	647	205	2274

* See footnote 22.

grouping of the schools was that commonly used in the Iowa Every-Pupil Testing Program.[22] The division of the scores was made by starting at the bottom and "stepping off" successive standard deviation "distances" (approximately). The distribution of scores was positively skewed; so it was necessary to combine the two upper score categories to provide satisfactorily large frequencies in the top row of cells.

The theoretical frequencies for Table 14 were computed in the usual way when the hypothesis of independence is under test. The computations required the use of three column sums and four row sums, as well as the value of $N (=2274)$. This made a total of 8 restrictions; so the number of $df = 20 - 8 = 12 = (5 - 1)(4 - 1)$. The value of χ^2 was computed as follows:

$$\chi^2 = \frac{(36 - 23.5)^2}{23.5} + \frac{(76 - 60.2)^2}{60.2} + \cdots$$
$$+ \frac{(88 - 83.5)^2}{83.5} + \frac{(75 - 47.3)^2}{47.3} = 99.038.$$

With $df = 12$, this value is highly significant and leads to a rejection of the hypothesis that the scores obtained on the test were independent of school size.

A comment should be made concerning the theoretical frequency in the upper right-hand cell of the table. Its value is 8.9. Ordinarily, a value

[22] The enrollment categories were: A, greater than 400; B, 126–400; C, 66–125; D, less than 66.

this small should not be used in obtaining an estimate of χ^2. In this case, however, the other 19 values of F_t are satisfactorily large, and the inclusion of one theoretical frequency that is less than 10 is permissible since an error in a single category will have slight effect on the resulting value of χ^2. The obtained value of χ^2 is so large that it makes no difference whether or not the small theoretical frequency is included in the calculations. It is only in situations of this general kind that one or two small theoretical frequencies may be retained. When the number of df is less than 4 or 5, and especially when $df = 1$, the use of theoretical frequencies of less than 10 should be strictly avoided.

Use of the Chi-square Test with Too Small Theoretical Frequencies

The studies of Lewis and Franklin (1944) and Lewis (1944). The commonest weakness in applications of the χ^2 test to contingency tables is the use of extremely small theoretical frequencies. This weakness is clearly present in most of the applications made in a paper by Lewis and Franklin (1944). The paper is concerned with the Zeigarnik effect (that is, with the relative amounts of recall of interrupted and completed tasks). In one experiment, 12 subjects were each presented with 18 problems, 9 of which were interrupted by the experimenter, the other 9 being completed without interruption. The ratio (RI/RC) of the number of interrupted tasks recalled to the number of completed tasks recalled is given for each subject in the top row of Table 15. In a previous study,

TABLE 15

Data upon the "Zeigarnik Effect" Presented by Lewis (1944), and Lewis and Franklin (1944)

Ratio of Recall Scores	Subjects													
	1	*2*	*3*	*4*	*5*	*6*	*7*	*8*	*9*	*10*	*11*	*12*	*13*	*14*
RI/RC—Group I (Indiv. Work Condition)	1.00	.80	.80	.80	.75	.71	.63	.57	.57	.44	.40	.25		
RI/RC—Group II (Coop. Work Condition)	1.67	1.25	1.00	1.00	1.00	1.00	1.00	1.00	1.00	1.00	.80	.75	.60	.14

Lewis (1944) had employed a cooperative work situation in which a coworker completed the tasks on which the subject was interrupted. The conditions were otherwise the same as in the later experiment by her and Franklin. Fourteen subjects were used. The RI/RC ratio for each subject is shown in the second row of Table 15. The median ratio for Group I was .67. The ratios for the two groups were divided (dichotomized) at

TABLE 16

Contingency Table Based upon Data in Table 15

	Group		
RI/RC	I	II	
Greater than 0.67	6 (8.3)	12 (9.7)	18
Less than 0.67	6 (3.7)	2 (4.3)	8
	12	14	26

this point. The 2×2 contingency table shown as Table 16 was the result. The hypothesis to be tested was that the conditions of the two experiments had no differential effects on the recall of interrupted and completed tasks. The theoretical frequencies are shown in parentheses in the table. The investigators calculated a value of χ^2 in the usual way. Finding it to be approximately 3.8 and to fall near the 5% level of confidence, they were inclined to reject the hypothesis.

Entirely aside from any of the conclusions reached by Lewis and Franklin, it must be firmly stated that all four of the theoretical frequencies in Table 16 and especially the two that are less than 5, are too small to warrant an application of the χ^2 test. Furthermore, the other applications made in their paper, with one or two possible exceptions, involved theoretical frequencies that should be avoided (theoretical frequencies, for example, of 4.4, 6.0, 7.0, 7.6, etc.).

Kuenne's study of transposition behavior. An application, which is correct in principle but which must be regarded somewhat unfavorably because of the size of the F_t's, comes from a paper by Kuenne (1946). Kuenne made a study of transposition behavior in four groups of young children, the ages for the groups being 3, 4, 5, and 6 years. Some of the children displayed size transposition, others of them did not. Kuenne realized that, because of the small number of cases in each age group, she could not apply the χ^2 test to the data for the four groups considered separately. Consequently, she combined the results for ages 3 and 4, and those for ages 5 and 6. The children were divided into two categories —those who did and those who did not meet the transposition criterion. The resulting 2×2 contingency table is shown as Table 17. The hypothesis to be tested is that the occurrence of transposition behavior was independent of age. Theoretical frequencies were determined by using two of the border sums and the value of N. It will be seen that two of these frequencies are less than 10. Because they are fairly close to 10, many

TABLE 17

Contingency Table Based upon Kuenne's (1946) Transposition Data

Age-Groups	Transposition	Non-Transposition	Total
3–4 years	3 (9.4)	15 (8.6)	18
5–6 years	20 (13.6)	6 (12.4)	26
	23	21	44

investigators would proceed as Kuenne did, and make a χ^2 test of the hypothesis of independence. In fact, there are textbooks in statistics which place 5 as the minimum value for the theoretical frequencies. A value of 5 is believed to be too low. In any event, it is the smallest value that should be used even when there are several other theoretical frequencies that are far greater than 10.

The value of χ^2 computed from Table 17 is 15.434. With 1 df, this value falls close to the 0.1% level. It is only because the value is so large that confidence can be placed in the conclusion that transposition behavior was related to age. In view of the smallness of all four of the theoretical frequencies, very great doubt would have remained if the χ^2 value had fallen at a border-line level of confidence. Whenever small theoretical frequencies enter into calculations of χ^2, the experimenter has no sound basis either for accepting or rejecting a hypothesis except when the value is quite extreme.

Yates' correction for continuity. No mention has as yet been made of a correction proposed by Yates (1934) which reduces the value of χ^2 to compensate for errors which may arise as a result of one of the approximations made in deriving the formula for the χ^2 distribution. It will be recalled that three approximations are made in this derivation. One of the three involves the substitution of an integral for a summation of discrete quantities. The approximation introduces an error (an error of overestimation) that is of consequence when values of F_t are small. The correction is justified only when the number of df is 1.[23] It provides for the reduction of all differences between observed and theoretical frequencies by 0.5. For example, all of the differences between observed and theoretical frequencies in Table 17 are 6.4. These are reduced to 5.9 if Yates' correction is applied. The calculation, using the correction, would be as follows:

[23] The correction should not be made if several values of χ^2 are to be summated. The additive principle does not apply to corrected values.

$$\chi^2 = (5.9)^2 \left(\frac{1}{9.4} + \frac{1}{8.6} + \frac{1}{13.6} + \frac{1}{12.4} \right) = (34.81)(.3768) = 13.1.$$

With 1 df, this value is still highly significant and leads to a rejection of the hypothesis. But the use of Yates' correction does not remove the objection to theoretical frequencies that are less than 10.

Questionable or Incorrect Categorizing

Lewis' (1944) study of recall interrupted and completed tasks. Another weakness which is sometimes present in applications of the χ^2 test to contingency tables is that the categorizing is done on either a questionable or a clearly incorrect basis. An illustration of incorrect categorizing is found in the paper by Lewis (1944) discussed above. The RI/RC ratios for 14 subjects were obtained in a "cooperative work experiment." The ratios are the ones given in the second row of Table 15. On the assumption that the recall of interrupted and completed tasks should have been the same, Lewis writes: ". . . we should have an equal number of ratios above 1.00 and below 1.00. . . . The expected distribution of ratios should, therefore, be 7 below 1.00 and 7 at 1.00 or above. The obtained distribution of ratios is 4 below 1.00 and 10 at 1.00 or above." The categorizing is plainly wrong; there is no more reason for placing an obtained ratio of 1.00 in the upper category than for placing it in the lower category. A better procedure would have been to divide the 8 ratios of 1.00 equally between the two categories. This flaw in Lewis' division of the ratios is remindful of the belief of many graduate students in psychology that it is quite permissible to set up several different sets of dichotomy lines, compute a value of χ^2 for each set, and finally select the dichotomies that yield a χ^2 value to support the experimenter's own point of view. In any investigation where the χ^2 test is to be applied, the categories must be established in a logically defensible and reliable manner—before the data are collected, if possible.

Anastasi and Foley's (1944) study of drawings of normal and abnormal subjects. The whole problem of categorizing may be brought clearly before the reader by taking an illustrative case from a study by Anastasi and Foley (1944). These two investigators had each of 340 normal subjects and 340 abnormal subjects draw a picture which depicted danger. The pictures were then divided into the 20 subject-matter categories listed in Table 18. The application of the χ^2 test yields a value of 99.603 which, with 19 df, falls far beyond the 0.1% level of confidence. This leads to a rejection of the hypothesis that the subject matter of the drawings was independent of the two "kinds" of subjects.

Let it again be emphasized that the criticism here, as elsewhere in the paper, is not directed at any of the conclusions reached by the in-

TABLE 18

Data from Anastasi and Foley's (1944) Study of Drawings of Normal and Abnormal Subjects

Subject-Matter Categories	F_o (Abnormal)	F_o (Normal)	F_t	χ^2
1. Traffic	105	124	114.5	1.576
2. Conventional sign or signal	36	37	36.5	.014
3. Skating, Ice	5	10	7.5	1.667
4. Falling	21	17	19.0	.421
5. Drowning, sinking, flood	8	10	9.0	.222
6. Falling objects, explosion	6	6	6.0	.000
7. Arms and explosives	10	22	16.0	4.500
8. War	3	7	5.0	1.600
9. Fire	26	37	31.5	1.921
10. Lightning, electricity	6	15	10.5	3.857
11. Animals	3	16	9.5	8.895
12. Abstract or conventionalized symbolism	3	8	5.5	2.273
13. Fantastic compositions	26	0	13.0	26.000
14. Several discrete objects	9	2	5.5	4.454
15. Scribbling or scrawl	6	0	3.0	6.000
16. Writing only	12	1	6.5	9.308
17. Miscellaneous	23	26	24.5	.184
18. Recognizable object not representing danger	8	1	4.5	5.444
19. Refusal to draw	14	1	7.5	11.267
20. No data for other reasons	10	0	5.0	10.000
	340	340		99.603

vestigators. But the illustration provides a very satisfactory basis for discussing the fundamental problem of categorizing. In the published article, the principles adopted in classifying the pictures are not explicitly stated. Furthermore, evidence is not presented regarding the reliability of the categories. Two generalizations may be offered. The first is that, *whenever possible, categories for frequency data should be established on the basis of completely external criteria* (for example, criteria that have been used or proposed by some other investigator) and should be set up independently of the data under study. Such a procedure frees a person from any charge of bias and guards against tendencies to juggle data. A second generalization is that *information on the reliability of categories should be offered,* and this is the case whether or not the categories have stemmed from an independent source.

A study of Table 18 shows rather quickly that the value of χ^2 for the drawings depicting danger would have been quite different if the categorizing had been different. For example, the amount 26.000 was contributed to the value of χ^2 by the frequencies of category 13 alone. As seen, this is the category "fantastic compositions." The decision which established this category and the judgments which placed 26 of the drawings of the abnormal subjects in the category and none of the drawings of the normal subjects in it, should have been explicitly justified and a precise statement concerning reliability should have been included. The discrepancy between the frequencies in category 13 (along with the discrepancies in such categories as "refusal to draw" and "no data for other reasons") required that there be discrepancies in one or more of the other categories. Unreliability at one point in a multicelled table automatically produces unreliability elsewhere.

It is well to emphasize, by reiteration, that when the χ^2 test is to be applied to a collection of data, the categories should be established independently of the data and, once established, should never be modified on the basis of the way the data happen to fall. Categories should usually, if not always, be established before the data have been scrutinized.

APPLICATIONS: III. THE GOODNESS OF FIT OF FUNCTIONS IN WHICH FREQUENCY IS THE DEPENDENT VARIABLE

The frequency (or relative frequency) of occurrence of a response is sometimes used as the dependent variable in psychological experiments. For example, in psychophysical investigations based on the method of constant stimuli, the number or proportion of judgments in a given direction serves as the dependent variable, while in studies of the conditioned response, the frequency of occurrence of the CR is often taken as the dependent variable. If a mathematical function is fitted to data of this type, it is sometimes possible to apply the χ^2 test in evaluating the goodness of the fit. However, care must be taken to insure a correct application.

An application of χ^2 to percentages (Grant & Norris, 1946). A recent application of the χ^2 test to percentages (or relative frequencies) is taken from a paper by Grant and Norris (1946.) These investigators were concerned with the influence of different amounts of dark-adaptation on the sensitization of the beta-response of the human eyelid to light.[24] One of their measures of degree of sensitization was the frequency of occurrence of the response. A subject looked straight ahead into a small box-like

---

[24] The beta-response is one of two reflexes displayed by the eyelid when the eye is light-stimulated.

enclosure which was painted flat black inside and out. The stimulus was a small circle of light emitted from a circular milk-glass plate, 10 cm. in diameter, located at the back of the enclosure. When illuminated, the plate had a surface brightness of 241 millilamberts. The duration of the stimulus was about 750 milliseconds.

Four experimental conditions were employed, all subjects serving in each condition. The conditions differed in the amount of dark-adaptation present in the subjects. The amount of dark-adaptation depended upon the total length of time the subjects spent in darkness. The measure of amount of adaptation was the product It, where I was the surface brightness of the stimulus plate and t was the number of seconds spent in darkness. The It products for the four conditions were 28,920; 187,980; 347,040; and 506,100. The corresponding values of t were 120, 780, 1414, and 2100 seconds.

Thirty-three subjects participated in the experiment. A subject was first dark-adapted for 120 seconds. The stimulus light was then presented four times, with a "control" trial between the first two and last two presentations. The control trial served as a check on possible conditioning. The four stimulus trials were separated by dark intervals of 35 sec. on the average. The first four stimulus trials (coming after 120 secs. in darkness) were the trials for condition 1. Condition 2, in which the stimulus light was again presented four times along with a control trial, came after the subject had spent a total of 780 secs. in darkness. Conditions 3 and 4, with the stimuli presented in the same general fashion, came after 1414 and 2100 secs. in darkness.

It should be noted that each of the thirty-three subjects was given four stimulus trials in each of the four dark-adapted conditions. A count was made of the number of beta-responses occurring in each subject in each condition. The results were then combined to provide frequencies of occurrence of the response for each condition. These frequencies were used in computing percentages. The four observed percentages are shown in the first row of Table 19. They indicate the *relative* frequency of occurrence of the beta-response in each of the experimental conditions. In condition 1, for example, 13.2% of the 132 possible responses of the eye-

TABLE 19
Data from the Experiment of Grant and Norris (1946)

	Conditions			
	1	*2*	*3*	*4*
P_o	13.2	30.0	45.6	51.5
P_t	11.05	36.10	44.30	49.35

lid displayed the beta-response. The other three percentages may be similarly interpreted.

A logarithmic function was fitted to the data. The two variables were It, the amount of dark-adaptation, and P_o, the relative frequency of occurrence of the beta-response. The fitted function was: $P_t = 13.38 \log_e It - 126.39.$[25] Its solution for the empirical values of It yielded the values of P_t given in the bottom row of Table 19.

A value of χ^2 was calculated as follows:

$$\chi^2 = \frac{(13.2 - 11.05)^2}{11.05} + \frac{(30.0 - 36.10)^2}{36.10} + \cdots + \frac{(51.5 - 49.35)^2}{49.35} = 1.579.$$

Two constants were estimated from the data, leaving 2 df. With this number of df, the obtained value of χ^2 falls near the 40% level of confidence. This led the investigators to conclude that the data could be satisfactorily represented by a logarithmic function.

There are four mistakes in this application of the χ^2 test: two computational mistakes and two "theoretical" mistakes. The computational mistakes will be discussed first. The calculated value of χ^2 was not corrected to take account of the use of percentages instead of frequencies. It should have been multiplied by the ratio 132/100. The second computational mistake was a failure to take account of the frequency of non-occurrence of the beta-response in each of the four conditions. In condition 1, for example, the beta-response occurred 13.2% of the time and failed to occur 86.8% of the time. This latter percentage played no part in the calculation of χ^2. All four percentages of non-occurrence should have been employed. The value of χ^2 for Table 19, when correctly computed, is 3.081. The number of df is still 2.

The other two mistakes were more basic. The use of χ^2 to test goodness of fit was not warranted, for two reasons. In the first place, there were linkages within conditions. Each of the subjects was given four trials in each of the four conditions of dark-adaptation. The results for the subjects were pooled. There were undoubtedly individual differences in capacity to display the beta-response; so there must have been linkages within conditions. In applying the χ^2 test, the investigators assumed, in effect, that the 132 trials per condition were given to 132 instead of 33 subjects. The test was inapplicable because the assumption could not justifiably be made.

The second reason that the χ^2 test should not have been used arises from the lack of independence from condition to condition. As already stated, a fundamental requirement for the use of the χ^2 test is independence between individual measures. The differences between indi-

---

[25] With common logarithms, the function becomes:

$$P_t = 30.81 \log_{10} It - 126.39.$$

vidual subjects that manifested themselves in any condition were certain to be maintained in the other conditions; so it cannot be assumed that the values of P_o in Table 19 are unrelated.

APPLICATIONS OF THE χ^2 TEST TO NON-FREQUENCY DATA

Some investigators are prone to use a χ^2 test of goodness of fit whenever a set of observed and theoretical values of any kind is available for comparison. This mistake apparently grows from a misinterpretation of the well-known formula for computing χ^2:

$$\chi^2 = \sum \frac{(F_o - F_t)^2}{F_t}. \tag{10}$$

Because this equation involves differences between observed and theoretical values, the conclusion is reached that the summation of the weighted squares of differences between observed and theoretical quantities yields a meaningful estimate of χ^2. Formulas superficially resembling equation (10) have been applied to non-frequency data, where theoretical values have been obtained from fitted curves.

Suppose that a study has been made of the amount of activity displayed by several groups of white rats deprived of food for differing lengths of time. Enough rats have been included in each deprivation group to provide means that are quite stable. These means, when plotted against time of food deprivation, show a systematic trend; so a curve is fitted to them. The equation for the curve permits the calculation of theoretical means which may be compared with the observed means. To test goodness of fit, differences between the observed and theoretical means are obtained. Each difference is squared and then divided by the appropriate theoretical mean. The sum of these weighted squared differences is taken as a meaningful estimate of χ^2.

Two of the fourteen papers (Chen & Irwin, 1946; Irwin & Chen, 1946) referred to in the opening paragraphs contain applications of this type. The fallaciousness of such applications becomes obvious when it is realized that values of χ^2 computed from non-frequency data vary in magnitude with the size of the units employed in measurement. Assume that two investigators have a common aim: To determine how the height of human males varies with age. They make measurements of the same individuals in various age groups, fit equations of the same form to their data, use these equations to compute theoretical values of height, and then obtain estimates of χ^2 in a manner similar to the one described above. One of the investigators has measured height in centimeters, the other in inches. Except for incidental discrepancies, the value of χ^2 calculated from the centimeter data will turn out to be 2.54 times the value calculated from the inch data.

The χ^2 Test of Linearity of Regression

A χ^2 test is recommended in certain textbooks (Peters & Van Voorhis, 1940, p. 319; Guilford, 1942, p. 237) as suitable for use, with non-frequency data, in evaluating linearity of regression. The formula, as usually presented, is as follows:

$$\chi^2 = \frac{\eta^2 - r^2}{1 - \eta^2} (N - k). \tag{19}$$

In this formula, η stands for correlation ratio; r for product-moment coefficient of correlation; N for the total number of measures; and k for the number of columns (groups) into which the measures have been divided. The number of df is $k - 2$. The formula yields a variable, the distribution of which approximates the χ^2 distribution, under certain conditions. It can be used with some degree of confidence provided that N is quite large and k is quite small, and provided also that the measures from column to column are independent and homoscedatic as well as normally distributed.

An exact test of linearity of regression, which is applicable whenever equation (19) is, can be made by computing an F-ratio. The formula is

$$F = \frac{(\eta^2 - r^2)(N - k)}{(1 - \eta^2)(k - 2)}. \tag{20}$$

The df's for this F are $(k - 2)$ and $(N - k)$. The reason that the χ^2 test, as defined by equation (19), can be substituted under any circumstances for the F test, as defined by equation (20), is that the distribution of F approximates the distribution of χ^2 when one of the df's for F is very small and the other is very large. The particular χ^2 distribution that is approximated is the one for the smaller df (that is, for $df = k - 2$). To state the point in another way: The sampling distribution of estimates of F, obtained with equation (20), approximates the sampling distribution of estimates of χ^2 obtained with equation (19), provided that k is very small relative to N.

Inasmuch as the χ^2 test, as represented by equation (19), is inexact, while the F test, as represented by equation (20), is exact, nothing is gained by using the χ^2 test.[26]

SPECIAL PROBLEMS: I. INDETERMINATE THEORETICAL FREQUENCIES

It sometimes happens, despite superficial indications to the contrary, that meaningful theoretical frequencies cannot be determined for a set of observed frequencies. This situation commonly arises from a lack of

---

[26] A fuller explanation of equation (20), together with a detailed discussion of other F tests of goodness of fit, is given by Lewis (1948).

independence between measures. Two illustrations will be presented to reveal some of the chief sources of difficulty.

First Illustration

The first illustration is concerned with coin-guessing data. Two hundred and forty university students each made a guess of heads or tails on four successive tosses of a coin. They recorded their guesses on individual record sheets. They were not told in advance the number of tosses that would be made, nor were they told how the coin turned up on the four tosses until the record sheets had been collected. The turns, in order of occurrence, were $H\ T\ T\ H$. The succession of guesses of each student could easily be compared with this succession of turns and the number of correct guesses for each could be tabulated. As expected, the number of correct guesses ranged from none to 4. The results are summarized in the first two columns of Table 20. As shown, 15 students made no correct

TABLE 20
Analysis of Coin-Guessing Data

Number Correct	F_o	$P(n)$	F_t	$\dfrac{(F_o - F_t)^2}{F_t}$	$P(n)'$	F_t'	$\dfrac{(F_o - F_t')^2}{F_t'}$
0	15	.0625	15	.000	.0490	11.76	.893
1	47	.2500	60	2.817	.1680	40.32	1.107
2	60	.3750	90	10.000	.3129	75.10	3.036
3	86	.2500	60	11.267	.3289	78.93	.633
4	32	.0625	15	19.267	.1412	33.89	.105
Σ	240	1.0000	240	43.351	1.0000	240.00	5.774

guesses; 47 made one correct guess; 60 made two correct guesses, etc. With frequency distributions of this type, it is common practice to apply the chi-square test after a binomial distribution function has been employed to calculate the required theoretical frequencies. The function for the present case, modeled after equation (3), would have the form

$$P(n) = \frac{4!}{n!(4-n)!}\left(\frac{1}{2}\right)^n \left(\frac{1}{2}\right)^{4-n} = \frac{4!}{n!(4-n)!}\left(\frac{1}{2}\right)^4, \tag{3'}$$

and would yield theoretical probabilities of obtaining n correct guesses in 4.

Values of $P(n)$ calculated with equation (3') are given in the third column of Table 20 with corresponding theoretical frequencies appearing in the fourth column. If these values of F_t were legitimate estimates, a value of χ^2 could be computed as illustrated in the fifth column of the table and used to test the hypothesis that the observed frequencies are distributed in accordance with equation (3'). In effect, this would be testing the hypothesis that each guess of every student was a purely chance occurrence, completely independent of every other guess, the probability of a guess of heads always being .50. The number of df is 4. If the hypothesis were true, the obtained value of χ^2, 43.351, would not be expected to arise in random sampling once in a million times.

The use of a binomial distribution function in calculating theoretical probabilities cannot be justified in this case. There is good reason for believing that the guess of a student on any of the last three tosses was not independent of previous guesses. The binomial distribution function is applicable only when there is a sound basis for assuming that every event under consideration is completely independent of every other event.

Theoretical frequencies for the empirical data in Table 20 are indeterminate—except as they might be estimated from probabilities yielded by extraneous empirical data. When hypothetical probabilities do stem from other empirical results, the hypothesis that can be tested may be quite different from the one that would be tested if a binomial distribution function yielded the theoretical values. For example, the data of Table 20 may be compared with somewhat similar findings published several years ago by Goodfellow (1938), who analyzed the patterning of the guesses on five successive tosses of a coin by a large number of radio listeners. These listeners participated in the "telepathic experiments" conducted in 1937–38 by the Zenith Foundation. The coin was tossed in the broadcasting studio in Chicago. A total of 5,687 members of the radio audience wrote down their guesses in order and mailed in their answer sheets. They were told in advance that they were to make five guesses, but were not told until two or three weeks after the broadcast how the coin had actually turned up each time. The conditions of guessing were not identical with those holding when the data in Table 20 were secured, but were similar enough to permit a comparison of the results.

Goodfellow (1938, Table 2) tabulated the results on the radio listeners in a way closely approximating that used in Table 8. In fact, the only difference of any consequence is that he tabulated percentages instead of frequencies. The "correct" pattern of $H\ T\ T\ H$ was checked against the 32 patterns in Goodfellow's table to obtain the proportions of individuals in the total group of 5,687 that hypothetically made 0, 1, 2, 3, and 4 correct guesses. These proportions are given in Table 20, in the column headed F_t'. The resulting value of χ^2, as shown in the last column of the table, is 5.774 with $df = 4$. The hypothesis that may now be tested is that the patterning of the guesses by the 240 students was the same as the patterning of the first four of five guesses by the large group of Zenith Foundation listeners. The obtained value of χ^2 falls near the 20%

level of confidence and provides no basis for rejecting this hypothesis. If the value of χ^2 *had been* large enough to justify a rejection of the hypothesis, it would not be possible to decide whether the patterning tendencies were basically different in the two groups or whether the differences in the conditions of guessing produced an apparent difference in patterning. Nevertheless, the test of the hypothesis, as stated, is an exact one. This is in sharp contrast to the indefiniteness which was present when a value of χ^2 was based on theoretical frequencies obtained with a binomial distribution function. Because of the strong likelihood of interdependence between the guesses of individual guessers, the highly significant value of χ^2 in the fifth column of Table 20 could be interpreted to mean that all guessers were biased, that some of the guessers were strongly biased while others were unbiased, that the probability of a guess of heads was not .50, that the probability of a guess of heads was .50 on some tosses but not on others, etc. The absence of independence and the consequent inability to obtain unequivocal theoretical (chance) frequencies made this application of the chi-square test a meaningless procedure.

It is not possible to provide rules-of-thumb for deciding whether theoretical frequencies are calculable or incalculable in particular situations. Decisions must ordinarily be based on careful logical analysis. However, *it is usually true that theoretical frequencies are incalculable if the observed frequencies are in any way related, and also if mutually contradictory assumptions can be made, with about equal justification, concerning the likelihood of occurrence or non-occurrence of the events (responses) that yielded the observed frequencies.*

SPECIAL PROBLEMS: II. THE NATURE OF IMPOSED RESTRICTIONS

All restrictions that are imposed during the determination of theoretical frequencies should be both linear and homogeneous. This limitation is seldom mentioned in either theoretical or practical treatments of the chi-square test, and even when mentioned is usually left unexplained. The main reason for the omission is that anything beyond a very superficial explanation cannot be given in other than mathematical terms.[27] It follows that only a few very general ideas can profitably be included here.

It is probably quite obvious to most readers why restrictions must be imposed—why the sums of the observed and theoretical frequencies, for example, must always be equalized. The value of ΣF_o is fixed for any set of empirical data. The value of ΣF_t cannot "wander around any place" without at times yielding entirely impossible cell frequencies.

[27] The excellent attempt at elementary explanation offered by Greenhood (1940, Chap. 3) is about as non-mathematical as it could conceivably be made, and yet requires a considerable amount of mathematical sophistication.

Any hypothesis must be tied down at some point to the sample data. There should be as much freedom as possible (for example, as much freedom as possible for fluctuations in the individual cell frequencies) and yet there must be enough restrictions to bring the over-all values within the same general area.

The one restriction that must always hold may be symbolized as follows:

$$\Sigma F_o = \Sigma F_t, \tag{21}$$

or

$$\Sigma(F_o - F_t) = 0. \tag{21a}$$

This restriction is clearly linear.

Some of the other restrictions that must be imposed in familiar applications of the chi-square test may be shown to be linear—and to be homogeneous also. Suppose, for example, that a normal curve is to be fitted to an array of observed frequencies. It is not enough to impose the single restriction specified by equation (21). The reason is that a multitude of different combinations of values of F_t, all arising from normal distributions, will each, when summated, equal ΣF_o. A second restriction must obviously be placed. It may be written

$$\Sigma F_o X = \Sigma F_t X, \tag{22}$$

or

$$\Sigma X(F_o - F_t) = 0, \tag{22a}$$

where X is the measure associated with the cell. Stated in more familiar verbal terms, the restriction is that the means of the observed and hypothetical arrays of measures shall be equal.

There is still too much freedom for values of F_t, if a normal curve is being fitted. The "scatter" of the hypothetical measures must also be restricted. In symbols,

$$\Sigma F_o X^2 = \Sigma F_t X^2, \tag{23}$$

or

$$\Sigma X^2(F_o - F_t) = 0. \tag{23a}$$

These equations, in effect, state that the variances of the observed and hypothetical arrays shall be the same.

The three restrictions represented by equations (21), (22), and (23) are imposed when a normal curve is fitted to a set of observed frequencies.[28] The equations are all linear *and* homogeneous in $(F_o - F_t)$. This is best seen from a comparison of equations (21a), (22a), and (23a).

---

[28] One exception to this generalization should be mentioned. If values of the hypothetical mean and hypothetical variance are not estimated from the empirical data but come from some extraneous source, the only restriction is that $\Sigma F_o = \Sigma F_t$.

CONCLUSIONS

Most readers will by now have correctly concluded that the chi-square test has a restricted usefulness. However, it usually cannot be replaced in those situations where it is applicable and it thus stands as a valuable research tool. Perhaps the chief trouble is that the test is too often applied without adequate prior planning; it is frequently "hit upon" and adopted after data have been collected and sometimes after other techniques of statistical analysis have been found unproductive. The aim of every investigator should be to plan, in advance, not only every detail of every experiment but every step in the analysis of the anticipated data. All contingencies cannot be foreseen; but if the chi-square test is to be employed, there is no good reason for failing to provide for independence among the measures and for frequencies of adequate size.

There should seldom if ever be any compromising on the requirement of independence.[29] There should usually be no compromising on the size of frequencies. There are occasions, of course, when it is very time-consuming and perhaps very expensive to add more cases to a mere handful. The best procedure under such circumstances is to try for an experimental design which utilizes each subject to the limit and leads to an analysis of data on an individual rather than a group basis. If it turns out that only a few subjects can be studied and the data on each one cannot be analyzed separately, it may still be possible to find a method of analysis which is more exact than the chi-square test. For example, if the data can be arranged into a 2×2 table and the individual cell frequencies are less than 10, the *exact treatment* proposed by Fisher (1946, pp. 96–97) is to be preferred to the chi-square test. The treatment is rather tedious to apply, but in view of its exactness there is no adequate excuse for avoiding it.

Many users and would-be users of the chi-square test gain erroneous impressions from what they read about limitations on the size of theoretical frequencies. A textbook says that frequencies of less than 10 are to be avoided. This statement is often interpreted to mean not that 10 is a limiting value to be exceeded whenever possible, but that 10 is a value around which the various theoretical frequencies may fall; and if an occasional frequency happens to be as low as 4 or 5, that is all right because other frequencies will be larger than 10 and everything will average out in the end. A textbook that gives 5 as the suggested

---

[29] Any investigator who applies the chi-square test to interdependent frequency data should always feel obligated to include, in published accounts of his findings, a full explanation of the procedures employed and a justification of them.

minimum tends to encourage the retention of impossibly small theoretical frequencies. And so does a test which states, in effect, that Yates' correction for continuity should be applied if the cell frequencies are 5 or less and precision is desired. This implies not only that frequencies of less than 5 are quite acceptable, but also that Yates' correction is an antidote for small frequencies. Both implications are fallacious.

The following excerpts from Yule and Kendall (1940, p. 422) may help to dispel false notions concerning the size of theoretical frequencies and also concerning the size of N:

> In the first place, N must be reasonably large. . . . . It is difficult to say exactly what constitutes largeness, but as an arbitrary figure we may say that N should be at least 50, however few the number of cells.
>
> No theoretical cell frequency should be small. Here again it is hard to say what constitutes smallness, but 5 should be regarded as the very minimum, and 10 is better.

Hoel (1947, p. 191), while giving 5 instead of 10 as the recommended minimal value of F_t, nevertheless emphasizes the importance of having a fairly large value of N by stating that if the number of cells or categories is less than 5, the individual theoretical frequencies should be larger than 5. Cramér (1946, p. 420 f.) firmly recommends a minimal value of 10 and says that if the number of observations is so few that the theoretical frequencies, even after grouping, are not greater than 10, the chi-square test should not be applied. In all but one of the illustrations used by Cramér, the theoretical frequencies are considerably larger than 10, and in the exceptional case, he admits (p. 440) that the frequencies are smaller "than is usually advisable." An investigator handicaps himself whenever he applies the chi-square test in relation to small theoretical frequencies.

There are a few applications of the χ^2 test which have not been described and illustrated in the present survey, either because they are quite specialized in character or because they provide only approximate solutions. One of the specialized applications which may be of interest to some readers is Bartlett's test of the homogeneity of variance (1937). Those who do not have access to Bartlett's original discussion will find a description of the test in Snedecor (1946, pp. 249-251). Another specialized application of interest is the use of χ^2 in setting the confidence limits for a population variance from a known sample variance. The procedure is explained by Hoel (1947, pp. 138-140) and need not be included here.

In general, any suggested applications of χ^2 which deviate from the well-established tests should be avoided except by those qualified to evaluate their full import or upon the advice of an expert.

References

Anastasi, A., & Foley, J. P., Jr. An experimental study of the drawing behavior of adult psychotics in comparison with that of a normal control group. *Journal of Experimental Psychology*, 1944, **34**, 169–194.

Arnold, M. B. Emotional factors in experimental neuroses. *Journal of Experimental Psychology*, 1944, **34**, 257–281.

Bartlett, M. S. Properties of sufficiency and statistical tests. *Proceedings of the Royal Society*, 1937, A **160**, 268–282.

Chen, H. P., & Irwin, O. C. Development of speech during infancy: Curve of differential percentage indices. *Journal of Experimental Psychology*, 1946, **36**, 522–525.

Cramér, M. *Mathematical methods of statistics*. Princeton: Princeton University Press, 1946.

Edwards, A. L. *Statistical analysis for students in psychology and education*. New York: Rinehart, 1946.

Fisher, R. A. *Statistical methods for research workers* (10th Ed.). Edinburgh: Oliver and Boyd, 1946.

Fry, T. C. The chi-square test of significance. *Journal of the American Statistical Association*, 1938, **33**, 513–525.

Goodfellow, L. D. A psychological interpretation of the results of the Zenith Radio experiments in telepathy. *Journal of Experimental Psychology*, 1938, **23**, 601–632.

Grant, D. A., & Norris, E. B. Dark adaptation as a factor in the sensitization of the beta response of the eyelid to light. *Journal of Experimental Psychology*, 1946, **36**, 390–397.

Greenhood, E. R., Jr. *A detailed proof of the chi-square test of goodness of fit*. Cambridge: Harvard University Press, 1940.

Guilford, J. P. *Psychometric methods*. New York: McGraw-Hill, 1936.

Guilford, J. P. *Fundamental statistics in psychology and education*. New York: McGraw-Hill, 1942.

Hoel, P. G. *Introduction to mathematical statistics*. New York: Wiley, 1947.

Holt, R. B. Level of aspiration: Ambition or defense? *Journal of Experimental Psychology*, 1946, **36**, 398–416.

Irwin, O. C., & Chen, H. P. Development of speech during infancy: Curve of phonemic types. *Journal of Experimental Psychology*, 1946, **36**, 431–436.

King, H. E., Landis, C., & Zubin, J. Visual subliminal perception where a figure is obscured by the illumination of the ground. *Journal of Experimental Psychology*, 1944, **34**, 60–69.

Kuenne, M. R. Experimental investigation of the relation of language to transposition behavior in young children. *Journal of Experimental Psychology*, 1946, **36**, 471–490.

Lewis, D. *Quantitative methods in psychology*. Ann Arbor: Edwards Brothers, 1948.

Lewis, H. B. An experimental study of the role of the ego in work. I. The role of the ego in cooperative work. *Journal of Experimental Psychology,* 1944, **34**, 113–126.

Lewis, H. B., & Franklin, M. An experimental study of the role of the ego in work. II. The significance of task-orientation in work. *Journal of Experimental Psychology,* 1944, **34**, 195–215.

Lindquist, E. F. *Summary report of results.* Iowa City: Sixth Annual Every-Pupil Testing Program, 1934.

Pearson, K. Experimental discussion of the (x^2, p) test for goodness of fit. *Biometrika,* 1932, **24**, 351–381.

Peatman, J. G. *Descriptive and sampling statistics.* New York: Harper, 1947.

Peters, C. C., & Van Voorhis, W. R. *Statistical procedures and their mathematical bases.* New York: McGraw-Hill, 1940.

Pronko, N. H. An exploratory investigation of language by means of oscillographic and reaction time techniques. *Journal of Experimental Psychology,* 1945, **35**, 433–458.

Seward, J. P., Dill, J. B., & Holland, M. A. Guthrie's theory of learning: A second experiment. *Journal of Experimental Psychology,* 1944, **34**, 227–238.

Shaw, F. J., & Spooner, A. Selective forgetting when the subject is not "ego-involved." *Journal of Experimental Psychology,* 1945, **35**, 242–247.

Smith, S. The essential stimuli in stereoscopic depth perception. *Journal of Experimental Psychology,* 1946, **36**, 518–521.

Snedecor, G. W. *Statistical methods* (4th Ed.). Ames, Iowa: Iowa State College Press, 1946.

Voelker, C. H. Phonetic distribution in formal American pronunciation. *Journal of the Acoustical Society of America,* 1934, **5**, 242–246.

Yates, F. Contingency tables involving small numbers and the x^2 test. *Supplement, Journal of the Royal Statistical Society,* 1934, **1**, 217–235.

Yule, G. U., & Kendall, M. G. *An introduction to the theory of statistics* (12th Ed.). London: Griffin, 1940.

2.2

The misuse of chi-square—
a reply to Lewis and Burke

Charles C. Peters

In the November, 1949, issue of the *Psychological Bulletin*,[1] Lewis and Burke discuss "The Use and Misuse of the Chi-square Test" in a long article in which they allege misuses in all but three out of 14 articles, and charge the Peters and Van Voorhis textbook with six misses out of seven tries. In our case every one of these six criticisms is invalid, although they could have made, but did not, a seventh criticism which would have been valid. If the same distribution of validities holds for the other writers (a possibility that I have not investigated), the situation may not be so bad as they make out. Lewis and Burke play up only some partial applications of chi-square at the expense of more fundamental and general considerations, thus frequently overshooting the mark in their generalizations. They make no mention of Karl Pearson's 1900 article in the London-Edinburgh-Dublin *Philosophical Magazine* (Vol. 50, pages 157–175), in which chi-square was first announced; yet no one can expect to have a basic understanding of this statistic without a thorough reading of this article and of Pearson's follow-up articles in volumes 14, 24 and 26 of *Biometrika*. We shall take up in turn some wrong allegations in the Lewis and Burke article.

1. That chi-square applies fundamentally only to frequencies. It is more general than that. The starting point for the derivation of the distribution function of chi and chi-square, which is always assumed for the "makings" of these statistics, is the normal distribution function for standard scores:

$$df = \frac{1}{\sqrt{2\pi}}\, e^{-z^2/2} dz \tag{1}$$

Charles C. Peters, "The Misuse of Chi-Square—A Reply to Lewis and Burke," *Psychological Bulletin,* **47**, 1950, 331–337. Copyright (1950) by the American Psychological Association, and reproduced by permission.

[1] Lewis, Don, & Burke, C. J. The use and misuse of the chi-square test. *Psychol. Bull.,* 1949, 46, 433–489.

By methods recently available, the distribution function for chi and for chi-square come rather easily from this normal distribution function. They are:

$$df(\chi) = \frac{1}{\dfrac{n-2}{2}! \, 2^{(n-2)/2}} \chi^{n-1} e^{-x^2/2} d(\chi) \tag{2}$$

$$df(\chi^2) = \frac{1}{\dfrac{n-2}{2}! \, 2^{n/2}} (\chi^2)^{(n-2)/2} e^{-x^2/2} d(\chi^2) \tag{3}$$

where n is the number of degrees of freedom.[2]

In this derivation no conditions whatever are imposed about the nature of z except that it shall be a deviation from a true value, that the z's shall be normally distributed about this central value in the population, and that the deviations of which they are composed be divided by the true variance. The deviations may be individual variates or frequencies or means or any other statistics, provided only that they are normally distributed in the population, are from a true central value, and are divided by the true variance. However, aside from the case of frequencies, we can not often know the true variance and (somewhat less rarely) the true central value; but sometimes we can.

2. That the sum of the theoretical frequencies must always equal the sum of the observed frequencies, and that our example reproduced in Table 1 of their article violates this principle. The generalized definition of chi-square is:

$$\chi^2 = \sum \frac{(\mu - x)^2}{\tilde{\sigma}^2} \tag{4}$$

where μ is the true central value and the tilde over the σ means that it is the true standard deviation. The true variance for the frequency in a cell is $N\tilde{p}\tilde{q}$, where $\tilde{p}$ is the true proportion belonging in the cell and $\tilde{q}$ equals $(1 - \tilde{p})$. So the generalized formula in the case of frequencies would be:

$$\chi^2 = \sum \frac{(f_t - f_o)^2}{N\tilde{p}\tilde{q}}. \tag{5}$$

This involves nothing whatever about what the frequencies shall sum to. In a setup such as ours of Table 1 in the Lewis and Burke article the f_o could be made to sum to equality with the f_t only by artificial doctoring, except in very rare cases. Yet, so long as the "cramping"

---

[2] Lewis and Burke do not give correctly the distribution function for chi-square. In their terminology, the exponent of chi-square should be $(r-2)/2$ instead of $(r+2)/2$.

features about to be discussed are not present, (5) is the basic formula to apply to get the correct chi-square.

But Pearson showed, in the original article referred to above, that *if* the condition is imposed upon a universe of samples that the total N shall always be the same, and the same for theoretical and observed frequencies so that the sum of the errors is zero, then a system of inter-correlations is introduced of which account must be taken in getting the values for chi-square. For under these restrictions, an excess of frequencies in one cell must necessarily be accompanied by a deficiency in others. He set up the system of intercorrelations for this condition along with the squared deviations, worked them through an elaborate and highly technical solution by determinants, and emerged with the now well-known formula:

$$\chi^2 = \sum \frac{(f_t - f_o)^2}{\tilde{p}N} = \sum \frac{(f_t - f_o)^2}{f_t}. \tag{6}$$

This is the formula that *must* be used when the limitation is imposed that the frequencies in a set shall always sum to N so that intercorrelations are involved; but, just as unequivocally, it must *not* be used otherwise.[3]

3. *That the frequencies of "non-occurrence" must always be present.* There is by no means always such necessity; it depends upon the nature of one's problem. One of the studies to which violation of this is charged is that by Grant and Norris, in which these investigators fitted a logarithmic curve and tested the agreement of the theoretical frequencies distributed to categories under it with the same aggregate of frequencies distributed according to observation. They applied the chi-square test in the customary manner for such frequency distribution testing. In such a procedure it would be entirely meaningless to set up for each category the number *not* in the category and compute chi-square elements for this second set of deviations to add to the ones conventionally used.

In their footnote on p. 435, Lewis and Burke also charge us with violation of this principle. In our example, statistics were given for the number of Italians, Russians, Poles, and Others who had been naturalized out of totals for those nationals in a community. The p was inferred from the sample, and the distribution tested for possible nationality differences. The chi-square was computed from only the row of "successes" by formula (6). We made the customary assumption

---

[3] We made a slip in using the customary formula, (6), in connection with our dice throwing example when we should have used formula (5). If formula (5) is used, chi-square is 12 with the same number of degrees of freedom. If Lewis and Burke had pointed this out, it would have been a valid criticism—the only one I acknowledge as legitimate.

that the total for the row is to be constant from sample to sample, and our generalization was about a universe of samples of which that description would hold. For our assumption of a fixed marginal total for the row, our procedure was in exact conformity with the conditions for which Pearson developed formula (6): a fixed N (the total number of naturalized) in a series of samples, f_t equal to f_o so that the sum of the errors in each sample is zero, a shifting incidence of the fixed number of naturalizations to the different nationalities in successive samples with the inevitable intercorrelations that would result, deviations of the frequencies in each cell from a theoretical number, and a test whether the deviations in this sample exceed those which could fall within the credible limits of sampling fluctuation. The degrees of freedom for this universe of samples are 3 and the p is .0024.

To install in this table a row of "non-occurrences" (the non-naturalized) would be essentially to establish a series of ratios, one for each nationality group, apply formula (6) to each with N being the number in that nationality group to get a chi-square for it, then sum across these two-cell columns for a composite chi-square for testing the homogeneity of these ratios. Karl Pearson says (*Biometrika*, Vol. 14, p. 418, and Vol. 24, p. 353) that a table thus set up is not a true contingency table to which the chi-square test is appropriate. At any rate, it automatically freezes the N_c's for the universe of samples, which is sometimes a natural and therefore desirable feature and sometimes is not. Instead of following blindly a rule regarding the nature of the universe about which one shall generalize, the researcher should make a choice of such universe according to what is most meaningful in his particular problem, then say what his universe is. In our naturalization problem, for example, we can state what the probability is that as great discrepancies in ratio as we found in our sample would arise merely by sampling fluctuation in a universe of samples of size N taken at random from the same or similar total communities; that $P(df = 7)$ would be .000138. Or we could state what would be the probability of the obtained discrepancies in a sub-universe of samples with always exactly the same numbers of Poles, Russians, etc. and exactly the same aggregate number of naturalizations and non-naturalizations; that $P(df = 3)$ is .000002. There is a whale of a difference, although here both, of course, are statistically highly significant refutations of the null hypothesis. In cases like this one the former generalization would seem to be much the more natural and meaningful. It requires, of course, that the values obtained from the sample be taken as acceptable estimates of the true values; but Pearson argued, very persistently and to this writer very convincingly, that these estimates are quite satisfactory and that the gain from using them is greater than the loss; and he supports his arguments with experimental evidence. An awkward and stilted generalization is too great a price to pay for excessive finicalness about true values. However, the fourfold-table (if used) is an exception, if one wishes outcomes that are comparable with those of differences between proportions or of tetrachoric correlations; for the standard error formulas of both of these statistics assume fixed marginal totals.

4. That chi-squares may not be summed when the same set of individuals is used in the series, because the sets are not independent; and that this principle is violated in our dice throwing example of their Table 1. That idea rests upon a misunderstanding of the nature of independence. When you throw a set of dice, they fall according to some law actually *within* them, within the limits of sampling fluctuation involved in that law. You also know the credible limits in the distribution of samples according to some *external* law which you suspect may be the law operating within the sample. You apply a chi-square test to see whether the behavior of the sample lies within the credible limits according to the external law. If it does not, you reject the hypothesis that your external law may be the law operating in the sample. You have made this test for one cell. Then you make a second throw and in it the dice behave again according to the law within them, quite independently of their behavior during the first throw. And so on with successive throws. The reason for summing these is merely to increase the reliability of the test and thus get a more dependable generalization for *these* dice. To do anything else than use the same dice, or the same subjects, would rob your investigation of all meaning for testing that hypothesis. But if on a second throw you picked up only *some* of the dice while allowing the others to lie and be counted in the next throw as they lay in the former throw, the sets would not then be independent. So if you are intending to generalize about the bias of *these* dice, or *these* subjects in a guess-heads or a choice-of-values experiment, it is correct to sum successive throws of the same dice, or successive responses of the same subjects. But if your purpose is to generalize about a wider universe of dice or of subjects, then a new sample of that wider universe must be chosen on each successive trial. Then you can generalize about the wider universe from which the samples have been randomly drawn. Our generalization was specifically about *these* dice.

With dice this procedure could scarcely go awry. But with human or animal subjects, if something happened between trials that might affect the next ones (as information about success), the chi-squares could not be meaningfully summed because we would have on the different trials not further samples from the same universe but each time a sample from a different universe. To study thus in a series of summed trials the bias of just *these* dice or *these* students may not be a very useful form of research, because the universe is too narrow to be, ordinarily, of much social importance. But it is sound statistically if one's purpose requires it.[4]

5. That the view that the number of degrees of freedom appropriate in testing the fit of a normal curve depends upon the hypothesis to be tested is erroneously attributed to Karl Pearson. It is not erroneously attributed to Pearson; it is correctly attributed to him, as a reading of *Biometrika*, Vol. 24, pp. 359 and 361 will show.

6. That certain writers, including us, recommend a chi-square test of goodness of fit of regression lines which is a rather poor one. This should have been attributed to R. A. Fisher, not to either of the two writers to whom reference is made. We merely passed on Fisher's scheme, though unfortunately with our blessing. But in a footnote on the same page we referred the reader to "a better test of the goodness of fit of regression lines" in a section of our book where a test is developed in terms of the unbiased correlation ratio, which is a statistic closely related to F.

Much of the difficulty with chi-square arises from a frenzy to extend its use into areas where it is not needed, because we have for them better statistics. All applications of chi-square with one degree of freedom should be cleared out with one sweep of the broom. For with one degree of freedom the distribution of chi is normal, so that nothing remains at that level that is distinctive for chi-square. Lewis and Burke say that for one *df* the square-root of chi-square (which is chi) is distributed as Student's t with N equal infinity. That is true because the distribution of Student's t with N equal to infinity *is* normal; but that fact does not follow from the state distribution function of Student's t, because that contains no χ. But it does follow easily from our formula (2); by substituting 1 for n and remembering that χ^0 equals 1 and that $(-\frac{1}{2})!$ (which equals gamma $\frac{1}{2}$) equals $\sqrt{\pi}$, formula (2) reduces to formula (1) (with, however, 2 in the numerator instead of 1, which does not affect the *shape* of the distribution and which compensates for the fact that chi's run only from zero to plus infinity instead of from minus infinity to plus infinity). If one applies in a four-fold table the well-known technique of a difference of proportions divided by the conventional standard error of a difference of proportions and goes to the normal curve for the interpretation of the probability, he will get the same result as by the chi-square technique to the hundredth decimal

---

[4] The example in our dice throwing of Table 1 was not, of course, the report of a research. It was intended merely as a schematic example for illustrating the meaning of deviations from a true value, worked up in the conventional manner. excessively simplified for pedagogical purposes. The objection that it would be a poor research, and that the frequencies are too small, seems under the circumstances to be picayune.

place—provided he is testing the same hypothesis and remembers that there belong in the standard error formula the "true" values of p and q. Even amateur statisticians know that you may not carry ratios from proportion to the normal curve and claim exactness for your probability in researches with such N's as we ever get in practice. But these same persons will go blithely and confidently ahead with chi-square, not knowing that none of the chi-square table entries hold strictly for N's short of infinity, and that this principle is violated just as badly in chi-square as it is in proportion.[5]

Tetrachoric correlation, which can be applied to any four-fold table where the variates make a continuous distribution, also has a constructive meaning which shows the nature and the strength of the law that is present (while chi-square shows only whether *some* law is present without indicating either its nature or its strength). Tetrachoric r is, by the cosine-pi formula, as easily computed as chi-square; it assumes a normal distribution only over the table as a whole, not in the separate cells; and it has a good and reasonably simple test of reliability.

There are also available now more convenient tests than chi-square for testing the goodness of fit of the normal curve (because simpler), and for testing the goodness of fit of all regression lines. There remains little more for chi-square than the test of multiple contingency tables with greater complexity than four-fold. If research workers would not, in a mad rush to use chi-square, raise an unnecessary dust, they would not need to complain that they can not see.

This note has involved criticism of six allegations made by Lewis and Burke in their article. But, in spite of these particular criticisms and possibly some others that might have been made, their article contains many sound and useful ideas, and serves a good purpose. But it should be read with critical alertness, not taken offhand as gospel.

---

[5] There is much talk about what minimum number of frequencies is required in a cell before using it in a chi-square calculation: whether five or ten or twenty, or what. The number required for *exactness* in fulfilling the assumptions is infinite; for *any* number short of infinity the determination is only rough, but good enough. In the binomial $(p + q)^N$, where $p + q$ equals 1, it is only as N (the frequency in the sample) approaches infinity that the distribution approaches the normality assumed for chi-square. When p (the proportion of the total N of the whole sample that belongs in the cell) and q are unequal, the distribution remains markedly skew far beyond the number usually mentioned as the minimum; but for any *definite p* it always approaches normality (around p) as the N of its exponent approaches infinity (and hence as the f in each cell approaches infinity). The damaging fact is not so much the intrinsic smallness of the f_t of the cell as it is the smallness of the p representing the proportion of the total number in the sample that belongs in the cell.

2.3

Some comments on "the use and misuse of the chi-square test"

Nicholas Pastore

Lewis and Burke in their recent article in this journal[1] rightly call attention to the necessity for familiarity and understanding of the assumptions underlying the chi-square test. It is unfortunate, however, that some of the crucial points in this rather lengthy article are either incorrect or confusing. The purpose of this comment is to indicate some of these incorrect or confusing statements.

1. There is a typographical error in the statement of the equation of the distribution function of chi-square (p. 439, equation 8). The numerator of the exponent of χ^2 should be $(r-2)$ instead of $(r+2)$.

2. On p. 434f. Lewis and Burke discuss a number of errors which were made by Peters and Van Voorhis in illustrating a chi-square problem. In this discussion Lewis and Burke fail to present or indicate a possibly superior method for dealing with the problem. Such a method would involve the determination of the proportion of the fourteen throws which yield exactly 0, 1, 2, . . . , 12 aces. The proportions are the successive terms of the expansion of $14(q+p)^{12}$, where $p = 1/6$ and $q = 5/6$. The number of degrees of freedom would then be 11. (In the Peters and Van Voorhis example, however, the chi-square test is not applicable, as Lewis and Burke themselves point out, because the theoretical frequencies are too small.)

3. In connection with the previous example, Lewis and Burke raise another objection to the procedure followed by Peters and Van Voorhis: ". . . the observed frequencies lack independence. They lack indepen-

Nicholas Pastore, "Some Comments on 'The Use and Misuse of the Chi-Square Test'," *Psychological Bulletin,* **47**, 1950, 338–340. Copyright (1950) by the American Psychological Association, and reproduced by permission.

[1] Lewis, Don, & Burke, C. J. The use and misuse of the chi-square test. *Psychol. Bull.* 1949, **46**, 433–489.

dence because the same twelve dice were thrown each time" (p. 434). This is an irrelevant objection. The usual assumption is that the wear and tear on a set of twelve dice are unrelated to the probability of the specified event. Furthermore, if the purpose is to determine whether a particular set of dice is biased, the thing to do is to experiment with the given set of dice. Lewis and Burke imply that the set of dice should be changed with each successive throw.

4. On pp. 441–443 the authors present and discuss a presumably "incorrect" application of the additive property of chi-square. Actually the application of the additive property of chi-square to this example is correct.

Ninety-six students were each asked to guess the fall of a coin five times. The actual turn of the coin was never revealed. "The hypothesis under test is that the guesses of students *on five successive tosses* of a coin are purely chance occurrences, with the probability of a guess of heads (by any student on any toss) equal to the probability of a guess of tails" (p. 442).[2] In other words, the authors propose to study the response-tendencies of students in order to determine whether such response-tendencies represent the same type of occurrence as tossing a set of five unbiased coins 96 times. It is assumed that each set of coins is properly shuffled before each toss. On the basis of the hypothesis, and the assumed independence of each toss, it is possible to calculate the expected number of heads for each set of 96 tosses. Since the coin is assumed to be unbiased the expected number of heads is 48. On the basis of this theoretical picture it is possible to calculate the probability of obtaining either a set of observed frequencies, or a set of more divergent frequencies. This probability can be calculated for each of the five coins. The authors calculate the chi-square value for each coin and then determine the appropriate probability with one degree of freedom. (See Table 3, p. 442). The relevant question is whether the five values of chi-square can be added, as well as the degrees of freedom, in order to determine the over-all probability for the set of five chi-squares. The authors claim that this addition is incorrect because "between tosses, the five guesses of each student were undoubtedly interrelated" (p. 443). Actually, however, the addition is quite permissible. It is correct that the additive property of chi-square assumes that the individual chi-squares are independent. This assumption correctly characterizes the theoretical model against which the observed frequencies are compared. If the probability corresponding to the sum of the chi-squares is small (less than five per cent or less than one per cent) then we may infer either that an unusual event has occurred, or that the hypothesis underlying the theoretical model is incorrect, or that some condition underlying the chi-square test was violated.

[2] Author's italics.

Although the authors deny the applicability of the additive property of chi-square to the particular example, they do affirm the applicability of chi-square to the first toss of the coin because "on a single toss, the guess of each student was independent of the guess of the other students" (p. 443). Following the thinking of the authors, the chi-square test should not be even applicable to the first toss because the guess of one student is not necessarily independent of the guesses of the other students. There could be a general cultural bias in favor of heads on the first toss. In addition, in the same sense in which the second guess of a student may be related to his first, the first guess may be related to some psychological tendency of the student.

5. On pp. 478–480 the authors present an example in detail in order to clarify the meaning of "independence between measures." The discussion of this example is quite confusing.

Each of 240 students was asked to guess the fall of four successive tosses of a coin. The authors use the binomial formula in order to determine the probability distribution of the number of hits. It is assumed that the response of a subject is just as likely to be "head" as "tail." Since the probability of a hit is $\frac{1}{2}$, the theoretical number of 0, 1, 2, 3, 4 hits can be calculated. This distribution is then compared with the observed distribution of hits. The value of chi-square in this problem is equal to 43.351, with four degrees of freedom (see Table 20, p. 479). Since the probability associated with this value of chi-square indicates the occurrence of a very unusual event, one would expect that the authors would reject the hypothesis. Since the original hypothesis was that the probability of a hit was equal to $\frac{1}{2}$, an alternative hypothesis would be one for which the probability of a hit is greater than $\frac{1}{2}$. The practical inference is that either the individuals responded with some knowledge of the actual fall of the coin or that the average response patterns of the subjects happened to coincide with the fixed pattern of the four coins ($H\ T\ T\ H$).

Rather than reject the hypothesis or question the associated conditions, the authors come to the remarkable conclusion that "The absence of independence and the consequent inability to obtain unequivocal theoretical (chance) frequencies made this application of the chi-square test a meaningless procedure" (p. 480). By the phrase "absence of independence" the authors mean that there is some psychological linkage between the responses of a subject to the four successive tosses of a coin. Thus it is that the authors reject the application of the binomial formula in his particular problem (p. 479). It should be noted, however, that the application of the binomial formula assumes independence in a statistical sense, viz., that the joint probability of two random events is equal to the product of the respective probabilities of the two events. The binomial formula can be correctly applied to the problem on hand because the probability of getting a hit is not affected by the psychological tendencies of the subject. Of course, since the pattern of the four coins was fixed, $H\ T\ T\ H$, it could happen that psychological tendencies of the subject may produce a disproportionate number of hits. To obviate this factor it is necessary to toss the four coins for each trial. The thing which is at fault, if there is any fault, in this application is not the binomial formula but the structure of the experiment.

6. In a criticism of a problem cited from the literature, Lewis and Burke fail to mention the significant point that the sum of the probabilities is greater than unity (viz., 1.7), whereas the sum of the probabilities must be equal to unity. Moreover, the authors' own alternative analysis of the same problem contains the same error, the sum of the probabilities being somewhat larger than 1.8.

2.4

On "the use and misuse of the chi-square test" —the case of the 2x2 contingency table

Allen L. Edwards

In their recent article in this journal, Lewis and Burke (1949) provide an excellent guide for the research worker in psychology in terms of what to do and what not to do in applying the χ^2 test of significance. It is, of course, obvious that there are conditions which will invalidate any test of significance. In the case of the χ^2 test, authorities tend to agree that one of these conditions is the presence of small theoretical frequencies. That Lewis and Burke are seriously concerned about what they believe to be violations of this principle by psychologists is indicated by the fact that warnings against the use of small theoretical frequencies appear through their paper. For example:

The use of "small theoretical frequencies" is listed as one of the "principal sources of error" in applications of the χ^2 test by psychologists in articles published in the *Journal of Experimental Psychology* over a three-year period (1949, pp. 433–434).

"The commonest weakness in applications of the χ^2 test to contingency tables is the use of extremely small theoretical frequencies" (1949, p. 460).

If we have but a single *df*, as in the case of the 2×2 contingency table, "the use of theoretical frequencies of less than 10 should be strictly avoided" (1949, p. 460).

"Whenever small theoretical frequencies enter into calculations of χ^2, the experimenter has no sound basis either for accepting or rejecting a hypothesis except when the value (of χ^2) is quite extreme" (1949, p. 462).

Allen L. Edwards, On "The Use and Misuse of Chi-Square Test"—The Case of the 2×2 Contingency Table, *Psychological Bulletin*, **47**, 1950, 341–346. Copyright (1950) by the American Psychological Association, and reproduced by permission.

Several applications of the χ^2 test by Lewis and Franklin (1944) are judged not acceptable because "they involved extremely small theoretical frequencies" (1949, p. 433).

In this paper we shall be concerned only with applications of the χ^2 test to the 2×2 contingency table. More specifically, we shall be concerned with what constitutes a small theoretical frequency in this application.

1. There is no quarrel with the idea that a large number of well-controlled observations is more satisfying than a small number of well-controlled observations. However, if only a limited number of well-controlled observations can be made, this may be more satisfying than a large number of poorly-controlled observations. At the same time, with other conditions remaining constant, a small number of observations will yield smaller theoretical frequencies than a large number of observations. Let us grant, then, that a large N is a good thing, if observations are well-controlled, and that we like to have it when possible. This, in turn, means that we like to avoid small theoretical frequencies. There is no disagreement on this point.

Disagreement does occur when we ask the pertinent question: How large must a theoretical frequency be before it is considered not small? Or, to turn the question around: How small may a theoretical frequency be before it is considered not large? The problem of adequately defining smallness, in the case of the χ^2 test, is, as Yule and Kendall (1947, p. 422) have pointed out, not a simple one.

Cramér (1946), according to Lewis and Burke, "firmly recommends a minimal value of 10" (1949, p. 487), and Lewis and Burke are quite explicit in stating that in their opinion: "A value of 5 is believed to be too low" (1949, p. 462). Yule and Kendall (1949, p. 422) take a more moderate view. They regard 5 as a minimal value, but add that 10 is better. Hoel (1947, p. 191) suggests that 5 is satisfactory if the number of cells or categories is equal to or greater than 5. But this is not the case with the 2×2 contingency table and, in this instance, Hoel suggests that it is better to have theoretical frequencies "somewhat greater than 5" (1947, p. 191). Fisher (1936, p. 87, 97), on the other hand, seems content with the rule of 5.

The position taken in this paper is that psychologists (and other research workers) are not necessarily "misusing" the χ^2 test when they apply it to the 2×2 contingency table when a theoretical cell frequency is as low as 5.

2. The exact probability for any set of frequencies in a 2×2 table can be obtained by direct methods of calculation as given by Fisher (1936, pp. 100–102). At the same time, it is important to emphasize, as Fisher has, that the χ^2 test applied to the same data "can only be of approximate accuracy" (1936, p. 97), and that its usefulness lies in "the comparative simplicity of the calculations" (1936, p. 100).

The "approximate accuracy" of the χ^2 test can be judged by the extent to which the probability associated with the obtained value of χ^2 is in accord with the obtained by the direct method. Agreement between the two values of P is likely to be best when theoretical frequencies are not small. The reasons for this are well stated by Lewis and Burke (1949, pp. 439–440). This does not mean, however, that χ^2 is not useful for its "approximate accuracy" is estimating the exact probabilities in 2×2 tables involving theoretical frequencies as small as 5. Nor is the application of the χ^2 test in this instance a "misuse" of χ^2.

Consistently rejecting a null hypothesis when P is .05 or less and accepting the hypothesis when P is greater than .05 is a practice followed by many research workers. It is suggested here that the research worker will not be led badly astray in evaluating the 2×2 contingency table with theoretical frequencies as small as 5 (1) if he consistently follows the P equal to or less than .05 standard (or some other standard which is acceptable); (2) if his attitude toward the null hypothesis is based upon the value of χ^2 corrected for continuity; and (3) if he calculates the exact probabilities for those cases where the probability value obtained by the χ^2 test is of borderline significance.

In practice, as Fisher (1936, p. 83) states with his usual common sense attitude, our interest is not primarily in the exact value of P for any given hypothesis, but rather in whether or not the hypothesis is open to suspicion.

3. Let us examine the two sets of data given by Lewis and Burke involving a 2×2 table with small theoretical frequencies. Table 1, given below, is their Table 16, and is based upon the data of Lewis (1944) and Lewis and Franklin (1944). With respect to this table, Lewis and Burke state that "all four of the thoretical frequencies . . . are too small to warrant an application of the χ^2 test" (7, p. 461).

TABLE 1

Classification of Subjects Working Under Different Conditions on the Basis of the Ratio (RI/RC) of the Number of Interrupted Tasks Recalled to the Number of Completed Tasks Recalled*

RI/RC	Individual Work Condition	Cooperative Work Condition	Total
Greater than 0.67	6	12	18
Less than 0.67	6	2	8
Total	12	14	26

* This table corresponds to Table 16 in the Lewis and Burke article (1949, p. 461). The data presented are based upon experiments by Lewis (1944) and Lewis and Franklin (1944).

Introducing Yates' correction for continuity in one of the standard computational formulas for χ^2, we have

$$\chi^2 = \frac{N\left(bc - ad - \dfrac{N}{2}\right)^2}{(a + b)(a + c)(b + d)(c + d)},\tag{1}$$

where the letters a, b, c, and d correspond to the cell frequencies and N represents the total number of observations. Calculating χ^2 for the data of Table 1, we obtain a value of 2.374. And since χ, the square root of χ^2 with 1 df, is distributed as a normal deviate z, we find from the table of the normal curve that when $z = 1.54$, P is .0618. The corresponding value of P for χ^2 will be (2) (.0618) = .1236.[1] What is the "approximate accuracy" of this value in terms of the probability obtained by direct methods?

Assuming the constancy of the marginal totals, as is assumed in calculating χ^2 also, the probability of any observed set of frequencies is given by the product of the factorials of the four marginal totals, divided by the product of the factorials of the grand total and the four cell entries (3, pp. 100–102). In terms of a formula[2]

$$P = \left(\frac{(a + b)!(a + c)!(b + d)!(c + d)!}{N!}\right)\left(\frac{1}{a!b!c!d!}\right).\tag{2}$$

The desired probability, however, involves not only the arrangement of frequencies as given in Table 1, but all other possible arrangements (deviations) which are more extreme. Thus we need the probabilities for the following arrangements:

$$
\begin{array}{cc@{\qquad}cc@{\qquad}cc}
6 & 12 & 5 & 13 & 4 & 14 \\
6 & 2 & 7 & 1 & 8 & 0 \\
\end{array}
$$

The probability desired will then be the sum of the probabilities for the three arrangements. Direct calculation shows this to be .0612, a value which may be compared with the probability of 0.618 for χ. We also have (2)(.0162) = .1224 with which we may compare the probability of .1236 for χ^2. Not only would no conclusions concerning significance be changed, if χ^2 is calculated with the correction for continuity, but the "approximate accuracy" of the χ^2 test seems quite good in this particular instance.

Table 2 reproduces Kuenne's (1946) data as given by Lewis and

---

[1] The reason why we take $2P$ for the probability associated with χ^2 is discussed in Edwards (1950) and Goulden (1939).

[2] A table of the logarithms of factorials such as may be found in Pearson's tables (1914) or the *Mathematical Tables from Handbook of Chemistry and Physics* (1941) facilities the necessary calculations.

Burke in their Table 17 and for which they are also concerned about the application of the χ^2 test because of small theoretical frequencies. From formula (1) we find χ^2 to be 13.1585, and χ to be equal to 3.627. From the table of the normal curve, in the manner previously described, we find that the value of P corresponding to χ is .0001433, and for χ^2 we have (2) (.0001433) = .0002866. Direct calculation, in the manner described earlier, shows that the corresponding values of P are .00000985 and (2) (.0000985) = .0001970, respectively. Again no conclusions con-

TABLE 2

Classification of Subjects in Two Age Groups on the Basis of Those Showing Transposition and Non-Transposition*

Age Groups	Transposition	Non-Transposition	Total
3–4 years	3	15	18
5–6 years	20	6	26
Total	23	21	44

* This table corresponds to Table XVII in the Lewis and Burke article (1949, p. 462). The data are based upon an experiment by Kuenne (1946).

cerning significance would be changed, regardless of whether the data are evaluated by means of χ^2 or by the direct method. As to the "approximate accuracy" of the χ^2 test, it may be observed that the discrepancy between the two values of P for χ^2 and for the direct method is .0000896.

A few additional examples taken from Goulden (1939) and Yates (1934) illustrate the degree of "approximate accuracy" which χ^2 may give, even when theoretical frequencies are smaller than the value of 5. In these examples the probability obtained by the direct method has been given by the source from which the data are taken. χ^2 has in all cases been corrected for continuity. For simplicity, the value of P is given for χ and as obtained by the direct method. The value P for χ^2 and the corresponding value for the direct method may be obtained by doubling the reported values. Examples 1, 2, and 3 are real in that they are based upon actual data, and Example 4 is simply an exercise taken from Goulden.

Example 1—Hellman's data cited by Yates (1934).

	Normal Teeth	Malocclusion	Total
Breast-fed	4	16	20
Bottle-fed	1	21	22
Total	5	37	42

$\chi = 1.068$ $P = .1427$
Direct method: $P = .1435$

Example 2—Grant's data cited by Goulden (1939).

	Group Blood 0	Blood Group Not 0	Total
Fond du lac Indians	18	11	29
Chipewyan Indians	13	1	14
Total	31	12	43

$\chi = 1.75$: $P = .0401$
Direct method: $P = .0349$

Example 3—Mainland's data on the position of polar bodies in the ova of the ferret cited by Goulden (1939).

	Similar	Different	Total
10μ apart	5	1	6
More than 10μ apart	1	6	7
Total	6	7	13

$\chi = 1.93$: $P = .0268$
Direct method: $P = .025$

Example 4—Exercise given in Goulden (1939).

	Recovered	Died	Total
Animals inoculated	7	3	10
Animals not inoculated	3	9	12
Total	10	12	22

$\chi = 1.68$: $P = .0465$
Direct method: $P = .0456$

As will be apparent, each of the examples cited involves theoretical frequencies of less than 5, yet the degree of "approximate accuracy" of the χ^2 test seems satisfactory. We should not expect, however, that all 2×2 tables with theoretical frequencies as small as these will yield P's, when tested by means of the χ^2 test, which are as close to those obtained by the direct method.

4. It is to be emphasized that there is no magic involved in the rule of 5 as a minimal theoretical frequency—any more than there is in the rule of 10. It will be true, however, that if we calculate the value of P directly whenever the theoretical frequency is under 10, we shall be involved in roughly about twice as much labor as would be the case if we accept 5 as the minimal theoretical frequency.

The procedure suggested here is to apply the rule of 5 and to calculate χ^2, keeping in mind the correction for continuity. If the obtained value of χ^2, in such cases, is of borderline significance, then calculate the value of P by the direct method. If theoretical frequencies of less than 5 are involved, the desired probability may be found, not too inconveniently, by the direct method.

References

Cramér, H. *Mathematical methods of statistics.* Princeton: Princeton University Press, 1946.

Edwards, A. L. *Experimental design in psychological research.* New York: Holt, Rinehart & Winston, Inc., 1950.

Fisher, R. A. *Statistical methods for research workers* (6th Ed.). Edinburgh: Oliver and Boyd, 1936.

Golden, C. H. *Methods of statistical analysis.* New York: Wiley, 1939.

Hoel, P. G. *Introduction to mathematical statistics.* New York: Wiley, 1947.

Kuenne, M. R. Experimental investigation of the relation of language to transposition behavior in young children. *Journal of Experimental Psychology* 1946, **36**, 471–490.

Lewis, D., & Burke, C. J. The use and misuse of the chi-square test. *Psychological Bulletin*, 1949, **46**, 433–489.

Lewis, H. B. An experimental study of the role of the ego in work. I. The role of the ego in cooperative work. *Journal of Experimental Psychology*, 1944, **34**, 113–126.

Lewis, H. B., & Franklin, M. An experimental study of the role of the ego in work. II. The significance of task-orientation in work. *Journal of Experimental Psychology*, 1944, **34**, 195–215.

Pearson, K. *Tables for statisticians and biometricians.* Cambridge: Cambridge University Press, 1914.

Yates, F. Contingency tables involving small numbers and the χ^2 test. *Supplement, Journal of the Royal Statistical Society*, 1934, **1**, 217–235.

Yule, G. U., & Kendall, M. G. *An introduction to the theory of statistics.* (13th Ed.). London: Griffin, 1947.

Mathematical tables from handbook of chemistry and physics (7th Ed.) Cleveland: Chemical Rubber Publishing Co., 1941.

2.5

Further discussion of the use and misuse of the chi-square test

Don Lewis
C. J. Burke

The articles by Peters (1950), Pastore (1950), and Edwards (1950), objecting in one way or another to our paper on chi-square (1949), indicate either that the paper was not carefully studied or that our exposition was less clear than we had supposed. Since Peters and Pastore raise several points in common, we are forced to conclude that the fault was ours to a considerable extent. A clarification of the points in question will be attempted.

We should perhaps say, parenthetically, that we are confident we could have written a clear, concise paper in mathematical language which would have left little or no room for disagreement between Peters, Pastore, and Edwards, on the one hand, and ourselves, on the other. Because the paper was aimed at the average user of statistics among psychologists, we rejected a mathematical type of exposition in favor of proceeding largely by example. We are disappointed but not surprised to discover that our treatment has been found obscure in a few places.

REPLY TO PETERS

Before the writing of the paper was begun, we carefully surveyed the basic literature on chi-square, including several papers by Pearson, but could see no good reason for basing our discussion on the earliest proofs. Full credit is due Pearson for originating the chi-square test. His proofs are adequate for establishing the general validity of the statistic. However, advancements have been made in the mathematical

Don Lewis and C. J. Burke, "Further Discussion of the Use and Misuse of the Chi-Square Test," *Psychological Bulletin*, **47**, 1950, 347–355. Copyright (1949) by the American Psychological Association, and reproduced by permission.

methods applied to statistical problems; mathematicians have been able to improve upon many of the early proofs. Such recent treatises as Cramér's *Mathematical Methods of Statistics* (1946) provide a more complete background for the χ^2 distribution function and for the tests associated with it, and they serve as a suitable point of departure for any discussion of χ^2. In making this statement, we intend no disparagement of Pearson. He falls easily among the four or five most eminent statisticians. It is a commonplace in science that later investigators, often persons of lesser ability, using the insights of a great innovator, frequently improve upon his work.

The specific points made by Peters will be discussed in the order in which he presents them.

1. That we claim chi-square applies fundamentally only to frequencies. An examination of our equations (1), (2), and (8) and of the comments in footnote 8, page 439, and on page 487 will show that we do no such thing. Our equation (8) is the exact equivalent of Peter's equation (iii), except for the error in the exponent of χ^2 which Peters and Pastore both mention. Several persons have written since November calling our attention to this error, and we discovered it ourselves soon after it appeared in cold ineradicable print. We will be greatly indebted to anyone who can provide a foolproof method of avoiding errors of this kind.

What we do claim is that the formula

$$\chi^2 = \sum \frac{(F_o - F_t)^2}{F_t} \qquad (10)^1$$

can be legitimately applied only to raw frequency data. Applications involving frequencies are the ones of greatest interest to most psychologists, so it seems appropriate to emphasize these applications and to enumerate certain fundamental assumptions and restrictions through a discussion of the binomial and multinomial distribution functions. We are well aware of the fact (stated in our footnote 10) that the chi-square distribution function may be derived from the joint normal distribution function in n variables, and aware of the additional fact (stated several times in the paper) that the chi-square statistic plays an important role in the study of variances.

2 and 3. That the sum of the theoretical frequencies need not equal the sum of the observed frequencies, and that the frequencies of non-occurrence need not always enter into the calculations. These two points can best be discussed together, as they are intimately related.

Cramér (1946, p. 426 ff.) rigorously proves an underlying theorem on

---

[1] The formula carries this number in the original paper.

chi-square and indicates how the proof of a generalization of the theorem would be handled. The generalized theorem is broad enough to include as special cases all applications of chi-square which are based on formula (10). It encompasses the tests of fixed probabilities, tests of goodness of fit, and tests of independence, and shows clearly the conditions which must prevail for other tests of the same general kind. It includes the case covered by Pearson as well as the new derivation requested by Peters in his third footnote, and establishes the fact that *the number of degrees of freedom is the same in both cases.* In Cramér's proof, it is necessary to assume that the frequency associated with every possible outcome of the experiment is used and that the sum of the theoretical frequencies is made equal to the sum of the observed frequencies. It should be emphasized that these restricting assumptions hold only for applications of chi-square based on formula (10).

Peters states that he and Van Voorhis erred in one of their examples (1940, Table XXXV) by using his formula (vi) instead of (v). This is simply an alternative way of wording one of our objections. We show in our paper (1949, pp. 437–438) that using his formula (v) is exactly equivalent to using our forumla (10) and taking account of frequencies of non-occurrence after the sums of the observed and theoretical frequencies have been equated.[2] What Peters does, in effect, is to accept our criticism but insist that it be stated in another way.

Peters maintains that the application of chi-square which he and Van Voorhis made in relation to numbers of naturalized citizens among several nationality groups is a correct one. Since a careful examination of this application is instructive, we give in Table 1, Part A, observed frequencies (of occurrence and non-occurrence of naturalization) adapted from Peters and Van Voorhis' Table XXXVI. In Part B of the table are the theoretical frequencies as calculated by Peters and Van Voorhis on the assumption of homogeneity of nationality groups with respect to the proportion naturalized. Note that the marginal totals in the two parts of the table are identical. The theorem given by Cramér provides unimpeachable proof that the quantity obtained by using formula (10) and summing over all eight categories will have a distribution approximating the χ^2 distribution with 3 df provided that the hypothesis of homogeneity is true. In this example, as can be readily verified, the value of χ^2 based on the eight categories is **29.72**. Peters and Van Voorhis used formula (10) also, but excluded the "remainder" categories and summed over only the top four categories, to obtain a value of **14.52**.

---

[2] In this particular case by using (v) Peters obtains a correct value of 12, as he reports in his second footnote (1950, p. 332). Using equation (10), we obtained the correct value of 12 and so stated in our fourth footnote (1949, p. 435). This agreement is not a coincidence.

TABLE 1

Observed and Theoretical Frequencies from Table 36 in Peters and Van Voorhis (1940)

	A. Observed Frequencies				
	Italians	*Russians*	*Polish*	*Others*	*Totals*
Number Naturalized	161	82	20	32	295
Remainder	205	34	32	22	293
Totals	366	116	52	54	588
	B. Theoretical Frequencies				
Number Naturalized	183	58	26	28	295
Remainder	183	58	26	26	293
Totals	366	116	52	54	588

Since their computations included only half of the categories, they would obtain for any similar sample a value considerably smaller than we would obtain, regardless of the truth or falsity of the hypothesis. The value computed from all eight categories is known to have, approximately, the χ^2 distribution with 3 *df*, when the hypothesis is true. It follows, of course, that the quantity they computed cannot have the distribution they attribute to it because it is systematically too small.[3] Thus, there is no logic for interpreting their statistic by means of the χ^2 distribution with 3 *df*. The fact that the decision reached in this particular case would be the same for both computational methods is completely beside the point.

Peters seems to imply that there are no differences between an application of this kind and the tests of goodness of fit for ordinary distribution functions. Where, we ask, in testing the goodness of fit of a normal curve, do categories of "success" and "failure" ever exist over the same class interval? The difference between the applications is really quite profound.

4. *That no difficulties arise in applying chi-square in situations where the presence of individual differences may bring about a lack of independence.* Here Peters misunderstands what we mean by *independence*, a misunderstanding that often arises because the term is used with at least four distinct meanings in connection with chi-square tests. We have in mind the well-known distinction between the Bernoulli-De Moivre theorems on the large sample binomial distribution (based on constant probabilities) and the Poisson generalization of these theorems (including variable and linked probabilities). A somewhat extreme, but

---

[3] On the same grounds, we reaffirm our statement that Grant and Norris erred in omitting frequencies of non-occurrence.

simple, example will clarify our objection to the dice-throwing illustration used by Peters and Van Voorhis.

Suppose that we have 20 coins which are of three kinds. With two of them, the probability of obtaining either a head or a tail is $\frac{1}{2}$; but none of the coins have heads on both sides while the remaining nine have tails on both sides. If we toss the 20 coins together we will observe either 9, 10 or 11 heads, the remainder being tails. From such a toss, the value of χ^2 computed with formula (10) with theoretical frequencies of 10 must be either 0.0 or 0.2, with a probability of $\frac{1}{2}$ associated with each. If we were to toss the coins five times and combine the results as Peters and Van Voorhis did in their dice-throwing illustration, we would obtain for the computed quantity the following exact distribution:

χ^2—0.0   0.2   0.4   0.6   0.8   1.0
P—1/32   5/32   10/32   10/32   5/32   1/32

If a quantity having this distribution were interpreted in accordance with the χ^2 distribution, the hypothesis that the coins are unbiased would never be rejected since, with 5 df, a χ^2 value larger than 1.0 has a probability of occurrence of over 0.95. And yet, we know that the coins are biased.

Too much emphasis should not be placed on the incorrectness of adding χ^2 values in a situation of this kind. The results would not be much better if the hundred turns (20 coins each tossed 5 times) were treated as a unit. The use of formula (10) with the pooled results, the theoretical frequencies now being 50, would yield a quantity having the following exact distribution:

χ^2—0.0   0.04   0.16   0.36   0.64   1.0
P—126/512   210/512   120/512   45/512   10/512   1/512

Again, the hypothesis that the coins are unbiased would never be rejected through the use of the χ^2 distribution because, with 1 df, the probability of obtaining a value of χ^2 larger than 1.0 by chance is over 0.3.

In order to make a meaningful study of the coins, we would be obliged to specify (if we could) some event which had a uniform and not a varying probability of occurrence. With a population of coins of the three specified types, we might test the hypothesis that the probability of selecting a coin at random, tossing it, and observing a head is $\frac{1}{2}$. To do this, we would need to select different coins for each repetition of the experiment (a procedure which apparently surprises Pastore).

We would not object quite so strongly to the dice-throwing illustration of Peters and Van Voorhis if there were not a simple modification in procedure which eliminates the difficulty. If each die were tossed a sufficient number of times to permit the computation of separate values of χ^2 to represent the "behavior" of the separate dice and the χ^2 values were then summed over the set of dice, the deviations due to individual differences would appear. In this design, there could be no compensating effects from one die to another. The same general procedure could be used for generalizing to a larger collection of coins.

The objection may be raised that our coin example is too bizarre to be taken seriously. (Who ever heard of a coin with heads or tails on both sides?) Our answer is that, in any situation when individual differences in the underlying probabilities exist, the same effect is present although usually not so conspicuously. We are inclined to believe that individual differences among coins and dice may actually contribute to their ways of turning. We insist that with subjects in psychological experiments, individual differences cannot be ignored.

Peters finds us picayune in our discussion of one of his examples, and his attitude is not entirely unwarranted. However, for better or worse, we are forced to judge his book partly by the uses our students and colleagues make of it. Some psychologists are prone to take any example in an authoritative text as a model of experimental design and to copy it in their own work. The text by Peters and Van Voorhis is widely used, and justifiably so; in many respects, most notably in its exposition of mathematical backgrounds, it is an admirable book. This wide use imposes certain obligations on the authors. We suspect that they may be partly responsible for experimental designs which they would certainly not sanction. This is admittedly a poor way to judge books, but in the field of statistics we have no other recourse. Because of the unexpected uses to which examples may be put, we believe that correctness should never be sacrificed to pedagogical simplicity.

5. *That we erred in not attributing to Karl Pearson the notion that a quantity computed according to fixed rules can have two different distributions depending on the choice of the experimenter.* We readily concede that Pearson discusses this point in almost the same way that Peters and Van Voorhis do. And yet, we find a certain ambivalence in his position. At times, he writes as if he regarded his recommendations as approximations which are acceptable only on empirical grounds. It comes down to a matter of interpretation, and we doubt that anything can be proved one way or the other. We retract the statement to which Peters objects as being too firmly worded and admit that his interpretation of Pearson's views may be the correct one. However, the retraction does not alter the fact that the number of degrees of freedom for χ^2 in situations of the type discussed by Pearson and Peters is unaffected by the way the investigator phrases the hypothesis under test.

6. *That the χ^2 test of linearity of regression is adequate in some circumstances, but can be replaced by a better test.* This is precisely the view taken in our paper. We might have attributed the χ^2 linearity test

to Fisher, as Peters says we should have done, but it was not our aim to designate the originator of every technique we mentioned. For our purposes, the best points of departure were the formulas and recommendations in the tests and reference works most widely used by psychologists.

We share many of the views expressed by Peters in his closing paragraphs but are less sanguine than he concerning the likelihood of finding suitable replacements for χ^2.

REPLY TO PASTORE

Several of Pastore's criticisms have been answered either directly or indirectly in our reply to Peters. Those which remain will now be discussed, with the numbering kept the same as his.

2. He contends that we fail to offer a better method for handling the dice-throwing problem of Peters and Van Voorhis. We actually present a correct method (1949, pp. 446–447) indicating that separate values of χ^2 should be obtained for the individual dice and that these values could then be summed. This procedure is without flaws. The method proposed by Pastore is also a correct one and would avoid the difficulties posed by the possibility of individual differences, but a large number of throws (much larger than 14) would be required. Also, because there are 13 categories, there would be 12 df instead of 11, as stated by Pastore—provided, of course, that enough throws were made to yield theoretical frequencies of acceptable size in all categories.[4] With the technique we suggest, the particular dice (if any) which were biased could be spotted; with Pastore's technique this would not be possible.

4. Pastore maintains that values of χ^2 can be combined even when they are known to be interdependent, provided that there is independence in the underlying model. It is made quite clear in our paper (1949, p. 442 ff.) that we are discussing a case in which linkages between guesses on separate trials are empirically known to be present. Frankly, we are rather surprised at being asked to defend the thesis that a test should not be based on a demonstrably false assumption, and wonder why anyone would wish to use a model which is already known to be inappropriate. However, we shall present an artificial example of a rather extreme type to clarify the major issue.

Suppose that we run a coin-guessing experiment of the kind described in our original paper, but someone instructs the subjects to

---

[4] The probability of 12 aces on a single throw would be about .000000000459, so it would take around 20 billion throws to give a theoretical frequency of 10 in that category.

alternate their guesses throughout the experiment. We happen to select a group of 100 subjects such that, on the first guess, 55 responses are heads and 45 are tails. The resulting value of χ^2 is 1. On the second guess, the responses are 45 heads and 55 tails, and the value of χ^2 is again 1. After 20 guesses, we would have a composite χ^2 value of 20, with 20 df. Since a value larger than this would have a probability greater than 0.4 of occurring by chance, we would retain the hypothesis that each guess was mediated entirely by chance factors. But in this situation, the only random event is the first guess. We wonder if Pastore would wish to consider this set of guesses as a purely chance affair throughout.

If the experiment were repeated under the same conditions, 43 guesses of heads and 57 guesses of tails might be obtained on the first trial. The resulting value of χ^2 would be 1.89. Larger values than this have a probability roughly equal to 0.15 associated with them. The difference between 1.89 and 1.00 is not large and little importance would be attached to it. After 20 guesses, however, we would have a composite value of 37.8, significanct at the 1% level. Thus, under these conditions, first-trial differences which are unimportant can lead, after summation, to rather extreme differences in interpretation.

Whenever interdependencies exist, effects of the type just illustrated will be present. Pastore's extension of our argument to the first guess is clearly illegitimate. Any deviation of the frequencies from the theoretical values to be tested can be unambiguously interpreted and no difficulties arise. Also, if a different sample of subjects is used on each guess in obtaining a composite value of χ^2, there are no interdependencies to complicate the result.

5. It is asserted that, in one of our examples illustrating indeterminate theoretical frequencies, the test which we reject is a correct one. Since a detailed discussion of this point turns upon interdependencies of the kind already treated in detail and would merely be a repetition of an argument previously given, we shall confine ourselves to a few general remarks. (If one considers what would happen with a group of subjects who were systematically alternating their guesses, the difficulties become quite clear.)

Pastore may be justified in saying that the obtained value of χ^2 ($=43.351$, with 4 df) is large enough to warrant a rejection of the hypothesis without further question. But here again, he shows a willingness to use a model that is known to be inapplicable. What can be gained from such a procedure? If the hypothesis is rejected, nothing is gained because it was already known to be false. If it is not rejected, it still cannot be retained in view of its known falseness.

Pastore writes as if a statistic can be correct for an experiment even though the structure of the experiment is faulty. As far as we are concerned, the experimental design and the statistical treatment constitute a unit and must be treated as such. Incidentally, we should like to learn Pastore's technique for testing the hypothesis that a probability is greater than ½.

6. Pastore objects because we include (1949, pp. 481–482) two sets of probabilities which sum to more than unity. Considered in their proper setting, they are obviously conditional probabilities. The sum of the corrected probabilities is unity in each case, since the sum of the observed frequencies is equal to the sum of the theoretical frequencies. Perhaps this point can be made clear by an example which is a close parallel to the case under consideration.

Suppose that someone offers us a gift which is hidden under one of three upturned cups. The gift is to be ours if we can guess the cup under which it lies. The probability is ⅓ that we will guess right. Our first guess is wrong, but we are invited to try again. The cups and gift remains as they were. The probability of a correct guess is now ½. Following our second incorrect guess, a third one is proffered. Our memory is sufficiently good—under ordinary circumstances—to make the third probability 1.0. The sum of this set of probabilities is 1.83.

REPLY TO EDWARDS

Edwards' paper (1950) seems less a criticism of ours than an attempted advance based on a new collection of evidence; and we doubt that there are any serious disagreements between him and us. Our paper, although strongly recommending a minimal value of 10 for theoretical frequencies, allows for smaller values in the 2×2 table and proposes the use of Fisher's exact treatment whenever they occur. Edwards argues for frequencies as low as five and for the use of χ^2 as a sort of screening device and the use of the exact treatment as a check. The difference in viewpoint is not very great.

We may have overstated our case. At least Edwards' examples are quite convincing as far as they go, and his recommendations seem reasonable.[5] Nevertheless, we are inclined to take a position of "wait-and-see." In many cases, as shown by Edwards' examples, chi-square tests based on small theoretical frequencies are not far off, and there appears to be the possibility of proving that such tests will never be far off. Until such proof is forthcoming, however, the possibility must be accepted that combinations of numbers in fourfold tables may exist

[5] With reference to his reworking of the frequency data obtained by Lewis, Lewis and Franklin, and Kuenne, we disavow in our paper any criticism of the experimental conclusions reached by these investigators.

for which the divergences could not be neglected. In this connection, we will be quite happy to see final proof that we are wrong.

As a final word, we wish to express the sincere hope that everything we have written on chi-square will (to borrow a phrase from Peters) "be read with critical alertness, not taken offhand as gospel."

References

Cramér, H. *Mathematical methods of statistics*. Princeton: Princeton Univ. Press, 1946.

Edwards, A. L. On "The use and misuse of the chi-square test"—The case of the 2 × 2 contingency table. *Psychological Bulletin*, 1950, **47**, 341–346.

Lewis, D., & Burke, C. J. The use and misuse of the chi-square test. *Psychological Bulletin*, 1949, **46**, 433–489.

Pastore, N. Some comments on "The use and misuse of the chi-square test." *Psychological Bulletin*, 1950, **47**, 338–340.

Peters, C. C. The misuse of chi-square—A reply to Lewis and Burke. *Psychological Bulletin*, 1950, **47**, 331–337.

Peters, C. C., & Van Voorhis, W. R. *Statistical procedures and their mathematical bases*. New York: McGraw-Hill, 1940.

2.6

Letter to the editor
on Peters' reply
to Lewis and Burke

C. J. Burke

In connection with the discussion of chi-square in the July 1950 issue of the *Psychological Bulletin*, the paper submitted by Peters differs in an important respect from the copy to which Dr. Lewis and I wrote our reply. In our discussion on p. 348 we have a reference to Peters' third footnote. In his published criticism of our original paper, the footnote in question was omitted, with the consequence that our remark is left dangling. Further consequences are that our reply to Peters' point 3 on pp. 333–334 of his article is inadequate, since this reply was based upon the missing footnote. The omitted footnote is crucial for our reply and serves better than anything we could write to illustrate Peters' confusions with respect to the chi-square test. The omitted footnote, numbered 3 in Peters' original manuscript, reads as follows:

> There may be needed a re-examination of the applicability of Pearson's formula *vi* to the new conditions for which he did not intend it. He developed the formula on the assumption of freely floating marginal totals, only the total N being fixed. When R. A. Fisher, for the sake of using true values for a narrower universe rather than estimated ones for a broader universe, proposed constant marginal totals with a consequent change in the number of degrees of freedom, he does not appear to have re-examined Pearson's fundamental derivation under these new conditions. With freely floating marginal totals and only a limitation on the total N, a certain system of intercorrelations among the cells arose of which he took account. Under the new conditions a new system of intercorrelations would arise which might very well not

be the same. A new derivation with the new assumptions might not come through with the same simple formula *vi*. I am not myself able to carry this through; but invite the attention of expert mathematicians to it.

Our comment on page 348 indicates that the derivation Peters requests is already available in one of the best known statistical sources and that he is mistaken in his notion that the degrees of freedom is changed.

> The editor is entirely responsible for the error which caused Drs. Lewis and Burke to criticize a footnote in Dr. Peters' original manuscript which he subsequently deleted from the galley proof of the article. Dr. Peters wrote to the editor before his paper was sent to Drs. Lewis and Burke for comment, and asked that the footnote in question be deleted. He stated that if the paper had gone to press he would take the note out of the galley proof. Through an oversight, the editor did not delete the footnote from the manuscript, nor did he read the galley proof of the article. Galley proofs and subsequent editorial work are now handled in the central office of the American Psychological Association.

Peters' comments on his point 3 are part of the same confusion. Without the footnote just quoted our reply does not cover the points he raises and adequate clarification would require considerable discussion. With the footnote, the discussion as given in our reply is easy and adequate.

2.7

Some methods for strengthening the common χ^2 Test[*]

William G. Cochran

1. INTRODUCTION

Since the χ^2 tests of goodness of fit and association in contingency tables are presented in many courses on statistical methods for beginners in the subject, it is not surprising that χ^2 has become one of the most commonly-used techniques, even by scientists who profess only a smattering of knowledge of statistics. It is also not surprising that the technique is sometimes misused, e.g. by calculating χ^2 from data that are not frequencies or by errors in counting the number of degrees of freedom. A good catalogue of mistakes of this kind has been given by Lewis and Burke (1949).

In this paper I want to discuss two kinds of failure to make the best use of χ^2 tests which I have observed from time to time in reading reports of biological research. The first arises because χ^2 tests, as has often been pointed out, are not directed against any specific alternative to the null hypothesis. In the computation of χ^2, the deviations $(f_i - m_i)$ between observed and expected frequencies are squared, divided by m_i in order to equalize the variances (approximately), and added. No attempt is made to detect any particular pattern of deviations $(f_i - m_i)$ that may hold if the null hypothesis is false. One consequence

William G. Cochran, "Some Methods for Strengthening the Common Tests," *Biometrics*, **10**, 1954, 417–451. The following sections have been deleted from the original article: 6.3, Comparison of Mean Scores; 6.4, Step by Step Comparisons of the p_i; 7.1. Score Methods; 7.2, Use of Analysis of Variance; 9.0. The Effect of Extraneous Variation, and the Appendix.

* Work assisted by a contract with the Office of Naval Research, U. S. Navy Department. Dept. of Biostatistics paper no. 278.

is that the usual χ^2 tests are often insensitive, and do not indicate significant results when the null hypothesis is actually false. Some forethought about the kind of alternative hypothesis that is likely to hold may lead to alternative tests that are more powerful and appropriate. Further, when the ordinary χ^2 test does give a significant result, it does not direct attention to the way in which the null hypothesis disagrees with the data, although the pattern of deviations may be informative and suggestive for future research. The remedy here is to supplement the ordinary test by additional tests that help to reveal the significant type of deviation.

In this paper a number of methods for strengthening or supplementing the most common uses of the ordinary χ^2 test will be presented and illustrated by numerical examples. The principal devices are as follows:

1.   Use of small expectations in computing χ^2.
2.   Use of a single degree of freedom, or a group of degrees of freedom, from the total χ^2.
3.   Use of alternative tests.

Most of the techniques have been available in the literature for some time: indeed, most of them stem from early editions of Fisher's "Statistical Methods for Research Workers." Research which has clarified the problem of subdividing χ^2 in contingency tables is more recent, and still continues.

In the hope of avoiding some confusion, the symbol of X^2 will be used for the quantity that we calculate from the sample in a chi-square test. The symbol χ^2 itself will refer to a random variate that follows the distribution in the χ^2 tables, and will sometimes be used in the phrase "the χ^2 test."

2. USE OF SMALL EXPECTATIONS IN COMPUTING X^2

In order to prove that the quantity $X^2 = \Sigma (f_i - m_i)^2 / m_i$ is distributed as χ^2 when the null hypothesis is true, it is necessary to postulate that the expectations m_i are large: in fact, the proof is strictly valid only as a limiting result when the m_i tend to infinity. For this reason many writers recommend that the m_i be not less than 5 when applying the test in practice, and that neighboring classes be combined if this requirement is not met in the original data. Some writers recommend a lower limit of 10 for the m_i.

It is my opinion that these recommendations are too conservative, and that their application may on occasion result in a substantial loss of power in the test. I give this as an opinion, because not enough research has been done to make the situation quite clear. However, the exact distribution of $\Sigma (f_i - m_i)^2/m_i$, when the expectations are small, has been worked out in a number of particular cases by Sukhatme (1938), Neyman and Pearson (1931) and Cochran (1942, 1936). These results indicate that the χ^2 tables give an adequate approximation to the exact distribution even when some m_i are much lower than 5.

Loss of power from following a rule that $m_i \geq 5$ occurs because this rule tends to require grouping of classes at the tails or extremes of the distribution. These are often the places where the difference between the alternative hypothesis and the null hypothesis stands out most clearly, so that the grouping may cover up the most marked difference between the two hypotheses. Information about the extent of the loss of power is unfortunately very scanty, because the power function of X^2 is known only as a limiting result when the m_i are large. The following illustration suggests that the loss can be large.

Suppose that we have a sample of $N = 100$. The null hypothesis is that the data follow a Poisson distribution with mean m known to be 1, when actually the data follow the negative binomial distribution $(q - p)^{-n}$, where $n = 2$, $q = 1.5$, $p = 0.5$. What is the chance of rejecting the null hypothesis at the 5% level of significance?

The expected frequencies of 0, 1, 2, 3, and 4 or more occurrences on the two hypotheses are shown in Table 1.

If an m_i as low as 1.90 is allowed, we can use 5 classes in the χ^2 test, with 4 degrees of freedom since the m_i are known. To make all $m_i \geq 5$, we must pool the last two classes and have 3 degrees of freedom; and to make all $m_i \geq 10$, we must pool the last three classes and have 2 degrees of freedom. In order to obtain an approximation to the powers of these three ways of applying the χ^2 test, we shall use the asymptotic

TABLE 1

Expected Frequencies on the Null and Alternative Hypothesis

	Expected frequencies		$\dfrac{(m_i' - m_i)^2}{m_i}$		
				Grouped	
i	Poisson m_i	Neg. Bin. m_i'	Ungrouped	$m_i \geq 5$	$m_i \geq 10$
0	36.79	44.44	1.590	1.590	1.590
1	36.79	29.63	1.393	1.393	1.393
2	18.39	14.82	0.693	0.693	
3	6.13	6.58	0.033	$\Big\}1.181$	$\Big\}0.009$
4+	1.90	4.53	3.640		
Totals $N =$	100.00	100.00	$\lambda = 7.349$	$\lambda = 4.857$	$\lambda = 2.992$

result for the power function. This is a non-central χ^2 distribution, with parameter λ of non-centrality, where

$$\lambda = \sum \frac{(m_i' - m_i)^2}{m_i}$$

The larger the value of λ, the higher is the power. The contributions to λ from each class are shown in the right-hand columns of Table 1. For the ungrouped case, note that the extreme class 4+ is much the largest contributor to λ. Grouping the last two classes considerably reduces this contribution, while grouping the last three classes diminishes it almost to zero. The approximate probabilities of rejecting the null hypothesis may be read from Fix's tables (1949) of the non-central χ^2 distribution. These probabilities are 0.56 for the ungrouped case, 0.43 for $m_i \geq 5$ and 0.32 for $m_i \geq 10$. The loss in sensitivity from grouping is evident. Perhaps a more revealing comparison is to compute from Fix's tables the sizes of samples N that would be needed in the two grouped cases in order to bring the probabilities of rejection up to 0.56, the value for the ungrouped case. The results are $N = 136$ when the last two classes are grouped and $N = 191$ when the last three classes are grouped, as against $N = 100$ in the ungrouped case.

This example is only suggestive, and probably favors the ungrouped case slightly, because the computations are based on large-sample results. However, the losses in power from grouping, measured in terms of equivalent sample sizes, are impressive.

2.1 Recommendation about Minimum Expectations

Elsewhere, (1952), I have given recommendations about the minimum expectation to be used in χ^2 tests. These working rules may be summarized, in slightly revised form, as follows:

(a) *Goodness of fit tests of unimodal distributions* (*such as the normal or Poisson*). Here the expectations will be small only at one or both tails. Group so that the minimum expectation at each tail is at least 1.

(b) *The 2×2 table.* Use Fisher's exact test (i) if the total N of the table < 20, (ii) if $20 < N < 40$ and the smallest expectation is less than 5. Mainland (1948) has given useful tables of the exact test for these cases. If $N > 40$ use X^2, corrected for continuity.

(c) *Contingency tables with more than 1 d.f.* If relatively few expectations are less than 5 (say in 1 cell out of 5 or more, or 2 cells out of 10 or more), a minimum expectation of 1 is allowable in computing X^2.

Contingency tables with most or all expectations below 5 are harder to prescribe for. With very small expectations, the exact distribution of X^2 can be calculated without too much labor. Computing methods

[handwritten margin note: Contrast Snedecor & Cochran p 217.]

have been given by Freeman and Halton (1951). If X^2 has less than 30 degrees of freedom and the minimum expectation is 2 or more, use of the ordinary χ^2 tables is usually adequate. If X^2 has more than 30 degrees of freedom, it tends to become normally distributed, but when the expectations are low, the mean and variance are different from those of the tabular χ^2. Expressions for the exact mean and variance have been given by Haldane (1939). Compute the exact mean and variance, and treat X^2 as normally distributed with that mean and variance.*

Further research will presumably change these recommendations, but I do not believe that the recommendations will lead users far astray.

Succeeding sections will deal with some of the common applications of the χ^2 test to goodness of fit problems and to contingency tables. The alternative or supplementary tests to be presented are those that seem most often useful, but they by no means exhaust the possibilities. The important guiding rule is to think about the type of alternative that is likely to hold if the null hypothesis is false, and to select a test that will be sensitive to this kind of alternative.

3. THE GOODNESS OF FIT TEST OF THE POISSON DISTRIBUTION

This is the test already referred to in Table 1, except that in practice the parameter m must usually be estimated from the data. We have

$$X^2 = \sum \frac{(f_i - m_i)^2}{m_i}$$

where f_i is the observed and m_i the expected frequency of an observation equal to $i(i = 0, 1, 2 \ldots)$.

If the data do not follow the Poisson distribution, two common alternatives are as follows:

(1) The data follow some other single distribution, such as the negative binomial or one of the "contagious" distributions. Another way of describing this case is to postulate that the individual observations, say x_j, follow Poisson distributions, but with means that vary from observation to observation so as to follow some fixed frequency distribution. For instance, as has been shown, the negative binomial distribution can be produced by assuming that these means follow a Pearson type III distribution.

(2) The means of the observations x_j follow some systematic pattern. With data gathered over several days, the means might be

---

* In a previous paper (1952) I recommended this procedure only when the degrees of freedom in X^2 exceed 60. Some unpublished research suggests that 30 is a better division point.

constant within a day, but vary from day to day, or they might exhibit a slow declining trend.

3.1 Test of the Variance

In both cases (1) and (2), a comparison between the observed variance of the observations x_j and the variance predicted from Poisson theory will frequently be more sensitive than the goodness of fit test. The variance test is made by calculating

$$X_v^2 = \sum_{j=1}^{N} \frac{(x_j - \bar{x})^2}{\bar{x}},$$

or if the calculation is made from the frequency distribution of the x_j,

$$X_v^2 = \sum_i \frac{f_i(i - m)^2}{m},$$

where $m = \bar{x}$ is the sample mean. The quantity X_v^2 is referred to the χ^2 tables with $(N - 1)$ degrees of freedom. The variance test is an old one: it was introduced by Fisher in the first edition of "Statistical Methods for Research Workers" under the heading "Small samples of the Poisson series," because it can be used when the sample is too small to permit use of the goodness of fit test.

The increased power of the variance test over the goodness of fit test was strikingly shown in some sampling experiments conducted by Berkson (1940), in a situation in which the data followed a binomial distribution, which has a *smaller* variance than the Poisson. A rough calculation for the example in Table 1 gives 0.76 as the probability of rejecting the Poisson hypothesis by the variance test, as compared with 0.56 for the best of the goodness of fit tests. In practice, I have often found the variance test significant when the goodness of fit test was not. Berkson (1938) presents some data that illustrate this point.

3.2 Subdivision of Degrees of Freedom in the Variance Test

If it is suspected that the means of the observations x_j may change in some *systematic* manner, as in case (2) above, more specific tests of significance can be obtained by selecting certain degrees of freedom from χ_v^2. The ordinary rules of the analysis of variance are followed in subdividing $\Sigma(x_i - \bar{x})^2$, and the denominator $\bar{x}$ is used to convert the partial sum of squares approximately to χ^2. A few examples will be given. (The formulas presented are intended to make clear the structure of the χ^2 components, but they are not always the speediest formulas to use in computing the components.)

3.3 Test for a Change in Level

To test for an abrupt change in the mean of the distribution, occurring after the first N_1 observations, we take

$$X^2 = \frac{N_1 N_2}{(N_1 + N_2)} \frac{(\bar{x}_1 - \bar{x}_2)^2}{\bar{x}} \qquad \text{(1 d.f.)}$$

where $N_2 = (N - N_1)$; $\bar{x}_1$, and $\bar{x}_2$ are the sample means in the two parts of the series and $\bar{x}$ is the overall mean. This test may be extended to compare a group of means.

3.4 Test for a Linear Trend

We may anticipate that the means will follow a linear regression on some variate z_j (frequently a time-variable). In this case

$$X^2 = \frac{[\Sigma(x_j - \bar{x})(z_j - z)]^2}{\bar{x}\Sigma(z_j - \bar{z})^2} \qquad \text{(1 d.f.)}$$

3.5 Detecting the Point at which a Change in Level Occurs

This problem has been illustrated by Lancaster (1949) in an experiment in which increasing concentrations of disinfectant are poured on a series of suspensions of a bacterial culture, each suspension having a constant amount of bacteria. The observations x_j represent numbers of colonies per plate. The problem is to find the value of j for which the disinfectant is strong enough to begin reducing the number of colonies. To this end we compare each observation with the mean of all previous observations, looking for the first value of j at which X^2 becomes large owing to a drop in the number of colonies. We thus obtain a set of independent single degrees of freedom:

$$X_1^2 = \frac{(x_1 - x_2)^2}{2\bar{x}}; \qquad X_2^2 = \frac{(x_1 + x_2 - 2x_3)^2}{6\bar{x}};$$

and in general

$$X_r^2 = \frac{(x_1 + x_2 + \cdots + x_r - rx_{r+1})^2}{r(r + 1)\bar{x}}$$

Note. The above set of X^2 values will add up to the total variance X_v^2, but as Lancaster has pointed out, there are other natural ways of subdividing X_v^2 in which the separate X^2 do not add up to X_v^2. When comparing x_1 with x_2, we might decide to disregard the remainder of the observations and compute X_1^2 as

$$X_1'^2 = \frac{(x_1 - x_2)^2}{2\bar{x}_{12}}$$

where $\bar{x}_{12}$ is the mean of x_1 and x_2. A set of successive X^2 values computed in this way will not add up to X_v^2, because the denominator changes from term to term, whereas X_v^2 carries the denominator $\bar{x}$.

The practice of computing X^2 components by using only those parts of the data that are immediately involved has something in its favor (despite the non-additive feature), at least if the total X_v^2 has already been shown to be significant. For in that event we have already concluded that the data as a whole do not follow a single Poisson distribution, and the overall mean $\bar{x}$ is of dubious validity as an estimate of the Poisson variance for a part of the data. On the other hand, if the total X_v^2 is not significant, but we suspect that some component is, the additive partition is convenient and should be satisfactory in a preliminary examination.

3.6 Single Degrees of Freedom in the Goodness of Fit Test

In the goodness of fit comparison of the observed frequency f_i with the expected frequency m_i of i occurrences, we can test any linear function of the deviations

$$L = \Sigma g_i(f_i - m_i)$$

where the g_i are numbers chosen in advance.

In the case in which the mean of the Poisson, and hence the m_i, are given in advance, the variance of L is

$$V(L) = \sum g_i^2 m_i - \frac{(\Sigma g_i m_i)^2}{N} \tag{1}$$

In the more common situation in which the Poisson mean m is estimated from the data,

$$V(L) = \sum g_i^2 m_i - \frac{(\Sigma g_i m_i)^2}{N} - \frac{[\Sigma g_i m_i(i - m)]^2}{Nm} \tag{2}$$

where the sums are over the values 0, 1, 2, . . . of i. In either case

$$X_1^2 = \frac{L^2}{V(L)} \tag{3}$$

is approximately distributed as χ^2 with 1 d.f. I plan to publish justification for formula (2), which appears to be new. By appropriate choice of the g_i, a test specific for a given pattern of deviations is obtained.

In particular, to test any single deviation $(j_i - m_i)$ when m is estimated from the data, we take

$$L = (f_i - m_i) : V(L) = m - \frac{m_i^2}{N}\left\{1 + \frac{(i - m)^2}{m}\right\} \tag{4}$$

As an illustration, the data in Table 2 are for a sample which gave a satisfactory fit to a Poisson distribution. However, in copying down

TABLE 2
Goodness of Fit Test for a Sample with a Gross Error

i	f_i	m_i	Contribution to X^2
0	52	47.65	0.40
1	67	77.04	1.31
2	58	62.28	0.29
3	52	33.56	10.13
4	7	13.56	3.17
5	3	4.39	0.44
6+	1	1.52	0.18
Total	240	240.00	15.92

the frequencies before fitting the Poisson, the frequency of 3 occurrences was erroneously written as 52 instead of 32.

The value of m is 1.6167. The total X^2, 15.92 with 5d.f., is significant at the 1 per cent level. The large contribution to X^2 from $i = 3$ excites notice. In order to test this deviation, we take

$$L = f_3 - m_3 = 52 - 33.56 = 18.44$$

$$V(L) = 33.56 - \frac{(33.56)^2}{240}\left\{1 + \frac{(3 - 1.6167)^2}{1.6167}\right\} = 23.31$$

$$X_1^2 = \frac{(18.44)^2}{23.31} = 14.59$$

This comparison accounts for the major part of the total X^2. It must be pointed out, however, that the X_1^2 test applies only to a deviation picked out before seeing the data. Thus the test can be applied validly, for $i = 0$, say, if we suspect beforehand that the data follow the Poisson distribution for $i \geq 1$, but that the frequency for $i = 0$ may be anomalous. If the test is applied, as here, to a deviation selected because it looks abnormally large, the significance P obtained is too low. I do not have an expression for the correct significance probability when we select the largest deviation. It appears intuitively that the correct probability lies between P and kP, where k is the number of classes in the goodness of fit test. Since P is about 0.00013 and k is 7, the upper limit is .00091, which is still highly significant statistically.

4. THE GOODNESS OF FIT TEST OF THE BINOMIAL DISTRIBUTION

For the binomial distribution, there is a series of tests analogous to those given for the Poisson distribution. A typical observation consists of the number of successes x_i out of n independent trials. We have a sample of N such observations. The ordinary goodness of fit test is made by recording the frequency f_i with which i successes occur in the sample, fitting a binomial to these frequencies, and calculating X^2 as

$$X^2 = \sum \frac{(f_i - m_i)^2}{m_i},$$

where m_i is the corresponding expected frequency.

As in the Poisson case, departures from the binomial frequently occur either because

(1) the data follow a different frequency distribution, usually with a larger variance (or the probabilities of success p_i show some kind of random variation from observation to observation).

(2) the probabilities p_i are affected by a systematic source of variation.

In both cases, a comparison of the observed and expected variances is likely to be more sensitive than the goodness of fit test.

4.1 Test of the Variance

The test criteria, all distributed approximately as χ^2, are as follows,

n constant, p given in advance:

$$X_v^2 = \frac{\sum_i (x_j - np)^2}{npq} = \frac{\sum_i f_i(i - np)^2}{npq}, \qquad N \text{ d.f.}$$

n constant, p estimated:

$$X_v^2 = \frac{\sum_i (x_j - \bar{x})^2}{n\hat{p}\hat{q}} = \frac{\sum_i f_i(i - n\hat{p})^2}{n\hat{p}\hat{q}}, \qquad (N-1) \text{ d.f.}$$

where $n\hat{p} = \bar{x}$, and $\hat{q} = 1 - \hat{p}$.

n varying, p given in advance:

In this case we cannot make a simple goodness of fit test (unless the sample is large enough to be divided into batches, each with n constant, so that the test can be made separately for each batch). If $p_j = x_j/n_j$ is the observed proportion of successes in the jth member of the sample,

$$X_v^2 = \frac{\sum n_j(p_j - p)^2}{pq} \qquad N \text{ d.f.}$$

n varying, p estimated:

$$X_v^2 = \frac{\Sigma n_j (p_j - \hat{p})^2}{\hat{p}\hat{q}} = \frac{\sum \frac{x_j^2}{n_j} - \frac{(\Sigma x_j)^2}{(\Sigma n_j)}}{\hat{p}\hat{q}}, \qquad (N-1) \text{ d.f}$$

where $\hat{p} = (\Sigma\ x_j)/(\Sigma\ n_j)$ is the estimate of p from the total sample.

There is another way of deriving the variance test. Arrange the data in a $2 \times N$ contingency table, as follows.

Successes	x_1	x_2		x_N
Failures	$n_1 - x_1$	$n_2 - x_2$		$n_N - x_N$
Total	n_1	n_2		n_N

Then the X^2 that is used to test for association in this $2 \times N$ table may be shown to be identical with X_v^2.

If N exceeds 30 and the expectations are small, it was pointed out in section 2 that X^2 in contingency tables tends to a normal distribution with a mean and variance somewhat different from those of χ^2. Use of Haldane's correct expressions for the mean and variance of X^2 was recommended in this case. The same procedure is recommended in the variance test if N exceeds 30 and the average n_i is less than 10. Haldane's expressions are rather complicated when the n_i vary. In the fairly common situation in which n is constant and p is estimated from the data, the following results suffice in almost all cases.

$$E(X_v^2) \doteq (N-1)\left(1 + \frac{1}{Nn}\right) \tag{5}$$

$$V(X_v^2) \doteq 2(N-1)\left(\frac{n-1}{n}\right)\left(1 - \frac{1 - 7\hat{p}\hat{q}}{Nn\hat{p}\hat{q}}\right) \tag{6}$$

The important correction term is that in $(n-1)/n$ in the variance: the terms in $1/Nn$ are usually small. These results will be used in the numerical example which follows.

The data in Table 3, taken from a previous paper (1936), illustrate the application of the goodness of fit test and the variance test to the same observations. The original data, due to Dr. J. G. Bald, consisted of 1440 tomato plants in a field having 24 rows with 60 plants in a row. For each plant it was recorded whether the plant was healthy or attacked by spotted wilt as of a given date. As one method of examining whether the distribution of diseased plants was random over the field, the plants were divided into 160 groups of 9, each group consisting of 3 plants $\times$ 3 rows. Thus $N = 160$, $n = 9$. The choice of $n = 9$ was of course arbitrary, and I do not know what would have been the best choice. The obvious alternative to a binomial distribution is that the values of p_i vary from one group of 9 to another, indicating a patchiness in distribution.

TABLE 3

Number of Diseased Plants in Groups of 9 Plants

i	f_i	m_i	Contr. to X^2
0	36	26.45	3.45
1	48	52.70	0.42
2	38	46.67	1.61
3	23	24.11	0.05
4	10	8.00	0.50
5	3		
6	1	2.05	4.25
7	1		
8	0		
N = 160		159.98	10.28

Allowing a minimum expectation of 1, we must pool the last 4 classes in Table 3. The value of X^2 is 10.28, with 4 d.f., since p is estimated; the significance P is 0.036. (If we pooled the last 5 classes in order to have a minimum expectation of 5, we would obtain a significance P of 0.046.)

For the variance test, we compute X_v^2 from the observed frequency distribution in Table 3. We have

$$\Sigma if_i = 261 = Nn\hat{p}, \quad \text{so that } \hat{p} = 0.18125.$$

$$X_v^2 = \frac{\Sigma i^2 f_i - (\Sigma if_i)^2/N}{n\hat{p}\hat{q}} = \frac{727 - (261)^2/160}{9(0.18125)(0.81875)}$$

$$= 225.55, \text{ with 159 d.f.}$$

Since $N = 160$ and $n = 9$, we use the normal approximation to the distribution of X_v^2, based on the correct mean and variance. In expressions (5) and (6), the terms in $1/Nn$ are negligible ($Nn = 1440$). Hence we take

$$E(X_v^2) = 159 : V(X_v^2) = 2(159)(8)/9 = 282.66$$

The approximate normal deviate is

$$\frac{225.55 - 159}{\sqrt{282.66}} = 3.96$$

This has a significance P less than 0.0001, much lower than that obtained from the goodness of fit tests.

Subdivision of the sum of squares for X_v^2, which may be useful in testing for systematic variation of the p_i, will be discussed in the section which deals with $2 \times N$ contingency tables.

4.2. Single Degrees of Freedom in the Goodness of Fit Test

Let

$$L = \Sigma g_i(f_i - m_i)$$

be a specified linear function of the deviations of observed from expected frequencies. If p is given, formula (1) holds for the variance of L. If p is estimated from the data,

$$V(L) = \sum g_i^2 m_i - \frac{(\Sigma g_i m_i)^2}{N} - \frac{[\Sigma g_i m_i(i - n\hat{p})]^2}{N n \hat{p} \hat{q}} \qquad (7)$$

For a single deviation, $(f_i - m_i)$, selected in advance,

$$V(L) = m_i - \frac{m_i^2}{N}\left\{1 + \frac{(i - n\hat{p})^2}{n\hat{p}}\right\} \qquad (8)$$

Then $L^2/V(L)$ is approximately distributed as χ^2 with 1 d.f.

5. THE GOODNESS OF FIT TEST OF THE NORMAL DISTRIBUTION

When the normal distribution is fitted to a body of data, both the mean and the variance are estimated from the sample: consequently, no variance test is possible. However, the variance test is just an application of the general procedure in which we compare the lowest moments (or cumulants) in which the sample can differ from the theoretical distribution that is being fitted. In this sense, the analogue of the variance test is the test for skewness (as given e.g. in Fisher's "Statistical Methods for Research Workers," §14), which will often detect a departure from normality that escapes the goodness of fit test. The test for kurtosis is also useful in this connection.

As with the Poisson and binomial distributions, a specified linear function L of the deviations in the goodness of fit test can be scrutinized. The formula for the variance of L is somewhat more complicated, because the observed frequencies are subject to 3 constraints. If

$$L = \Sigma g_i(f_i - m_i)$$

then

$$V(L) = \sum g_i^2 m_i - \frac{(\Sigma g_i m_i)^2}{N} - \frac{(\Sigma g_i d_i m_i)^2}{N s^2} - \frac{[\Sigma g_i m_i(d_i^2 - s^2)]^2}{2 N s^4}$$

where

$$d_i = \text{(midpoint of } i\text{th class)} - \text{(sample mean)}$$
$$s^2 = \text{sample estimate of variance}$$

In order to apply this test, construct 3 additional columns containing the quantities $g_i m_i$, d_i and $(d_i^2 - s^2)$, respectively. $V(L)$ is then easily computed.

For testing a *single* deviation, $V(L)$ simplifies to

$$V(L) = m_i - \frac{3m_i^2}{2N} - \frac{m_i^2 d_i^4}{2Ns^4}$$

The series of supplementary tests described above for the Poisson, binomial and normal distributions can be extended to other distributions. In particular, variance and skewness tests for the negative binomial distribution have been developed by Anscombe (1950) and further illustrated by Bliss (1953).

6. SUBDIVISION OF DEGREES OF FREEDOM IN THE 2 × N CONTINGENCY TABLE

This section describes some useful ways in which the total X^2 for a $2 \times N$ contingency table may be subdivided. The notation, which continues that already used for the binomial distribution, is as follows.

| | Number of | | | Proportion of Successes |
	Successes	Failures	Total	
	x_1	$n_1 - x_1$	n_1	$p_1 = x_1/n_1$
	x_2	$n_2 - x_2$	n_2	$p_2 = x_2/n_2$
	x_N	$n_N - x_N$	n_N	$p_N = x_N/n_N$
Totals	T_x	$T - T_x$	T	$\hat{p} = T_x/T$

For many purposes, a formula due to Brandt and Snedecor is useful in interpreting the total X^2. The formula is:

$$X^2 = \frac{\sum_{j=1}^{N} n_j(p_j - \hat{p})^2}{\hat{p}\hat{q}} \qquad (9)$$

Thus the total X^2 is seen to be a weighted sum of squares of the deviations of the individual proportions of success p_j from their mean, with weights $n_j/\hat{p}\hat{q}$. Consequently, if we subdivide this weighted sum of squares into a set of independent components by the rules of the analysis of variance, we obtain a corresponding subdivision of X^2 into independent components.

6.1 Test for a Change in the Level of p

In order to test whether the value of p is different in the first N_1 rows from that in the subsequent N_2 rows $(N = N_1 + N_2)$, we may subdivide X^2 into the following 3 components.

	d.f.
Difference between p's in first N_1 and last N_2 rows	1
Variation among p_j within the first N_1 rows	$(N_1 - 1)$
Variation among p_j within the last N_2 rows	$(N_2 - 1)$

For the following example I am indebted to Dr. Douglas P. Murphy. A group of women known to have cancer of the uterus and a corresponding 'control' group (primarily from a dental clinic and several women's clubs) were selected. From each of a defined set of relatives of the selected person (the proband), data were secured about the presence of cancer. A higher proportion of cancer cases among relatives of the cancer proband would indicate some kind of familial aggregation of the disease, perhaps of genetic origin (1952).

In one table, data were presented separately for those relatives who were of the same generation as the proband (e.g. sister) and for those relatives who were one generation earlier (e.g. mother). Some breakdown of this kind is advisable, because cancer attacks mainly in middle or old age, and the 'cancer' and 'control' groups might be found to differ in the proportions of young relatives which they contained. The data, which are a small part of a much more intensive investigation, appear in Table 4.

There are several ways of computing the total X^2 in a $2 \times N$ table. One form of the Brandt-Snedecor formula is

$$X^2 = \frac{\Sigma x_j p_j - \hat{p} T_x}{\hat{p}\hat{q}} \tag{10}$$

This expression is useful when there is some question of a systematic variation in the p_j, because we will want to compute the p_j in order to have a look at them. When the p_j have been computed, the numerator of X^2 can be obtained from formula (10) directly on the computing machine, without any intermediate writing down. The only disadvantage is that a substantial number of decimal place must be retained in the p_j.

If this method is to be used, first compute $\hat{p}$, $\hat{q}$ and the product $\hat{p}\hat{q}$ (bottom right of Table 4). Since $\hat{p}\hat{q}$ is about 1/20, the numerator of X^2 must be correct to at least 3 decimal places if we want X^2 correct to 2 decimal places. Further, since T_x is 313, we should have 6 decimal places in the p_j. (The symbol x_i should be assigned to the column with the *smaller* numbers: this makes the computations lighter and necessitates fewer decimals in the p_j.)

From inspection of the p_j in Table 4, a large difference in cancer

TABLE 4

Cancer among Relatives of 'Cancer' and 'Control' Probands

	Proband	With x_j	Without	Total n_j	Proportion with cancer p_j
		No. of relatives			
Earlier	Cancer	86	814	900	0.095556
Generation	Control	117	1038	1155	0.101299
Same	Cancer	49	1475	1524	0.032152
Generation	Control	61	1580	1641	0.037172
	Total	313	4907	5220	$\hat{p} = 0.059962$ $\hat{q} = 0.940038$ $\hat{p}\hat{q} = 0.056367$

rates between the two generations is evident. Within each generation, the differences in rates between the cancer and control groups appear tiny. To illustrate the methods, the total X^2 will be partitioned into the 3 relevant components. All that is necessary is to subdivide the sum of squares in the numerator, and then divide each component by $\hat{p}\hat{q}$. For the numerator of the total X^2, we have from formula (10),

$$\text{Total } S.S. = (86)(0.095556) + \cdots + (61)(0.037172) - (313)(0.059962)$$
$$= 5.1447$$

For the comparison of the two generations, we form the auxiliary 2×2 table.

	With x_j	Without	Total n_j	Proportion p_j
Earlier generation	203	1852	2055	0.098783
Same generation	110	3055	3165	0.034755
Total	313	4907	5220	0.055962

The same formula can be used for this table.

$$\text{Generations } SS = (203)(0.098783) + (110)(0.034755)$$
$$- (313)(0.055962)$$
$$= 5.1079.$$

For the comparisons between cancer and control groups within generations, we have

$$\text{First generation } SS = (86)(0.095556) + (117)(0.101299)$$
$$- (203)(0.098783) = 0.0169.$$

TABLE 5
Subdivision of χ^2 into Components (2 × 4 Table)

Component	d.f.	S.S.	X^2	From 2 × 2 tables
First vs. later generation	1	5.1079	90.62	90.62
Cancer vs. Control: first gen.	1	0.0169	0.30	0.19
Cancer vs. Control: later gen.	1	0.0199	0.35	0.59
Total	3	5.1447	91.27	91.40

The second generation SS, obtained similarly, gives 0.0199. In Table 5 the results are summarized and converted to X^2 values on division by $\hat{p}\hat{q}$.

There is no indication of any difference in cancer rates between the cancer and control groups within either generation. In order to complete this analysis, we should make a combined test of Cancer vs. Control from the two generations. Methods for making tests of this kind are discussed in section **7**.

As mentioned previously in connection with the corresponding subdivision for the Poisson distribution, separation of X^2 into *additive* components is convenient for a preliminary examination of the data. But if differences in p are found between groups of rows, the X^2 values for comparisons made *within* groups need to be recomputed. This is clear in the present example. The additive method requires the assumption that the estimated variance of any p_j is $\hat{p}\hat{q}/n_j$. However, the huge X^2 value of **90.62** between generations shows that the combined $\hat{p}$ cannot be regarded as a valid estimate of the proportion of cancer cases within either of the individual generations.

The procedure is to recompute the two 'within-generation' X^2, each from its own 2 × 2 table. These values, which no longer add to the original total X^2, are given in the right hand column of Table 5. The interpretation of the data is not altered. The X^2 values computed from individual parts of the table seldom differ greatly from the additive X^2, but they can do so in certain circumstances and are worth looking at, as a precaution, in analyses of this type.

In this example the rows were divided into **2** groups. The same methods may be applied to test the variation in p among any number of groups, and also within each group. To obtain an additive separation, we subdivide the numerator of X^2 as indicated in the example. Alternatively, for a non-additive separation, we can form an auxiliary table in which each row is a group total and obtain the X^2 for this table, and a further X^2 for each group considered by itself.

6.2. Test for a Linear Regression of p

In some contingency tables we may expect that the p_j will bear a linear relation to a variate z_i that is defined for each of the N rows of the table. In others, where the rows fall into a natural order, it is not unreasonable to assign scores z_j to the rows, in an attempt to convert the ordering into a continuous scale, with which the p_j may show a linear relation. Since p_j is assigned a weight $n_j/\hat{p}\hat{q}$, the regression coefficient b of p_j on z_j is obtained by the standard formula for weighted regressions:

$$b = \frac{\Sigma n_j(p_j - \hat{p})(z_j - \bar{z}_w)}{\Sigma n_j(z_j - \bar{z}_w)^2} \tag{11}$$

where $\bar{z}_w$ is the weighted mean of the z_j.

For computing purposes, the numerator and denominator of b are where $\bar{z}_w$ is the weighted mean of the z_j. conveniently expressed as follows (note that $n_j p_j = x_j$):

$$\text{Num.} = \sum x_j z_j - \frac{T_x(\Sigma n_j z_j)}{T} \tag{12}$$

$$\text{Den.} = \sum n_j z_j^2 - \frac{(\Sigma n_j z_j)^2}{T} \tag{13}$$

The X^2 for regression, with 1 d.f., is

$$X^2 = \frac{(\text{Num.})^2}{\hat{p}\hat{q}(\text{Den.})} \tag{14}$$

As an illustration, the data in Table 6, for which I am indebted to the Leonard Wood Memorial, (American Leprosy Foundation) are taken from an experiment on the use of drugs (sulfones and strepto-mycin) in the treatment of leprosy. The rows denote the *change* in the overall clinical condition of the patient during 48 weeks of treatment: the columns indicate the degree of infiltration (a measure of a certain type of skin damage) present at the beginning of the experiment. The question of interest is whether patients with much initial infiltration progressed differently from those with little infiltration. Patients did not all receive the same drugs, but since no difference in the effects of drugs could be detected, it was thought that the data for different drugs could be combined for this analysis.

The total X^2 is 6.88, with 4 d.f. (P about 0.16). However, the p_j (the proportions of patients with severe infiltration) decline steadily from 39% in the "markedly improved" class to 8% in the "worse" class. This suggests that a regression of the p_j on the clinical change might furnish a more sensitive test.

The data are typical of many tables in that the rows (clinical

TABLE 6

196 Patients Classified According to Change in Condition and Degree of Infiltration

Clinical change		Score z_j	Degree of infiltration 0 − 7	Degree of infiltration 8 − 15 x_j	Total n_j	$p_j = x_j/n_j$ (in %)	$n_j z_j$
Improve-ment	Marked	3	11	7	18	39	54
	Moderate	2	27	15	42	36	84
	Slight	1	42	16	58	28	58
	Stationary	0	53	13	66	20	0
	Worse	−1	11	1	12	8	−12
Total			144	52	196	0.26531	184
				T_x	T	$\hat{p}$	

changes) are ordered. In order to compute a regression, this ordering must be replaced by a numerical scale. I have supposed that scores 3, 2, 1, 0, −1, as shown in Table 6, have been assigned to the five classes of clinical change. Such scores are to some extent subjective and arbitrary, and some scientists may feel that the assignment of scores is slightly unscrupulous, or at least they are uncomfortable about it. Actually, any set of scores gives a *valid* test, provided that they are constructed without consulting the results of the experiment. If the set of scores is poor, in that it badly distorts a numerical scale that really does underlie the ordered classification, the test will not be sensitive. The scores should therefore embody the best insight available about the way in which the classification was constructed and used. In the present example, I considered an alternative set 4, 2, 1, 0, −2, on the grounds that the doctor seemed to deliberate very carefully before assigning a patient to the "markedly improved" or to the "worse" class, but I decided that the presumption in favor of this scale was not strong enough.

To compute the regression X^2, the only supplementary column needed is that of the products $n_j z_j$, shown on the right of Table 6. From equations (12) and (13) for b,

$$\text{Num.} = \sum x_j z_j - \frac{T_x(\Sigma n_j z_j)}{T}$$

$$= (7)(3) + (15)(2) + \cdots + (1)(-1) - \frac{(52)(184)}{196}$$

$$= 17.1837 \tag{15}$$

$$\text{Den.} = \sum n_j z_j^2 - \frac{(\Sigma n_j z_j)^2}{T}$$

$$= (54)(3) + (84)(2) + \cdots + (-12)(-1) - \frac{(184)^2}{196}$$

$$= 227.2653 \tag{16}$$

$$X^2 \text{ for regression} = \frac{(17.1837)^2}{(227.2653)\hat{p}\hat{q}} = 6.666 \tag{1 d.f.}$$

This is significant at the 1% level. The total X^2 has now been subdivided as follows.

	d.f.	X^2
Regression of p_j on z_j	1	3.67
Deviations from regression	3	0.21
Total	4	6.88

7. THE COMBINATION OF 2 × 2 CONTINGENCY TABLES

Suppose that we are comparing the frequencies of some occurrence in two independent samples, and that the whole procedure is repeated a number of times under somewhat differing environmental conditions. The data then consist of a series of 2×2 tables, and the problem is to make a combined test of significance of the difference in occurrence rates in the two samples. The data obtained in comparing the effectiveness of two agents in dosage-mortality experiments are a typical example, in which the repetitions of the experiment are made under a series of different dosage levels. My concern here, however, is with cases where there is no variate corresponding to dosage level, and no well-established theory of how to combine the data.

One method that is sometimes used is to combine all the data into a single 2×2 table, for which X^2 is computed in the usual way. This procedure is legitimate only if the probability p of an occurrence (on the null hypothesis) can be assumed to be the same in all the individual 2×2 tables. Consequently, if p obviously varies from table to table, or we suspect that it may vary, this procedure should not be used.

Another favorite technique is to compute the usual X^2 separately for each table, and add them, using the fact that the sum of g values of χ^2, each with 1 d.f., is distributed as χ^2 with g d.f. This is a poor method. It takes no account of the signs of the differences $(p_1 - p_2)$ in the two samples, and consequently lacks power in detecting a difference that shows up consistently in the same direction in all or most of the individual tables.

An alternative is to compute the X values, and add them, taking account of the signs of the differences. Since X is approximately normally distributed with mean 0 and unit S.D., the sum of g independent X values is approximately normally distributed with mean 0 and S.D. $\sqrt{g}$. Hence the test criterion,

$$\frac{\Sigma X}{\sqrt{g}}$$

is referred to the standard normal tables.

This method has much to commend it if the total N's of the individual tables do not differ greatly (say by more than a ratio of 2 to 1) and if the p's are all in the range 20–80%. For the following illustrative data I am indebted to Dr. Martha Rogers. The comparison is between mothers of children in the Baltimore schools who had been referred by their teachers as presenting behavior problems, and mothers of a comparable group of control children who had not been so referred. For each mother, it was recorded whether she had suffered any infant losses (e.g. stillbirths) previous to the birth of the child in the study. The comparison is part of a study of possible associations between behavior problems in children and complications of pregnancy of the mother. Since these loss rates increase with later birth orders, and since the samples might not be comparable in birth orders, the data were examined separately, as a precaution, for 3 birth-order classes (see Table 7). The two groups of children are referred to as 'Problems' and 'Controls.'

Note that the loss rate is higher in the 'Problems' sample in all 3 tables. Since the N's in the separate tables lie within a 2:1 ratio, and the p's are between 18% and 48%, addition of the X values is indicated. The individual X values are, respectively, 0.650, 0.436, 1.587, all being

TABLE 7
Data on Number of Mothers with Previous Infant Losses

Birth Order		No. of mothers with		Total	% Loss
		Losses	None		
2	Problems	20	82	102	19.6
	Controls	10	54	64	15.6
		30	136	$166 = N_1$	18.1
3–4	Problems	26	41	67	38.8
	Controls	16	30	46	34.8
		42	71	$113 = N_2$	37.2
5+	Problems	27	22	49	55.1
	Controls	14	23	37	37.8
		41	45	$86 = N_3$	47.7

given the same sign since the difference is in the same direction. For this test, the X values are computed without the correction for continuity. Hence the test criterion is the approximate normal deviate

$$\frac{0.650 + 0.436 + 1.587}{\sqrt{3}} = 1.54$$

The P value is just above 0.10. Addition of the X^2 values gives 3.131, with 3 d.f., corresponding to a P of about 0.38.

If the N's and p's do not satisfy the conditions mentioned, addition of the X's tends to lose power. Tables that have very small N's cannot be expected to be of much use in detecting a difference, yet they receive the same weight as tables with large N's. Where differences in the N's are extreme, we need some method of weighting the results from the individual tables. Further, if the p's vary from say 0 to 50%, the difference that we are trying to detect, if present, is unlikely to be constant at all levels of p. A large amount of experience suggests that the difference is more likely to be constant on the probit or logit scale. As a further complication, the term pq in the variance of a difference will change from one 2×2 table to another.

Perhaps the best method for a combined analysis is to transform the data to a probit or logit scale. Examples of this type of analysis are given by Winsor (1948) and Dyke and Patterson (1952): it is recommended if the data are extensive enough to warrant a searching examination. As an alternative, the following test of significance in the original scale will, I believe, be satisfactory under a wide range of variations in the N's and p's from table to table.

For the ith 2×2 table, let

$$n_{i1}, n_{i2} = \text{sample sizes}$$
$$p_{i1}, p_{i2} = \text{observed proportions in the two samples}$$
$$\hat{p}_i = \text{combined proportion from the margins}$$
$$d_i = p_{i1} - p_{i2} = \text{observed difference in proportions}$$
$$w_i = \frac{n_{i1}n_{i2}}{n_{i1} + n_{i2}}: \qquad w = \Sigma w_i$$

Then we compute the weighted mean difference

$$d = \frac{\Sigma w_i d_i}{w}$$

This has a standard error

$$\text{S.E.} = \frac{\sqrt{\Sigma w_i \hat{p}_i \hat{q}_i}}{w}$$

The test criterion is

$$\frac{\bar{d}}{\text{S.E.}} = \frac{\Sigma w_i d_i}{\sqrt{\Sigma w_i \hat{p}_i \hat{q}_i}}$$

TABLE 8
Mortality by Sex of Donor and Severity of Disease

Degree of disease	Sex of donor	Number of		Total	% deaths
		Deaths	Surv.		
None	M	2	21	23	8.7
	F	0	10	10	0.0
	Total	2	23	$33 = N_1$	$6.1 = \hat{p}_1$
Mild	M	2	40	42	4.8
	F	0	18	18	0.0
	Total	2	58	$60 = N_2$	$3.3 = \hat{p}_2$
Moderate	M	6	33	39	15.4
	F	0	10	10	0.0
	Total	6	43	$49 = N_3$	$12.2 = \hat{p}_3$
Severe	M	17	16	33	51.5
	F	0	4	4	0.0
	Total	17	20	$37 = N_4$	$45.9 = \hat{p}_4$

This is referred to the tables of the normal distribution. As explained in the Appendix, which gives supporting reasons for this criterion, the criterion was constructed so that it would be powerful if the alternative hypothesis implies a constant difference on either the probit or the logit scale. The form of the criterion is not one that I would have selected intuitively, and the reader who feels the same way should consult the Appendix.

The test will be illustrated from data published by Diamond et al. *See* Allen et al., 1949. Erythroblastosis fetalis is a disease of newborn infants, sometimes fatal, caused by the presence in the blood of an $Rh+$ baby, of anti-Rh antibody transmitted by his $Rh-$ mother. One form of treatment is an "exchange transfusion," in which as much as possible of the infant's blood is replaced by a donor's blood that is free of anti-Rh antibody. In 179 cases in which this treatment was used in a Boston hospital, the rather startling finding was made that there were no infant deaths out of 42 cases in which a female donor was used, but 27 infant deaths out of 137 cases in which a male donor was used. Since there seemed no *a priori* reason why there should be less hazard with female donors, a statistical investigation was made and reported in the reference.

One possibility was that male donors had been used primarily in the more severe cases. Consequently, the data were classified according to the stage of disease at birth, giving the four 2×2 tables shown in Table 8.

The N's do not vary greatly, but the p's range from 3 to 46%. The combined test of significance is made from the supplementary data in Table 9.

TABLE 9
Computations for the Combined Test

Stage	d_i	$\hat{p}_i$	$\hat{p}_i\hat{q}_i$	$w_i = \dfrac{n_{i1}n_{i2}}{n_{i1} + n_{i2}}$
None	+8.7	6.1	573	7.0
Mild	+4.8	3.3	319	12.6
Moderate	+15.4	12.2	1071	8.0
Severe	+51.5	45.9	2483	3.6

$$\frac{\bar{d}}{\text{S.E.}} = \frac{\Sigma w_i d_i}{\sqrt{\Sigma w_i \hat{p}_i \hat{q}_i}} = \frac{429.88}{\sqrt{25{,}537}} = 2.69$$

The significance probability is 0.0072. In data of this kind, a non-significant result would indicate that the surprising phenomenon can be explained by differences in the selection of cases. A significant result must be interpreted with caution, since conditions were not necessarily comparable for male and female donors even within cases of a given degree of severity. However, a significant result does encourage further study, e.g. by examining results in other hospitals.

The expectations in table 9 are so low that one might well doubt the validity of a normal approximation. At least, I did. However, the exact distribution of $\Sigma \ w_i d_i$ can be worked out, with some labor, by writing down the probabilities of all possible configurations for each 2×2 table. The total numbers of configurations are 3, 3, 7 and 5, respectively, for the four tables, so that the total number of possible samples is 315, of which some have negligible probabilities. The value of the test criterion and the probability was worked out for each sample. The exact significance level was found to be 0.0095, as against the normal approximation of 0.0072. The degree of agreement is reassuring, considering the extreme smallness of the expectations. I am indebted to Mrs. Leah Barron for performing the computations.

References

Allen, F. H., Diamond, Z. K., & Watrous, J. B. Erythroblastosis fetalis v. The value of blood from female donors for exchange transfusion. *The New England Journal of Medicine*, 1949, **241**, 799–806.

Anscombe, F. J. Sampling theory of the negative binomial and logarithmic distributions. *Biometrika*, 1950, **37**, 358–382.

Berkson, J. Some difficulties of interpretation encountered in the application of the chi-square test. *Journal of the American Statistical Association*, 1938, **33**, 526–536.

Berkson, J. A note on the chi-square test, the Poisson and the binomial. *Journal of the American Statistical Association*, 1940, **35**, 362–367.

Bliss, C. I. Fitting the negative binomial distribution to biological data. *Biometrics*, 1953, **9**, 176–196.

Cochran, W. G. Statistical analysis of field counts of diseased plants. *Journal of the Royal Statistical Society, Supplement*, 1936, **3**, 49–67. (a)

Cochran, W. G. The χ^2 distribution for the binomial and Poisson series, with small expectations. *Annals of Eugenics*, 1936, **7**, 207–217. (b)

Cochran, W. G. The χ^2 correction for continuity. *Iowa State College Journal of Science*, 1942, **16**, 421–436.

Cochran, W. G. The χ^2 test of goodness of fit. *Annals of Mathematical Statistics*, 1952, **23**, 315–345.

Dyke, G. V., & Patterson, H. D. Analysis of factorial arrangements when the data are proportions. *Biometrics*, 1952, **8**, 1–12.

Fix, E. Tables of the noncentral χ^2, *University of California Publication on Statistics*, 1949, **1**, 15–19.

Freeman, G. H., & Halton, J. H. Note on an exact treatment of contingency, goodness of fit and other problems of significance. *Biometrika*, 1951, **38**, 141–149.

Haldane, J. B. S. The mean and variance of χ^2 when used as a test of homogeneity, when expectations are small. *Biometrika*, 1939, **31**, 346–355.

Lancaster, H. O. The derivation and partition of χ^2 in certain discrete distribution. *Biometrika*, 1949, **36**, 117–129.

Lewis, D., & Burke, C. J. The use and misuse of chi-square test. *Psychological Bulletin*, 1949, **46**, 433–498.

Mainland, D. Statistical methods in medical research. *Canadian Journal of Research, E*, 1948, **26**, 1–166.

Murphy, D. P. *Heredity in uterine cancer.* Cambridge, Mass.: Harvard University Press, 1952.

Neyman, J., & Pearson, E. S. Further notes on the χ^2 distribution, *Biometrika*, 1931, **22**, 298–305.

Sukhatme, P. V. On the distribution of χ^2 in small samples of the Poisson series. *Journal of the Royal Statistical Society, Supplement*, 1938, **5**, 75–79.

Winsor, C. P. Factorial analysis of a multiple dichotomy. *Human Biology*, 1948, **20**, 195–204.

3

Parametric Techniques

INTRODUCTION

In statistics courses much time is usually spent on the most common methods of statistical analysis, such as Student's *t* test and the analysis of variance. Therefore Chapter 3 does not belabor these techniques. Instead, it presents articles that cover topics very much related to the standard parametric techniques but not necessarily covered intensively in the normal course offering.

People conduct research to answer a question; often they get involved with this question and test their idea experimentally as quickly as possible. Sometimes the desired information results from the experiment and sometimes it does not. Depending upon how the experiment is executed, the results may be so confounded with factors other than the specific one under consideration that the experimenters cannot answer the question they set out to answer. A perfect example of an experiment that was executed incorrectly is provided by "Student" in "The Lanarkshire milk experiment." The experimenters asked a very simple question: Would milk in the school diet enhance children's growth? The experiment was conducted and "Student" reviews for us why the authors of the study should not have drawn the conclusions they drew. This is not a difficult article to read, but to appreciate it one must remember that though it was written in 1931, today one can still find the same experimental mistakes that "Student" discussed then.

It should be pointed out that "Student" was the pen name for W. S. Gosset. Mr. Gosset worked in a brewery and thus was required to publish in respectable journals under a pen name. W. S. Gosset was responsible for the development of what is now called the *t* test, correctly called Student's *t* statistic.

Often we collect experimental data that do not conform to the required assumptions of our statistical tests. M. S. Bartlett, in his article on transformations, presents techniques that may enable you to rescale your scores or measures in such a way that the assumptions can be satisfied.

For example, one may have violated the assumption of homogeneity of variance. Homogeneity of variance refers to the assumption that both samples have the same variance. In other words, we can say that we are expecting the difference between groups to be in their means, not their variability. If the treatment increases the heterogeneity of scores in an experimental group, the variance of the control and experimental groups will not be statistically equivalent. Suppose we end our experiment with data that does violate the assumption of homogeneity of variance." What could we do, besides run the study over again? Bartlett suggests several transformations, such as the square root or log, that will aid us in this situation. The particular transformation of scores into new scaled values, to satisfy the violated assumption, depends upon the specific data under consideration. It should be apparent, however, that Bartlett's article will provide a sound basis for the selection of the proper transformation should it become necessary in your own statistical calculations.

The last article, by W. G. Cochran, describes the analysis of covariance and its proper use. Analysis of covariance is a statistical technique that allows us to evaluate the response or dependent measures to some variable (y) with greater statistical precision by removing the contribution of another variable (x). Thus we may statistically control for the contribution of the second or other variables. This technique is particularly fitting for the social sciences, since we often employ observational studies and cannot experimentally control "disturbing variables" in the natural setting. For example, we may be concerned with a comparison of the mean weight of children from a ghetto and those not from the ghetto. We may know that age is an important factor in the determination of weight. Since we cannot experimentally control for age in our observational study, we can statistically control the contribution of age by using the analysis of covariance, and we will then have a more precise comparison of mean weights of ghetto and non-ghetto children. What we are trying to do is to evaluate and then eliminate the contribution of all but the variable under study, in this case place of residence. Analysis of covariance is a very useful technique, and the article by Cochran provides a good background for it.

3.1

The Lanarkshire
milk experiment

By "Student."

Biometrika, 1931, 23, 398-

In the spring of 1930* a nutritional experiment on a very large scale was carried out in the schools of Lanarkshire.

For four months 10,000 school children received ¾ pint of milk per day, 5000 of these got raw milk and 5,000 pasteurised milk, in both cases Grade A (Tuberculin tested); another 10,000 children were selected as controls and the whole 20,000 children were weighed and their height was measured at the beginning and end of the experiment.

It need hardly be said that to carry out an experiment of this magnitude successfully requires organisation of no mean order and the whole business of distribution of milk and of measurement of growth reflects great credit on all those concerned.

It may therefore seem ungracious to be wise after the event and to suggest that had the arrangement of the experiment been slightly different the results would have carried greater weight, but what follows is written not so much in criticism of what was done in 1930 as in the hope that in any further work full advantage may be taken of the light which may be thrown on the best methods of arrangement by the defects as well as by the merits of the Lanarkshire experiment.

The 20,000 children were chosen in 67 schools not more than 400 nor less than 200 being chosen in any one school, and of these half were assigned as "feeders" and half as "controls," some schools were provided with raw milk and the others with pasteurised milk, no school getting both.

This was probably necessary for administrative reasons, owing to the difficulty of being sure that each of as many as 200 children gets the right kind of milk every day if there were a possibility of their

---

* Department of Health for Scotland. *Milk Consumption and the Growth of School Children.* By Dr. Gerald Leighton and Dr. Peter L. McKinlay. (Edinburgh and London: H.M. Stationery Office, 1930.)

getting either of the two. Nevertheless, as I shall point out later, this does introduce the possibility that the raw and pasteurised milks were tested on groups of children which were not strictly comparable.

Secondly, the selection of the children was left to the Head Teacher of the school and was made on the principle that both "controls" and "feeders" should be representative of the average children between 5 and 12 years of age; the actual method of selection being important I quote from Drs. Leighton and McKinlay's Report: "The teachers selected the two classes of pupils, those getting milk and those acting as 'controls,' in two different ways. In certain cases they selected them by ballot and in others on an alphabetical system." So far, so good, but after invoking the goddess of chance they unfortunately wavered in their adherence to her for we read: "In any particular school when there was any group to which these methods had given an undue proportion of well fed or ill nourished children, others were substituted in order to obtain a more level selection." This is just the sort of after-thought that most of us have now and again and which is apt to spoil the best laid plans. In this case it was a fatal mistake, for in consequence the controls were, as pointed out in the Report*, definitely superior both in weight and height to the "feeders" by an amount equivalent to about 3 months' growth in weight and 4 months' growth in height.

Presumably this discrimination in height and weight was not made deliberately, but it would seem probable that the teachers, swayed by the very human feeling that the poorer children needed the milk more than the comparatively well to do, must have unconsciously made too large a substitution of the ill-nourished among the "feeders" and too few among the "controls" and that this unconscious selection affected, secondarily, both measurements.

Thirdly, it was clearly impossible to weigh such large numbers of children without impedimenta. They were weighed in their indoor clothes, with certain obvious precautions, and the difference in weight between their February garb and their somewhat lighter clothing in June is thus necessarily substracted from their actual increase in weight between the beginning and end of the experiment. Had the selection of "controls" and "feeders" been a random one, this fact, as pointed out in the Report*, would have mattered little, both classes would have been affected equally, but since the selection was probably affected by poverty it is reasonable to suppose that the "feeders" would lose less weight from this case than the "controls." It is therefore not surprising to find that the gain in weight of "feeders" over "controls," which includes this constant error, was more marked, relatively to their growth rate, than was their gain in height, which was fortunately not similarly affected.

* See footnote on p. 398.

[159]

Fourthly, the "controls" from those schools which took raw milk were bulked with those from the schools which took pasteurised milk.

Now with only 67 schools, at best 33 against 34, in a district so heterogeneous both racially and socially, it is quite possible that there was a difference between the averages of the pupils at 33 schools and those of the pupils at another 34 schools both in the original measurements and in the rate of growth during the experiment.

In that case the average "control" could not be used appropriately to compare with either the "raw" group or the "pasteurised" group.

This possibility is enhanced by the aforementioned selection of "controls" which can hardly have been carried out in a uniform manner in different schools.

Fortunately it would still be possible to correct this, for the figures for the different schools must still be available in the archives.

Diagrams 1 and 2 give the average heights of "controls," raw milk "feeders" and pasteurised milk "feeders" for boys and girls respectively. The heights at the beginning of the experiments are set out against a uniform age scale centering each group at the half year above the whole

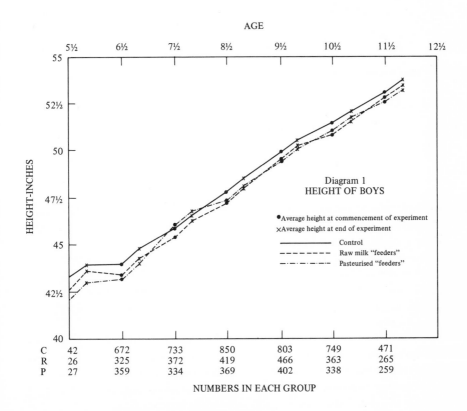

DIAGRAM 1   HEIGHT OF BOYS

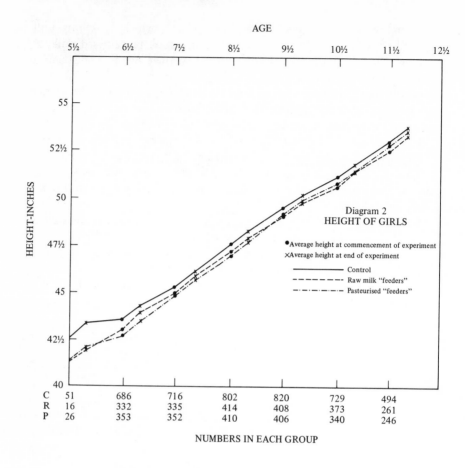

AGE

HEIGHT-INCHES

Diagram 2
HEIGHT OF GIRLS

•Average height at commencement of experiment
✕Average height at end of experiment

——————— Control
– – – – – – – Raw milk "feeders"
–·—·—·—·– Pasteurised "feeders"

	5½	6½	7½	8½	9½	10½	11½
C	51	686	716	802	820	729	494
R	16	332	335	414	408	373	261
P	26	353	352	410	406	340	246

NUMBERS IN EACH GROUP

DIAGRAM 2   HEIGHT OF GIRLS

number. This is doubtless accurate enough except for the first group aged
"5 and less than 6," which was very much smaller in numbers than the
other groups, either because only the older (or larger) children are sent
to school between 5 and 6 or because the teachers did not think that the
smaller children would be able to play their part. For this reason they
should probably be centred more to the right compared to the others. A
similar argument might lead us to centre the "11 and over" group a little
more to the left.

The average heights at the end of the experiment are of course set
out four months to the right of those at the beginning and it will be
noticed that except for the first group, which is clearly out of place, not
any of the points diverge very much from their appropriate line of
growth whether "controls," "raws" or "pasteuriseds."

The case is very different in Diagrams 3 and 4 which show the corresponding average weights. Here there is, after the first two ages, a very decided dip, especially in the later ages. The weights at the end of the experiment are too low. This might be accounted for by a tendency in older children to grow normally in height and subnormally in weight during the spring, but I think it much more likely that older children wear about 1 lb. more clothes in February than they do in June, while in the case of younger children a more limited wardrobe permits of fewer discards.

The authors have tried to show that the selection of the "controls" has not affected the validity of the comparisons, by computing the correlation coefficients between the original heights (and weights) and the growth during the experiment for each of the 42 age groups into which the measurements were divided. These they find to be quite small even though they are here and there significant, and they argue that the additional height and weight of the "controls" was without effect on the comparison of subsequent growth.

Now this might have been a perfectly good argument had the height and weight been selected directly, but if, as I have indicated was very likely the case, the selection was made according to some unconscious scale of well being, then it is surely natural to suppose that the relatively ill nourished "feeders" would benefit more than their more fortunate school mates, the "controls," would have done by the extra ¾ pint of milk per day.

That being so how are we to regard the conclusions of the Report*:

(1) "The influence of the addition of milk to the diet of school children is reflected in a definite increase in the rate of growth both in height and weight."

This conclusion was probably true; the average increase for boys' and girls' height was 8 per cent. and 10 per cent. over "controls" and for boys' and girls' weights was 30 per cent. and 45 per cent., respectively, and though, as pointed out, the figures for weights were wholly unreliable it is likely enough that a substantial part of the difference in height and a small part of that in weight were really due to the good effect of the milk. The conclusion is, however, shifted from the sure ground of scientific inference to the less satisfactory foundation of mere authority and guesswork by the fact that the "controls" and "feeders" were not randomly selected.

(2) "There is no obvious or constant difference in this respect between boys and girls and there is little evidence of definite relation between the age of the children and the amount of improvement. The results do not support the belief that the younger derived more benefit than the older children. As manifested merely by growth in weight and height the increase found in younger children through the addition of

---

* See footnote on p. 398.

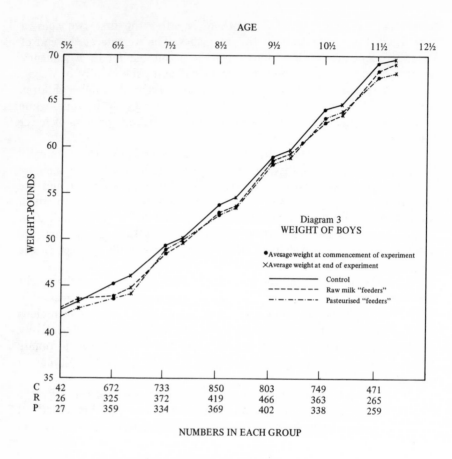

DIAGRAM 3   WEIGHT OF BOYS

milk to the usual diet is certainly not greater than, and is probably not even as great as, that found in older children."

Now from the authors' point of view, believing in the validity of their comparisons in weight, this is much understating the case, as the following table derived from Capt. Bartlett's condensed tables shows:

Age in years	Gain in weight in ozs. by Feeders over Controls		Gain in height in inches by Feeders over Controls		As % of control			
					Weight		Height	
	Boys	Girls	Boys	Girls	Boys	Girls	Boys	Girls
5, 6 and 7	1.13 ± .73	1.24 ± .72	.083 ± .011	.059 ± .011	9	13	11	8
8 and 9	3.15 ± .68	4.47 ± .67	.071 ± .011	.098 ± .010	30	51	10	14
10 and 11	5.21 ± .85	7.88 ± .79	.037 ± .012	.055 ± .012	78	73	5	8

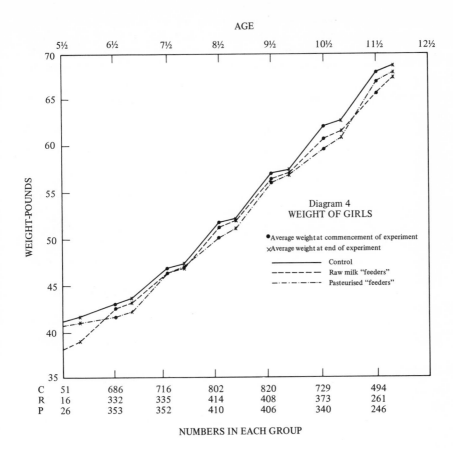

AGE

DIAGRAM 4   WEIGHT OF GIRLS

Note that the P.E.'s are calculated from Capt. Bartlett's tables and are subject, as his are, to his having interpreted the methods of the original Report correctly.

From this they might have concluded:

(*a*) That in the matter of weight older children, both boys and girls, derived more benefit than younger, while

(*b*) In height the younger boys did better than the older, though the difference is not quite significant, but that there was no regular tendency in the matter of girls' height.

In the light of previous criticism, however, we must be content to say that apparently the differential shedding of clothes between the "feeders" and the more fortunate "controls" is more marked with older children (and possibly with girls than with boys), and that there is some probability that younger boys gain in height more than older.

Finally, conclusion (3) runs: "In so far as the conditions of this investigation are concerned the effects of raw and pasteurised milk on growth in weight and height are, so far as we can judge, equal."

This conclusion has been challenged by Capt. Bartlett*, and by Dr. Fisher and Capt Bartlett†, who conclude that there is definite evidence of the superiority of raw over pasteurised milk in both height and weight.

Even they, however, point out that the raw and pasteurised milk were not supplied to the same schools, and their conclusion amounts to saying: "If the groups of children taking raw and pasteurised milk respectively were random samples from the same population, the observed differences would be decisively in favour of the raw milk."

Unfortunately they were not random samples from the same population; they were selected samples from populations which may have been different, and moreover the "controls" with which they were compared were not appropriate to either group; and so—again it is a matter of guess and authority—I would be very chary of drawing any conclusion from these small biased differences.

That is not to say that there is no difference between the effect of raw and pasteurised milk—personally I believe that there is and that it is in favour of raw milk—but that this experiment, in spite of all the good work which was put into it, just lacked the essential condition of randomness which would have enabled us to prove the fact.

This note would be incomplete without some constructive proposals in case it should be considered necessary to do further work upon the subject, and accordingly I suggest the following:

(1) If it should be proposed to repeat the experiment on the same spectacular scale,

(a) The "controls" and "feeders" should be chosen by the teachers in pairs of the same age group and sex, and as similar in height, weight and especially physical condition (i.e. well or ill nourished) as possible, and divided into "controls" and "feeders" by tossing a coin for each pair. Then each pair should be considered to be a unit and the gain in weight and height by the "feeder" over his own "control" should also be considered as a unit for the purpose of determining the error of the gain in weight or height.

In this way the error will almost certainly be smaller, perhaps very much smaller, than if calculated from the means of "feeders" and "controls."

If in addition the social status of each pair be noted (well to do, medium, poorly nourished or some such scale) further useful information will be available for comparing pasteurised and raw "feeders."

---

* "Nutritional Value of Raw and Pasteurized Milk," by Stephen Bartlett, M. C., B.Sc. (*Journal of the Ministry of Agriculture*, April, 1931).

† *Nature*, April 18th, 1931, p. 591, "Pasteurised and Raw Milk."

If this is found to be too difficult a perfectly good comparison can be made by adhering to the original plan of the 1930 experiment and drawing lots to decide which should be "controls" and which "feeders" (this is better than an alphabetical arrangement), but the error of the comparison is likely to be larger than in the plan outlined above.

(b) If it is at all possible each school should supply an equal number of raw and pasteurised "feeders," again by selection of similar children followed by coin tossing, but I fear that this is a counsel of perfection.

(c) Some effort should be made to estimate the weight of clothes worn by the children at the beginning and end of the experiment: possibly the time of year could be chosen so that there would be little change in this respect.

(2) If it be agreed that milk is an advantageous addition to children's diet—and I doubt whether any one will combat that view—and that the difference between raw and pasteurised milk is the matter to be investigated, it would be possible to obtain much greater certainty at an expenditure of perhaps 1–2 per cent. of the money* and less than 5 per cent. of the trouble.

For among 20,000 children there will be numerous pairs of twins; exactly how many it is not easy to say owing to the differential death rate, but, since there is about one pair of twins in 90 births, one might hope to get at least 160 pairs in 20,000 children. But as a matter of fact the 20,000 children were not all the Lanarkshire schools population, and I feel pretty certain that some 200–300 pairs of twins would be available for the purpose of the experiment.

Of 200 pairs some 50 would be "identicals" and of course the same sex, while half the remainder would be non-identical twins of the same sex.

Now identical twins are probably better experimental material than is available for feeding experiments carried out on any other mammals, and the error of the comparison between them may be relied upon to be so small that 50 pairs of these would give more reliable results than the 20,000 with which we have been dealing.

The proposal is then to experiment on all pairs of twins of the same sex available, noting whether each pair is so similar that they are probably "identicals" or whether they are dissimilar.

"Feed" one of each pair on raw and the other on pasteurised milk, deciding in each case which is to take raw by the toss of a coin.

Take weekly measurements and weight without clothes.

Some way of distinguishing the children from each other is necessary or the mischievous ones will play tricks. The obvious method is to take

---

* This is a serious consideration; the Lanarkshire experiment cost about £7500.

finger-prints, but as this is identified with crime in some people's minds, it may be necessary to make a different indelible mark on a fingernail of each, which will grow off after the experiment is over.

With such comparatively small numbers further information about the dietetic habits and social position of the children could be collected and would doubtless prove invaluable.

The comparative variation in the effect in "identical" twins and in "unlike" twins should furnish useful information on the relative importance of "Nature and Nurture."

To sum up: The Lanarkshire experiment devised to find out the value of giving a regular supply of milk to children, though planned on the grand scale, organised in a thoroughly business-like manner and carried through with the devoted assistance of a large team of teachers, nurses and doctors, failed to produce a valid estimate of the advantage of giving milk to children and of the difference between raw and pasteurised milk.

This was due to an attempt to improve on a random selection of the controls which in fact selected as controls children who were on the average taller and heavier than those who were given milk.

The hypothesis is advanced that this was due not to a selection of the shorter, lighter children as such to take the milk, but to an unconscious bias leading the teachers to pick out for this purpose the needier children whom the milk would be most likely to benefit.

This hypothesis is supported by the fact that while the advantage derived from the milk was only 8–10 per cent. of the gain in height, without much variation for age, it was 30–45 per cent. of the gain in weight, varying from 9–13 per cent. in the younger children (who do not seem to have shed much clothing in the summer) up to 73–78 per cent. in the older children—who obviously did.

Suggestions are made for the arrangement:

(1) Of a similar large scale experiment on random lines, and

(2) Of a much smaller and cheaper experiment carried out on pairs of twins of like sex.

The second is likely to provide a much more accurate determination of the point at issue, owing to the possibility of balancing both nature and nurture in the material of the experiment.

3.2

The use of transformations

M. S. Bartlett

Biometrics, 1947, 3, 39-52,

1. THEORETICAL DISCUSSION

The purpose of this note is to summarize the transformations which have been used on raw statistical data, with particular reference to analysis of variance. For any such analysis the usual purpose of the transformation is to change the scale of the measurements in order to make the analysis more valid. Thus the conditions required for assessing accuracy in the ordinary unweighted analysis of variance include the important one of a constant residual or error variance, and if the variance tends to change with the mean level of the measurements, the variance will only be stabilized by a suitable change of scale.

If the form of the change of variance with mean level is known, this determines the type of transformation to use. Suppose we write

$$\sigma_x^2 = f(m), \qquad (1)$$

where σ_x^2 is the variance on the original scale of measurements x with the mean of x equal to m. Then for any function $g(x)$ we have approximately[1]

$$\sigma_g^2 = (dg/dm)^2 f(m), \qquad (2)$$

so that if σ_g^2 is to be constant, C^2 say, we must have

$$g(m) = \int \frac{C dm}{\sqrt{f(m)}}. \qquad (3)$$

For example if the standard deviation σ_x tends to be proportional to the mean level m we have $f(m)$ proportional to m^2, and $g(m)$ proportion to log m; i.e., we should use the logarithmic scale. Appropriate scales of this kind for types of data often encountered in statistical analysis are discussed in sections 2, 3, and 4.

---

[1] For a more precise formulation, see Curtiss (1943).

However, a constant variance is not the only condition we seek, and precautions are still necessary when using analysis of variance with the transformed variate. In the ideal case (Bartlett & Vendall, 1946),

(a) The variance of the transformed variate should be unaffected by changes in the mean level (this is taken to be the primary purpose of the transformations of sections 2, 3, and 4).

(b) The transformed variate should be normally distributed.

(c) The transformed scale should be one for which an arithmetic average is an efficient estimate of the true mean level for any particular group of measurements.

(d) The transformed scale should be one for which real effects are linear and additive.

Although these conditions are to some extent related [for example, (a) and (b) and (d) together simply (c)], we obviously cannot necessarily expect to arrange for conditions (b), (c), and (d) to be satisfied if our scale has already been fixed by condition (a).

Fortunately, the transformation of scale to meet condition (a) often has the effect of improving the closeness of the distribution to normality, a correlation of variability with mean level on the original scale often implying excessive skewness which tends to be eliminated after the transformation. But the validity of any assumption of normality should be watched, for while moderate departures from normality are known not to be serious, any large departures in the region of the more outlying observations are likely to affect the validity of significance tests (Bartlett & Kendall, 1946).

Condition (c) is required because the estimates which arise in analysis of variance are of the simple arithmetic average type, and we want to know that such estimates are efficient. The contention is sometimes made that the original scale is the more relevant one for taking sums and averages, and more understandable.

While this argument has some force and is a warning against making transformations without good reason, it loses strength when we remember that if the variability in the data varies with the mean level for different blocks or groups an unweighted average of the *observed* treatment responses is not necessarily the best estimate of the *true* treatment response, and the average on the transformed scale will often be the better estimate when re-converted to the original scale.

Lastly, it is a more effective and a simpler procedure to be working on a scale for which treatment or other effects are linear and additive; this implies in a layout of the randomized block type that real treatment $\times$ block interactions will not inflate the error term, and reduce the apparent significance of the treatments; and relatedly, that for treatments of the factorial type, interactions between the treatments due merely to scale will not necessitate narrower and less powerful conclusions about the treatment effects. Now it is not always possible to choose a scale to cover conditions (a) and yet be most reasonable for (d), though it may happen that a choice of scale for (a) improves the scale to some extent as far as (d) is concerned. In some cases where sufficient information about the variability is known from the nature of the data, we may decide to abandon the advantages of a simple analysis of variance under condition (a), and choose our scale with sole regard to (d), weighting our observations with appropriate weights depending on the known variability. Something like this happens, for example, when we make use of the probit transformation (section 5; cf. also section 7), which is chosen to provide a rational linear scale for percentage mortalities or other analogous percentages.

2. THE SQUARE ROOT TRANSFORMATION

When statistical data consist of integers, i.e., whole numbers, such as number of bacterial colonies in a plate count, or number of plants in a given area, homogeneous conditions will often lead to variation in these numbers x following the Poisson distribution. Since for such a distribution the variance is exactly equal to the mean, we readily obtain from our general equation in section 1, that to stabilize the variance we must work on the square-root scale. We have seen that this is only an approximate result and the exact values for different values of the mean m seem worth quoting from my original paper (Bartlett, 1936b, Table 1), as they may be useful in any comparisons of the observed variance of our data with this theoretical value. I recommended the use of $\sqrt{(x + \frac{1}{2})}$ in place of $\sqrt{x}$ when very small numbers were involved (e.g., means in the range

TABLE 1
Variance of Poisson Variate on Transformed Scale

Mean m (on original scale)	$\sqrt{x}$	$\sqrt{(x + \frac{1}{2})}$
0.0	0.000	0.000
0.5	0.310	0.102
1.0	0.402	0.160
2.0	0.390	0.214
3.0	0.340	0.232
4.0	0.306	0.240
6.0	0.276	0.245
9.0	0.263	0.247
12.0	0.259	0.248
15.0	0.256	0.248

10 to 2, especially when zeros are occurring among the observed numbers), and the variance for this quantity is also shown.

In practice we often use an analysis of variance for data of the above integer type because we suspect heterogeneity of one kind or another to be present, especially if our data have been collected under field conditions. We do not then need to assume Poisson variation, but will still transform to the square-root scale if the variation of $\sqrt{x}$ appears stable. The following set of data (Bartlett, 1936b; Cochran, 1938), representing weed-infestation counts in one of a series of experiments on weed control in cereals, is an example of stability of variance on the square-root scale, even when the level of variability is far higher than expected on the assumptions of a Poisson distribution for x (Table 2).

TABLE 2

Plan Showing Layout of Experiment on Oats, and Numbers of Poppies (in 3¾ sq. ft. Areas)

Block A	(1)* 438	(4) 17	(2) 538	(5) 18	(3) 77	(6) 115
B	(3) 61	(2) 422	(6) 57	(1)* 442	(5) 26	(4) 31
C	(5) 77	(3) 157	(4) 87	(6) 100	(2) 377	(1)* 319
D	(2) 315	(1)* 380	(5) 20	(3) 52	(4) 16	(6) 45

* Control.

As a rule, a scale chosen to stabilize variance will be one on which arithmetic averages will provide efficient estimates, though this requirement, which I have called condition (c), is not altogether independent of (d). For data for which "homogeneity" represents some well-defined assumption such as Poisson variation, it is also useful to be sure that if the data were in fact homogeneous, such estimates are not throwing away too much information, and this condition is satisfied for the Poisson variate transformed to the square-root scale (cf. Bartlett, 1936b, p. 69, where it was noted that the minimum percentage efficiency in large samples was 88 percent for $\sqrt{x}$ and 96½ percent for $\sqrt{(x + \frac{1}{2})}$), even although the best estimate of the true mean of a perfectly homogeneous Poisson set of observations is actually the arithmetic mean on the original scale. In an interesting theoretical paper, Cochran (1940) has discussed the appropriate analysis of variance for data of a Poisson (or binomial) type, for which real block or other group differences extra to the treatment effects may be present provided they are assumed additive on the transformed scale. He shows that the direct analysis of variance on the transformed scale is then really a first approximation to a more exact analysis, in which any loss of efficiency in estimation is reduced to zero. This method, however, becomes irrelevent if the data do not belong to the exact distributional type assumed, and consequently its use would seem rarely justifiable in practice for field data of the type here considered.

3. LOGARITHMIC TRANSFORMATIONS

The stability of variance on the square-root scale in the case of the series of weed-control experiments was rather unexpected; since, if considerable heterogeneity in numbers is present, the variance is often found still to be correlated with the mean level on a square-root scale, and may only be stabilized if transformation is made to the logarithmic scale. The natural explanation of a variance greater than the mean is that the mean level itself fluctuates, so that

$$\sigma_x^2 = m + \sigma_m^2. \tag{4}$$

For biological populations, increases in numbers are often proportional to the numbers already present, giving rise to variations in mean from place to place themselves proportional to the mean. This illustrates how σ_m^2 might be proportional to m^2, so that we might expect

$$\sigma_x^2 = m + \lambda^2 m^2. \tag{5}$$

For λ large, or m large, this variance law implies the logarithmic transmation.

In some problems it is possible that λ could be estimated well enough to justify a more exact transformation corresponding to a variance of the type represented by equation (5). This transformation would be $\lambda^{-1} \mathrm{Sinh}^{-1} [\lambda \sqrt{x}]$, or equivalently $\lambda^{-1} \log \{ \sqrt{(1 + \lambda^2 x)} + \lambda \sqrt{x} \}$ (Beall, 1942). For example, it is known that under certain assumptions about the way m varies, the Poisson distribution becomes a "negative binomial" distribution, this distribution often fitting observational data which do not conform to the narrower Poisson type. For such data $\lambda^{-1} \mathrm{Sinh}^{-1} [\lambda \sqrt{x}]$ transformation would be appropriate.[2] For small $\sqrt{x}$ it becomes equivalent to the $\sqrt{x}$ transformation, and for small numbers the transformation $\lambda^{-1} \mathrm{Sinh}^{-1} [\lambda \sqrt{(x + \frac{1}{2})}]$ would seem somewhat better. For large $\lambda \sqrt{x}$ it becomes equivalent to the logarithmic transformation.

This transformation, however, has the disadvantage of requiring an

---

[2] Compare the $\mathrm{Sin}^{-1} \sqrt{x}$ transformation for the ordinary binomial (section 4).

approximate knowledge of λ, and the empirical transformation log $(1 + x)$, which has been suggested in place of log x as a logarithmic transformation for integers to avoid the difficulty with zeros in the case of log x, seems likely to prove good enough in many cases; (it shows an approximate linear relationship with $\text{Sinh}^{-1} [\lambda \sqrt{(x + \frac{1}{2})}]$ for values of λ which appear likely in practice). Beall (1942) has, however, suggested that, in entomological field experiments where an estimate of λ is required, two plots for each treatment should be included in each randomized block; for such experimental designs the $\lambda^{-1} \text{Sinh}^{-1} [\lambda \sqrt{x}]$ scale would naturally be used.[3]

The $\text{Sinh}^{-1} [\lambda \sqrt{x}]$ scale is also appropriate for the somewhat more general variance law.

$$\sigma_x^2 = \mu^2(m + \lambda^2 m^2). \tag{6}$$

As an example[4] of data for which the empirical law (6) might have been fitted, the following "leatherjacket" counts (Table 3) are cited. The figures refer to an experiment on the control of leatherjackets by the use of toxic emulsions (Bartlett, 1936a, p. 190).

TABLE 3
Leatherjacket Counts

Treatment	1 (Control)	2 (Control)	3	4	5	6
Block I	92	66	19	29	16	25
II	60	46	35	10	11	5
III	46	81	17	22	16	9
IV	120	59	43	13	10	2
V	49	64	25	24	8	7
VI	134	60	52	20	28	11

The original analysis was actually made for $\sqrt{(x + \frac{1}{2})}$ for the *treated* plots only, the numbers on the control plots being used merely to indicate the degree of control by the treatments (see Table 4 below).

A comprehensive analysis of variance including the control plot numbers is somewhat a matter of convenience when the control plot numbers differ considerably from the treated plot numbers. Here if it is desired to include control plot numbers in the analysis they might still reasonably be included in the square-root analysis, but the use of the more general variance law (6) would be safer.

[3] Beall (1942) gives a table for values of $k = \lambda^2$ from 0 to 1. Since this table is for $\lambda^{-1} \text{Sinh}^{-1} [\lambda \sqrt{x}]$ and not $\lambda^{-1} \text{Sinh}^{-1} [\lambda \sqrt{(x + \frac{1}{2})}]$, I would recommend the empirical correction of replacing zero values of x by $\frac{1}{4}$ (cf. section 4).

[4] See also the examples in Beall (1942).

TABLE 4
Summary of Results

	1 2	3	4	5	6	S.E.	Sig. diff. (P = .05)
Mean $\sqrt{(x + \frac{1}{2})}$		5.58	4.43	3.84	3.03	0.407	1.23
Mean No./plot	73.1	31.8	19.7	14.8	9.8		
o/o Control		56	73	80	87		

It has been noted above that when biological populations change, the change is often proportional to the mean, implying changes independent of the mean on the logarithmic scale. However, suppose the fraction of area covered by a species of plant is the measurement; there is then a factor limiting the amount of growth, the fractional area never exceeding 1. In such situations I have found the transformation to the scale log $\{x/(1 - x)\}$ useful (Bartlett, 1937a, p. 163).

It is of interest to add to our list the well-known transformation used for a sample correlation coefficient r to make its distribution less skew and more stable in variance; viz., $\frac{1}{2} \log \{(1 + r)/(1 - r)\}$. Since the variance of a correlation coefficient is approximately $(1 - \rho^2)^2/(n - 1)$, where p is the true value of the coefficient and n the number of observations in the sample, we obtain this transformation from our equation (3) if we wish to make the variance independent of ρ. It is, of course, rare to have to analyze a set of correlation coefficients by analysis of variance, but if the problem arose the above transformation would be the appropriate one.

A more important problem that does frequently occur is the analysis of variance of a set of sample variances or standard deviations. A detailed discussion of this problem has been given elsewhere (Bartlett & Kendall, 1946), and as I have already quoted the illustrative example once in this country,[5] I will not do so here, but merely note that the variance of a sample variance s^2 is proportional to $(\sigma^2)^2$, and hence the logarithmic transformation is suitable.

4. THE INVERSE SINE OR ANGULAR TRANSFORMATION

The inverse sine square-root transformation

$$g(x) = \text{Sin}^{-1} \sqrt{x} \qquad (7)$$

bears the same relation to estimated probabilities or proportions x with

[5] In a paper "Applications of Analysis of Variance," given at Princeton University on November 1, 1946.

binomial variance $p(1 - p)/n$, where n is the number of individuals in the sample, as the square-root transformation does to a Poisson variate. The approximately constant variance on the new scale is $821/n$, provided that the inverse sine, which denotes an angle, is measured in degrees. A table in this form is given in Fisher and Yates' *Tables* (Fisher & Yates, 1938, p. 42; see also Bliss, 1937). An alternative table in radians was given in Bartlett (1937b) and on this scale the variance is $0.25/n$. In Tables 5 and 6 below are quoted (Bartlett, 1937a, p. 167) the data

TABLE 5
Number of Dead Flies

Treatments	A	B	C	D	E	F	G*
1	24	(25)	17	17	18	23	1
2	25	(25)	15	17	25	25	1
3	24	(25)	12	17	24	23	1
4	21	(25)	20	22	16	23	10
5	25	(25)	21	13	22	23	4, 6

* Control (with one extra replication).

and analysis of results in one of a series of experiments for which this transformation was used in the routine analysis. It is not a particularly ideal "textbook" example, but is useful as an example of the rough evaluation of insecticides in contrast with detailed evaluations for which the probit transformation (see section 5) is more appropriate. The insecticides were here in the form of toxic sprays, and no exact dose for any insect is known.

TABLE 6
Summary of Results

	A	B	C	D	E	F	G	S.E.
o/o Kill	95	(100)	68	69	84	94	15	
Sin⁻¹ √x (radians)	1.41	(1.57)	0.98	0.99	1.22	1.34	0.37	0.082

In the above analysis no correction for "discontinuity" was used, since adding one-half to the observed numbers cannot consistently be carried through to the top end of the scale, near 100 percent kill. It was, however, pointed out in a footnote to my original discussion (Bartlett, 1937a, pp. 167–168) that an empirical but fairly useful correction is simply to write ¼ wherever 0 occurs (and $n - \frac{1}{4}$, for n), and leave the other integers unchanged. This correction has a similar effect in "smoothing" the jumps due to the data consisting of whole numbers, the most violent jumps on the transformed scale being from 0 to 1 (or from $n - 1$ to n).

In the theoretical discussion of Poisson and binomial variation by Cochran (1940), already referred to in section 2, Cochran has pointed out (p. 346) that in an exact analysis of percentages the above empirical correction would become replaced by special adjustments, but he also notes that such an analysis would only apply to binomial data. It thus appears that the empirical correction I have suggested will remain useful in practical applications. For example, in the series of insecticide experiments referred to above, the mean variance was of the order of 0.03, as against $\frac{1}{4} \times 1/25 = 0.01$ for binomial variation, so that the assumption of exact binomial variability would certainly not have been tenable.

References

Bartlett, M. S. Some notes on insecticide tests in the laboratory and in the field. *Journal of the Royal Statistical Society, Supplement,* 1936, **3,** 185. (a)

Bartlett, M. S. Square-root transformation in analysis of variance. *Journal of the Royal Statistical Society, Supplement,* 1936, **3,** 68. (a)

Bartlett, M. S. Some examples of statistical methods of research in agriculture and applied biology. *Journal of the Royal Statistical Society, Supplement,* 1937, **4,** 137. (a)

Bartlett, M. S. Subsampling for attributes. *Journal of the Royal Statistical Society Supplement,* 1937, **4,** 121 (b)

Bartlett, M.S. A modified probit technique for small probabilities. *Journal of the Royal Statistical Society, Supplement,* 1946, **7,** 113.

Bartlett, M. S., & Kendall, D. G. The statistical analysis of variance-heterogeneity and the logarithmic transformation. *Journal of the Royal Statistical Society, Supplement,* 1946, **7,** 128.

Beall, G. The transformation from entomological field experiments so that the analysis of variance becomes applicable. *Biometrika,* 1942, **32,** 243.

Bliss, C. I. The analysis of field experimental data expressed in percentages. (in Russian). *Plant Protection,* Leningrad, 1937, fasc. **12,** 67.

Cochran, W. G. Some difficulties in the statistical analysis of replicated experiments. *Empire Journal of Experimental Agriculture,* 1938, **6,** 157.

Cochran, W. G. The analysis of variance when experimental errors follow the Poisson or binomial laws. *Annals of Mathematical Statistics,* 1940, **11,** 335.

Curtiss, J. H. On transformation used in analysis of variance, *Annals of Mathematical Statistics,* 1943, **14,** 10.

Fisher, R. A., & Yates, F. *Statistical tables for biological, agricultural and medical research.* Edinburgh: Oliver and Boyd, 1938.

APPENDIX

SUMMARY OF TRANSFORMATIONS

Variance in terms of mean m	Transformation	Approximate variance on new scale	Relevant distribution
m	$\sqrt{x}$, (or $\sqrt{(x + \frac{1}{2})}$	0.25	Poisson
$\lambda^2 m$	for small integers)	$0.25\lambda^2$	Empirical
$2m^2/(n - 1)$	$\log_e x$	$2/(n - 1)$	Sample variances
$\lambda^2 m^2$	$\log_e x, \log_e (x + 1)$	λ^2	Empirical
	$\log_{10} x, \log_{10} (x + 1)$	$0.189\lambda^2$	
$m(1 - m)/n$	$\mathrm{Sin}^{-1} \sqrt{x}$, (radians)	$0.25/n$	Binomial
	$\mathrm{Sin}^{-1} \sqrt{x}$, (degrees)	$821/n$	
$m(1 - m)/n$	Probit	Not constant*	
$m(1 - m)/n$	$\log_e [x/(1 - x)]$	$1/[mn(1 - m)]$	
$\lambda^2 m^2(1 - m)^2$	$\log_e [x/(1 - x)]$	λ^2	Empirical
$(1 - m^2)^2/(n - 1)$	$\frac{1}{2} \log_e [(1 + x)/(1 - x)]$	$1/(n - 3)$	Sample correlations
$m + \lambda^2 m^2$	$\lambda^{-1} \mathrm{Sinh}^{-1} [\lambda \sqrt{x}]$, or	0.25	Negative binomial
	$\lambda^{-1} \mathrm{Sinh}^{-1} [\lambda \sqrt{(x + \frac{1}{2})}]$		
$\mu^2(m + \lambda^2 m^2)$	for small integers	$0.25\mu^2$	Empirical
—	To expected normal scores	1 for large n*	Ranked data

* See Fisher and Yates (1938) for exact values.

3.3

Analysis of covariance: its nature and uses[*]

William G. Cochran

1. INTRODUCTION

This paper is intended as an introduction to the subsequent papers in this issue. It discusses the nature and principal uses of the analysis of covariance, and presents the standard methods and tests of significance.

As Fisher (1932) has expressed it, the analysis of covariance "combines the advantages and reconciles the requirements of the two very widely applicable procedures known as regression and analysis of variance." This dual role can be illustrated by a two-way classification in which rows represent treatments, and columns represent blocks or replications. The typical mathematical model appropriate to the analysis of covariance is

$$y_{ij} = \mu + \tau_i + \rho_j + \beta(x_{ij} - x_{..}) + e_{ij} \tag{1}$$

Here y_{ij} is the yield or response, while x_{ij} is an auxiliary variate, sometimes called the *concomitant variate* or *covariate*, on which y_{ij} has a linear regression with regression coefficient β. The constants μ, τ_i and ρ_i are the true mean response and the effects of the ith treatment and jth replication, respectively. The residuals e_{ij} are random variates, assumed in standard theory to be normally and independently distributed with mean zero and common variance.[**]

From the viewpoint of analysis of variance, equation (1) may be rewritten as

$$y_{ij} - \beta(x_{ij} - x_{..}) = \mu + \tau_i + \rho_j + e_{ij} \tag{2}$$

William G. Cochran, "Analysis of Covariance: Its Nature and Uses," *Biometrics*, **13**, 1957, 261–281. Sections 6.0 (The Reduced Sum of Squares for Treatment) and 8.0 (Multiple Covariance) in the original article have been deleted here.

[*] Paper No. 319, Department of Biostatistics.

[**] The symbols $xz_{..}$, $y_{..}$ denote overall means, while $x_{i.}$, $y_{i.}$ denote treatment

In this form, (2) is the typical equation for an analysis of variance of the quantities

$$y_{ij} - \beta(x_{ij} - x_{..})$$

These are the deviations of y_{ij} from its linear regression on x_{ij}, or the values of y_{ij} after adjustment for this linear regression. In this setting, τ_i may be regarded as the true effect of the ith treatment on y_{ij}, after adjustment for the linear regression on the covariate x_{ij}. Thus the technique enables us to remove that part of an observed treatment effect which can be attributed to a linear association with the x_{ij}.

When the objective of the analysis is to fit a regression of y on x, the parameters τ_i and ρ_i in equation (1) represent "nuisance" parameters, included in the mathematical specification in order to make it realistic. In this way the analysis of covariance extends the study of regression relationships to data of complex structure in which the nature of the regression is at first sight obscured by structural effects like the τ_i and ρ_i.

2. PRINCIPAL USES

These may be grouped under several headings.

2.1 To Increase Precision in Randomized Experiments

This is probably the most frequent application. The covariate x is a measurement, taken on each experimental unit before the treatments are applied, which is thought to predict to some degree the final response y on that unit. The first illustration of the covariance method in the literature was of this type (Fisher [1932]). The variate x was the yield of tea per plot in a period preceding the start of the experiment, while y was the tea yield at the end of a period of application of treatments (in this illustration, the treatments were "dummy"). Adjustments of the responses y for their regression on x removes the effects of variations in initial yields from the experimental errors, insofar as these effects are measured by the linear regression. In this example these effects might be due either to inherent differences in the tea bushes or to soil fertility differences that were permanent enough to persist during the course of the experiment.

With a linear regression equation, the gain in precision from the covariance adjustment depends primarily on the size of the correlation coefficient ρ between y and x on experimental units (plots) that receive the same treatment. If σ_y^2 is the experimental error variance when no covariance is employed, the adjustments reduce this variance to a value which is effectively about

$$\sigma_y^2(1 - \rho^2) \left\{ 1 + \frac{1}{f_e - 2} \right\}.$$

where f_e is the number of error d.f. The factor involving f_e is needed to take account of errors in the estimated regression coefficient. If ρ is less than 0.3 in absolute value, the reduction in variance is inconsequential, but as ρ mounts towards unity, sizeable increases in precision are obtained. In Fisher's example ρ was 0.928, reflecting a high degree of stability in relative yield of a plot from one period to another. The adjustment reduced the error variance roughly to a fraction $\{1 - (0.928)^2\}$, or about one-sixth, of its original value. Some of the most spectacular gains in precision from covariance have occurred in situations like this, in which the covariate represents an initial calibration of the responsiveness of the experimental units.

In this use the function of covariance is the same as that of local control (pairing and blocking). It removes the effects of an environmental source of variation that would otherwise inflate the experimental error. When the relation between y and x is linear, covariance and blocking are about equally effective.* If instead of using covariance we can group the units into replications such that the x values are equal within a replication, this blocking also reduces the error variance to $\sigma_y^2(1 - \rho^2)$.

The potentialities of preliminary measurements as a means of increasing precision have frequently been recognized by experimenters. In animal feeding experiments the response is taken as gain in weight (final–initial weight) rather than final weight itself. Insulin may be assayed from the drop in blood sugar (initial reading—reading 3 hours after injection of insulin) instead of from the 3 hour reading. The weight of a treated muscle on the right side of the body may be taken as a percentage of the weight of the corresponding untreated muscle on the left side. Such adjustments make the best use of the covariate only when the relation between y and x is exactly that implied by the adjustment. In the animal feeding and insulin examples, the assumption is that $\beta = 1$; in the muscles, that y/x is independent of x and has constant variance. If these assumptions do not hold, the adjustment falls short of the optimum and sometimes is worse than no adjustment at all. By a covariance analysis, the experimenter can utilize his knowledge or speculations about the general nature of the relation between y and x, but still leave flexibility in the process by including parameters like β that are estimated from the data. Incidentally, he can verify from the covariance analysis whether a specific simple adjustment like the use of $(y - x)$ is good enough, as it sometimes is.

In a covariance analysis, the preliminary variate x may be measured on a completely different scale from that of the response y,—a situation in which the experimenter would have difficulty in creating a "home-

* See p. 281.

made" method of adjustment. Bartlett (1937) used a visual estimate of the degree of saltiness of the soil to adjust cotton yields. Federer and Schlottefeldt (1954) used the serial order (1, 2, . . . 7) of the plot within a replication as a basis for a quadratic regression adjustment of tobacco data, thereby removing the effects of an unexpected gradient in fertility within the replications. Similarly, the reading performances of children under different methods of instruction may be adjusted for variations in their initial I.Q.'s. Note also that x need not be a direct causal agent of y—it may, for instance, merely reflect some characteristic of the environment that also influences y.

When covariance is used in this way, it is important to verify that the treatments have had no effect on x. This is obviously true when the x's were measured before the treatments were applied. But sometimes the x variates are measured after treatments have been applied, as when plant number shortly before harvest is used to adjust crop yields for uneven growth, or as happened in the index of saltiness used by Bartlett. When the treatments do affect the x-values to some extent, the covariance adjustments take on a different meaning. They no longer merely remove a component of experimental error. In addition, they distort the nature of the treatment effect that is being measured. If the higher yields given by superior treatments are due mostly to their effects in increasing numbers of plants, a covariance adjustment, which attempts to measure what yields would be if plant numbers were equal for all treatments, may remove most of the real treatment effect. The F-test of treatments against error for the x-variate is helpful when there is doubt whether treatments have had some effect on x.

2.2 To Remove the Effects of Disturbing Variables in Observational Studies

In fields of research in which randomized experiments are not feasible, we may observe two or more groups differing in some characteristic, in the hope of discovering whether there is an association between this characteristic and a response y. Examples are differences in heights of urban and rural school children, differences in illness rates between tenants of public and slum housing, and differences in expenditures for luxuries between clerical and manual workers. In observational studies it is widely realized that an observed association, even if statistically significant, may be due wholly or partly to other disturbing variables $x_1, x_2 \ldots$ in which the groups differ. Where feasible, a common device, analogous to blocking in randomized experiments, is to match the groups for the disturbing variables thought to be most important. In the same way, a covariance adjustment may be tried for x-variables that have not been matched.

In a comparison of the heights of children from two different types of school, Greenberg (1953) found that the two groups differed slightly, though not significantly, in mean age. A covariance adjustment for age resulted in a more sensitive comparison of the heights. As a more complex example, Day and Fisher (1937) adjusted the log S. D. of leaf length (used as a measure of within-species variability) for fluctuations in length, breadth and thickness of leaves in comparing populations of *Plantago maritima* from different regions.

In observational studies covariance can perform two distinct functions. One is to remove bias. To illustrate, it follows from model (1) that the *unadjusted* difference $(y_{i.} - y_{i.})$ between the means of two groups is

$$y_{i.} - y_{j.} = \tau_i - \tau_j + \beta(x_{i.} - x_{j.}) + e_{i.} - e_{j.}.$$

If the two groups have not been matched for x, the difference $(x_{i.} - x_{i.})$ may reflect a real difference in their x-distributions, being much larger than can be accounted for by within-group variations. The term $\beta(x_{i.} - x_{i.})$ is then of the nature of a bias which if allowed to remain will render tests of significance and confidence limits invalid. If model (1) applies to the data at hand, a covariance adjustment removes their bias. Most users of covariance in observational studies would, I think, regard coping with bias as its primary function. However, even if there are no real differences between the x-distributions in the two groups, so that there is no danger of bias, covariance may still be used to increase the precision of the comparison as in the applications in section 2.1.

Unfortunately, observational studies are subject to difficulties of interpretation from which randomized experiments are free. Although matching and covariance have been skillfully applied, we can never be sure that bias may not be present from some disturbing variable that was overlooked. In randomized experiments, the effects of this variable are distributed among the groups by the randomization in a way that is taken into account in the standard tests of significance. There is no such safeguard in the absence of randomization.

Secondly, when the x-variables show real differences among groups—the case in which adjustment is needed most—covariance adjustments involve a greater or less degree of extrapolation. To illustrate by an extreme case, suppose that we were adjusting for differences in parents' income in a comparison of private and public school children, and that the private-school incomes ranged from \$10,000–\$12,000, while the public-school incomes ranged from \$4,000–\$6,000. The covariance would adjust results so that they allegedly applied to a mean income of \$8,000 in each group, although neither group has any observations in which incomes are at or even near this level.

Two consequences of this extrapolation should be noted. Unless the regression equation holds in the region in which observations are lacking, covariance will not remove all the bias, and in practice may remove only a small part of it. Secondly, even if the regression is valid in the no man's land, the standard errors of the adjusted means become large, because the standard error formula in a covariance analysis takes account of the fact that extrapolation is being employed (although it does not allow for errors in the form of the regression equation). Consequently the adjusted differences may become insignificant statistically merely because the adjusted comparisons are of low precision.

When the groups differ widely in x, these difficulties imply that the interpretation of an adjusted analysis is speculative rather than soundly based. While there is no sure way out of this difficulty, two precautions are worth observing.

(i) Consider what internal or external evidence exists to indicate whether the regression is valid in the region of extrapolation. Sometimes the fitting of a more complex regression formula serves as a partial check.

(ii) Examine the standard errors of the adjusted group means, particularly when differences become non-significant after adjustment. Confidence limits for the difference in adjusted means will reveal how precise or imprecise the adjusted comparison is.

2.3 To Throw Light on the Nature of Treatment Effects

This application is closely related to the previous one. In a randomized experiment, the effects of several soil fumigants on eelworms, which attack some English farm crops, were compared. After the treatments had been given time to exert their effects, the numbers of eelworm cysts per plot and the yields of the crop, spring oats, were both recorded. Significant effects were produced on both eelworms and oats. It would be of interest to discover whether the reductions in numbers of eelworms were the causal mechanism in producing the observed differences in oats yields. If the treatment effects on the oats (y) disappear after adjusting by covariance for differences in the numbers of eelworms (x), this suggests, at least at first sight, that the treatment differences are simply a reflection of the differences produced by the fumigants on eelworm numbers. There are numerous instances of this kind in which a concomitant variable might be in part the agent through which the treatments produce their effects on the principal response. A covariance analysis offers the possibility of exploring whether this is so.

Here again, however, there are difficulties which restrict the utility of this ingenious tool. . . . Since in most of these applications the treatments will have produced significant effects on x, there is the problem of extrapolation discussed in the previous section. In addition . . . the interpretation of the adjusted y averages requires careful study. Sometimes these averages have no physical or biological meaning of interest to the investigator, and sometimes they do not have the meaning that is ascribed to them at first glance.

2.4 To Fit Regressions in Multiple Classifications

The simplest situation, discussed in elementary text books, involves a single classification. By standard techniques we can (i) fit a separate regression of y on x within each class, (ii) test whether the slopes or positions of the lines differ from class to class, and (iii) if advisable, make a combined estimate of a common slope. As an example of a regression in a row $\times$ column classification, the regression of wheat yield on shoot height, number of plants and number of ears was worked out from a series of growth studies on wheat conducted in Britain (Cochran [1938]). Each quartet of observations (y, x_1, x_2, x_3) represented the mean over several plots at one station in one year. In order that soil, geographic and seasonal factors would be adequately sampled, data were obtained from 7 stations for each of 5 years, making 35 quartets. Consequently it was necessary to fit constants for the mean yield of each station and the mean yield of each year.

2.5 To Analyze Data when Some Observations Are Missing

An interesting by-product of the covariance method, first pointed out by Bartlett (1937), is that it may be used to compute the exact analysis of variance when some observations are missing. To each missing observation we assign any convenient value (*e.g.* 0, 5, or 100) and introduce a dummy x-variate that takes the value[1] for the missing unit, and 0 for all other units. The standard covariance computations then give the correct least squares estimates of the treatment means and the exact F- and t-tests. This method is probably slower than the insertion of a missing value by the Yates formula (1933), but it is useful (i) with unfamiliar classifications where the Yates formula has not been worked out and (ii) where exact F- and t-tests are important, since the Yates' method gives only approximate tests.

Covariance can also be used (Nair [1939]) to estimate individual yields of a group of plots whose produce has inadvertently been combined, so that only a total over the group is known.

3. THE STANDARD COMPUTATIONS

The computations, which will be reviewed briefly, are essentially the same for all mathematical models in which a single regression and a single residual variance are postulated. These cases include the simpler multiple classifications and experimental designs (randomized blocks, cross-over designs, latin squares) as well as balanced and partially balanced incomplete block designs without recovery of inter-block information. Separate discussion is required for hierarchical classifications involving more than one regression equation or more than one residual variance in the specification, as will split-plot designs or incomplete blocks when inter-block information is recovered.

The backbone of the standard procedure is an analysis of sums of squares and products into the Treatments and Error components. Table 1 shows the notation employed.

TABLE 1
Sums of Squares and Products

	D. f.	(x^2)	(xy)	(y^2)
Treatments	$(t - 1)$	T_{xx}	T_{xy}	T_{yy}
Error	f_e	E_{xx}	E_{xy}	E_{yy}
Sum	$t - 1 + f_e$	S_{xx}	S_{xy}	S_{yy}

A line is added giving the sums for treatments and error. Thus $S_{xx} = T_{xx} + E_{xx}$, etc.

The error s.s. for y is now divided into two parts: the s.s. for regression on x (1 d.f.) and the s.s. of deviations (Table 2). The same subdivision is made for the Sums.

TABLE 2
Partition of E_{xx} and S_{xx} into Components for Regression and for Deviations

	(y^2)	Regression D.f.	Regression S. s.	Deviations D.f.	Deviations S. s.	M. s.
Error	E_{xx}	1	E_{xy}^2/E_{xx}	$f_e - 1$	$E_{yy} - E_{xy}^2/E_{xx}$	s_e^2
Sum	S_{xx}	1	S_{xy}^2/S_{xx}	$t + f_e - 2$	$S_{yy} - S_{xy}^2/S_{xx}$	
Treatments (by subtraction)				$t - 1$	$T_{yy} - S_{xy}^2/S_{xx} + E_{xy}^2/E_{xx}$	s_t^2

The reduced s.s. for Treatments is obtained by subtracting the deviations s.s. for Error from that for the Sum (last line of Table 2).

The items of information most commonly wanted from this analysis are obtained as follows.

(i) The regression coefficient β. This is estimated from the Error line; $b = E_{xy}/E_{xx}$. The estimated standard of b is $s_e/\sqrt{E_{xx}}$, with $(f_e - 1)$ d.f.

(ii) The adjusted estimate of a treatment effect. In the simplest experimental designs, the *unadjusted* estimate for treatment i is simply the mean $y_{i.}$ of all observations having this treatment. The adjusted estimate is

$$y'_{i.} = y_{i.} - b(x_{i.} - x_{..}) \tag{1}$$

(iii) The estimated standard error of any linear function $L = \Sigma g_i y'_i$ of the adjusted treatment means is

$$s_L = s_s \sqrt{\frac{\Sigma g_i^2}{r} + \frac{[\Sigma g_i(x_{i.} - x_{..})]^2}{E_{xx}}} \tag{2}$$

with $(f_e - 1)$, d.f., where r is the number of replications. In particular the standard error of the difference between two adjusted treatment means is, putting $g_i = 1$, $g_j = -1$, and all other g's $= 0$,

$$s.e.(y'_{i.} - y'_{j.}) = s_e \sqrt{\frac{2}{r} + \frac{(x_{i.} - x_{j.})^2}{E_{xx}}} \tag{3}$$

(iv) For a test of the null hypothesis that all treatment effects are equal, we compute $F = s_t^2/s_e^2$, where s_t^2 and s_e^2 are the mean squares found in Table 2. This ratio has $(t - 1)$ and $(f_e - 1)$ d.f.

4. NATURE OF THE COVARIANCE ADJUSTMENT

The structure of the covariance adjustment, $-b(x_i - x_{..})$, is in accord with common sense: $(x_i - x_{..})$ measures the amount by which the x-value for this treatment exceeds the average x-value, while b measures the change in y expected to accompany unit change in x. In a specific application, the sizes and directions of the adjustments are determined by the data. In this respect a covariance adjustment differs markedly from the type of arbitrary adjustment that has sometimes earned a dubious reputation for the whole process of adjustment. It is not true, however, that a covariance adjustment is entirely objective, since the investigator must choose the type of regression equation (e.g. linear, quadratic, linear in log x) from which the adjustment is derived. Moreover, the x-variables do not always measure what we would like to think they measure. One occasionally meets extravagant claims for a covariance adjustment, as for example that data have been "adjusted to equalize socio-economic status," when what has actually been done is to adjust for a linear regression on a crude social rating of the father's occupation as reported by anyone who happens to be at home when the interviewer calls.

From equation (3), the estimated variance of the difference between two adjusted means, when averaged over all pairs of means, works out as

$$\frac{2s_e^2}{r} \left\{ 1 + \frac{t_{xx}}{E_{xx}} \right\} \tag{4}$$

where $t_{xx} = T_{xx}/(t-1)$ is the treatments *mean square* for x. If t-tests are to be made between several pairs of means, this expression may be used, as an approximation, for the variance of the difference between any pair (Finney [1946]). This saves the labor of computing a separate standard error for each pair, as is required by the exact formula (3). This device is not recommended when the treatments produce significant effects on x, because the variance of the difference may be substantially greater for some pairs than for others, so that the use of a single average variance becomes unsatisfactory.

More generally, the quantity

$$s_e^2 \left\{ 1 + \frac{t_{xx}}{E_{xx}} \right\} \tag{5}$$

may be regarded as the effective error variance *per unit* in a covariance analysis, where the term in brackets is an allowance for sampling errors in b. In a completed experiment, the gain in precision from covariance can be estimated by comparing (5) with $s_y^2 = E_{yy}/f_e$, the error mean square for y in its analysis of variance in Table 1. This comparison ignores the loss of 1 d.f. from the error which occurs with a covariance adjustment. The effect of this loss on the sensitivity of the t-tests is small even with only 5 d.f. in error.

5. THEORY OF THE TECHNIQUE

The theory for the simplest designs will be illustrated by the row $\times$ column classification, with treatments as rows and replications as columns. The model is

$$y_{ij} = \mu + \tau_i + \rho_j + \beta(x_{ij} - x_{..}) + e_{ij} \tag{6}$$

Following the method of least squares, the unknown parameters are estimated by minimizing

$$\sum_{i,j} \{ y_{ij} - m - t_i - r_j - b(x_{ij} - x_{..}) \}^2 \tag{7}$$

Since t_i need measure only the difference between the effect of the ith treatment and the general mean, we may assume

$$\sum_i t_i = 0; \qquad \sum_j r_j = 0.$$

The estimates. The algebra involved in finding the estimates can be reduced by introducing the variable

$$x_{ij}' = x_{ij} - x_{i.} - x_{.j} + x_{..} \tag{8}$$

Those familiar with the analysis of variance will recognize x_{ij}' as the contribution of the (i, j)th observation to the error of x in its analysis of variance. The properties of x'_{ij} that will be useful are

$$\sum_i x'_{ij} = 0; \qquad \sum_j x'_{ij} = 0 \tag{9}$$

$$\sum_{i,j} x'^2_{ij} = E_{xx}; \qquad \sum_{i,j} x'_{ij} y_{ij} = E_{xy} \tag{10}$$

The second relation in (10) is less familiar than the others, but is easily verified.

Now the given prediction equation

$$y_{ij} = m + t_i + r_j + b(x_{ij} - x_{..})$$

becomes identical with the prediction equation

$$y_{ij} = m' + t'_i + r'_j + bx'_{ij}$$

if the new estimates m', t', r'_j satisfy the relations

$$m = m'; \qquad t_i = t'_i - b(x_{i.} - x_{..}); \qquad r_j = r'_j - b(x_{.j} - x_{..}) \tag{11}$$

This may be verified by substitution. Further, since $\Sigma t_i = \Sigma r_j = 0$, it follows that $\Sigma t'_i = \Sigma'_{rj} = 0$.

Hence, instead of finding m, t_i, r_j and b so as to minimize (7), we can find m', t'_i, r'_i, and b to minimize

$$\sum_{i,j} (y_{ij} - m' - t'_i - r'_j - bx'_{ij})^2 \tag{12}$$

On expansion, (12) becomes

$$\sum_{i,j} (y_{ij} - m' - t'_i - r'_j)^2 - 2b \sum_{i,j} x'_{ij} y_{ij} + b^2 \sum_{i,j} x'^2_{ij}$$

Note that the omitted terms

$$\Sigma m' x'_{ij}, \qquad \Sigma t'_i x'_{ij}, \qquad \Sigma r' x'_{ij}$$

vanish because of relations (9).

Using (10), the quantity to be minimized is

$$\sum_{i,j} (y_{ij} - m' - t'_i - r'_j)^2 - 2bE_{xy} + b^2 E_{xx} \tag{13}$$

The advantage of this result is that b is disentangled from the other unknowns. Differentiation with respect to b gives

$$b = E_{xy}/E_{xx} \tag{14}$$

The other unknowns, m', t'_i, r'_j, must be chosen so as to minimize the first term in (13). But this is exactly the minimization involved in an ordinary analysis of variance of y without covariance. Hence, by the standard results for the analysis of variance,

$$m' = y_{..}; \qquad t'_i = y_{i.} - y_{..}; \qquad r'_j = y_{.j} - y_{..}$$

Finally, from (11), the least squares estimates for the covariance analysis are

$$
\left.
\begin{aligned}
m &= y_{..} \\
t_i &= y_{i.} - y_{..} - b(x_{i.} - x_{..}) \\
r_j &= y_{.j} - y_{..} - b(x_{.j} - x_{..})
\end{aligned}
\right\}
\tag{15}
$$

Since the t_i represent deviations from the overall mean, the estimate used in practice is

$$m + t_i = y_{i.} - b(x_{i.} - x_{..}) = y'_{i.}, \quad \text{say.} \tag{16}$$

Standard errors. From (13), the residual sum of squares may be written

$$E_{yy} - 2bE_{xy} + b^2E_{xx} = E_{yy} - \frac{2E_{xy}^2}{E_{xx}} + \frac{E_{xy}^2}{E_{xx}} = E_{yy} - \frac{E_{xy}^2}{E_{xx}} \tag{17}$$

The d.f. are $(f_e - 1)$, since 1 d.f. is subtracted for the regression. Hence s_e^2, the residual mean square, is computed as in Table 2.

To find the standard error of b, we may write

$$b = \frac{\Sigma x'_{ij} y_{ij}}{E_{xx}} = \frac{\Sigma x'_{ij}\{\mu + \tau_i + \rho_j + \beta(x_{ij} - x_{..}) + e_{ij}\}}{E_{xx}}$$

From the properties of the x'_{ij}, this reduces to

$$b = \beta + \frac{\Sigma x'_{ij} e_{ij}}{E_{xx}}$$

This equation expresses $(b - \beta)$ as a linear function of the random residuals e_{ij}. Hence

$$\sigma_b^2 = E(b - \beta)^2 = \frac{\sigma_e^2 \Sigma x'^2_{ij}}{E_{xx}} = \frac{\sigma_e^2}{E_{xx}}$$

In the same way we find, for an adjusted treatment mean,

$$y'_{i.} = m + t_i = \mu + \tau_i + e_{i.} - (x_{i.} - x_{..})\frac{\Sigma x'_{ij} e_{ij}}{E_{xx}}$$

For a linear comparison between the adjusted means, it follows that

$$L = \sum g_i y'_{i.} = \sum g_i(\mu + \tau_i) + \sum g_i e_{i.} - \left\{\sum g_i(x_{i.} - x_{..})\right\}\frac{\Sigma x'_{ij} e_{ij}}{E_{xx}}$$

By the properties of the x'_{ij}, the variate $\Sigma x'_{ij}, e_{ij}$ is uncorrelated with any of the means $e_{i.}$. This gives

$$\sigma_L^2 = \sigma_e^2 \left\{\frac{\Sigma g_i^2}{r} + \frac{[\Sigma g_i(x_{i.} - x_{..})]^2}{E_{xx}}\right\}$$

in agreement with the result in equation (2).

Note. The variate x'_{ij} can be used in the same way with all other designs (e.g. latin squares, incomplete blocks) to which the standard covariance theory applies. With an incomplete block design, for example, we define x'_{ij} as

$$x'_{ij} = x_{ij} - m_x - \hat{t}_{ix} - \hat{r}_{jx}$$

where $\hat{t}_{ix}$, $\hat{r}_{jx}$ are the estimates of the treatment and block effects, respectively, in the incomplete block analysis of variance of x. The rest of the algebraic development goes through without change.

Tests of significance. For the F-test, we quote the general theorem on the F-test in regression analysis, as applied to this problem. Let

$$D = \Sigma\{y_{ij} - m - t_i - r_j - b(x_{ij} - x_{..})\}^2$$
$$D'' = \Sigma\{y_{ij} - m'' - r''_i - b''(x_{ij} - x_{..})\}^2$$

where the constants are chosen in each case so as to minimize the corresponding sum of squares of deviations. Then if all τ_i are zero, the quantity

$$\frac{(D'' - D)}{(t - 1)} \bigg/ \frac{D}{(f_e - 1)}$$

is distributed as F with $(t - 1)$ and $(f_e - 1)$ d.f. For proofs see, e.g., Yates (1938), Anderson and Bancroft (1952).

From (17) the denominator $D/(f_e - 1)$ has been shown to be s_e^2. By the same approach, D'' is the sum of squares of deviations from the "error" regression when only replication effects are eliminated in the analysis of variance. But in that event the "error" will be equal to "treatments + error" in the present analysis. Hence the numerator of F, $(D'' - D)/(t - 1)$, is equal to s_t^2 as defined in Table 2.

6. ASSUMPTIONS REQUIRED FOR THE ANALYSES OF COVARIANCE

The assumptions required for valid use of the analysis of covariance are the natural extension of those for an analysis of variance, namely,

(i) Treatment, block and regression effects must be additive as postulated by the model,

(ii) The residuals e_{ij} must be normally and independently distributed with zero means and the same variance.

Although the effects of failures in these assumptions on the analysis of covariance as such do not appear to have been investigated, much of the related work on the analysis of variance carries over—for instance, that on the effects of non-normality or inhomogeneity of variance in the e_{ij}. The general precautions that have been given about the practical use of the analysis of variance should equally be observed in an analysis of covariance, see e.g. Cochran (1947).

Two assumptions that particularly involve the regression term in covariance should be noted. The treatment and the regression effects may not be additive. Bartlett (1937) pointed out that this danger might be present in the cotton-salt example already cited. On plots with a high salt content, the crop might be unable to respond to superior fertilizers. Thus, in an extreme case, the treatment effects may be zero if x lies above a certain value. If this happens, the covariance adjustment may still improve the precision, but (i) the meaning of the adjusted treatment effects become cloudy, and (ii) if covariance is applied in a routine way, the investigator fails to discover the differential nature of the treatment effects—a point that might be important for practical applications.

Secondly, the covariance procedure assumes that the correct form of regression equation has been fitted. Perhaps the most common error to be anticipated is that linear regressions will be used when the true regression is curvilinear. In a randomized experiment in which treatments do not affect x, the randomization ensures that the usual interpretations of standard errors and tests of significance are not seriously vitiated, although fitting the correct form of regression would presumably give a larger increase in precision. The danger of misleading results is greater when x shows real differences from treatment to treatment. Later investigations by Fairfield Smith in this issue suggest, however, that this disturbance is serious only in rather extreme situations.

References

Anderson, R. L., & Bancroft, T. A. Statistical theory in research. New York: McGraw-Hill, 1952, Chapter 14.

Bartlett, M. S. Some examples of statistical methods of research in agriculture. *Journal of the Royal Statistical Society, Supplement,* 1937, **4,** 137–183.

Cochran, W. G. Crop estimation and its relation to agricultural meteorology. *Journal of the Royal Statistical Society, Supplement,* 1938, **5,** 12–16.

Cochran, W. G. Some consequences when the assumptions for the analysis of variance are not satisfied. *Biometrics,* 1947, **3,** 22–38.

Day, B., & Fisher, R. A. The comparison of variability in populations having unequal means. An example of the analysis of covariance with multiple dependent and independent variates. *Annals of Eugenics,* 1937, **7,** 333.

Federer, W. T., & Schlottfeldt, C. S. The use of covariance to control gradients in experiments. *Biometrics,* 1954, **10,** 282–290.

Finney, D. J. Standard errors of yields adjusted for regression on an independent. *Biometrics,* 1946, **2,** 53.

Fisher, R. A. *Statistical methods for research workers* (4th ed.). Edinburgh, Oliver and Boyd, 1932.

Greenberg, B. G. The use of analysis of covariance and balancing in analytical surveys. *American Journal of Public Health*, 1953, **43**, 692–699.

Nair, K. R. The application of covariance technique to field experiments with missing or mixed-up yields. *Sankhya*, 1939, **4**, 581–588.

Yates, F. The analysis of replicated experiments when the field results are incomplete. *Empire Journal of Experimental Agriculture*, 1833, **1**, 129–142.

Yates, F. Orthogonal functions and tests of significance in the analysis of variance. *Journal of the Royal Statistical Society. Supplement*, 1938, **5**, 177–180.

4

Assumptions and Statistical Inference

INTRODUCTION

In order to use statistical techniques in any meaningful way, especially statistical tests of hypothesis, the user must understand the proper use of the statistical test he is employing, the assumptions underlying it, and how these are related to statistical inference. In Chapter 4 the reader will encounter articles designed to increase his sophistication with what might be called the "logic" of statistical inference.

This chapter can be viewed as having within it three subsections. The first includes the first seven articles. These articles, by four different authors, center around a debate on the use of one- or two-tailed tests of significance. M. R. Marks published the first article in this subsection in 1951. In it he provides a rationale for the delineation of two types of experiments, those based on one-tailed and those based on two-tailed tests of significance. The terms "one-tailed" and "two-tailed" refer to the nature of the alternative hypothesis, that is, whether we expect just any difference from the null hypothesis or whether we expect the difference to be in a specific direction, either smaller or larger. A two-tailed alternative hypothesis means we cannot predict the direction of the difference between our groups. The one-tailed test means that we can expect a difference to occur in a given direction.

"Tails" are the extremes of the sampling distribution. For example, if we use a two-tailed test of significance and we choose a significance level of 5 percent, we are saying that by chance we expect 2.5 percent of our statistics to be judged significant at each end of the distribution, a total of 5 percent. Five percent of the time we expect to reject the null hypothesis when in fact the obtained difference could have occurred by chance. The one-tailed test has the same significance, since we still are going to reject the null hypothesis incorrectly 5 percent of the time (using that significance level). However, in the case of the one-tailed test, the 5 percent is distributed at one end of the distribution. We have decided before conducting the experiment that if a difference between the means for our groups occurs in our experiment it will occur in one direction. The use of a one-tailed test usually increases the power of our test statistic, that is, its ability to signify differences when differences do in fact occur.

L. V. Jones, in the second article, also takes a definite stand on the use of one- and two-tailed tests, suggesting that we consider the use of the one-tailed test more than we have, and employ it where appropriate. Following the articles of Marks and Jones comes the article by W. E. Hick. Hick does not agree completely with Marks and provides some criticisms of Marks' article. Marks in turn answers Hick's criticisms step by step, hopefully clarifying the issues raised by Hick.

The next article, by C. J. Burke, provokes thought and takes issue with Jones. Burke suggests that Jones may mislead the social scientists into using one-tailed tests when they shouldn't and, further, that the use of one-tailed tests can create consequences for social science research not of a statistical nature. He cites for example, the possibility that "discovery of new psychological phenomena will be hindered" since the one-tailed test mediates against unexpected outcomes.

Jones' reply to Burke, aside from technical answers, comes to the conclusion that Burke's argument is more of a plea for careful research than it is an indictment of one-tailed tests. The last article in this subsection is a reply by Burke to Jones that also is a summary of the preceding arguments and a final plea for the use of two-tailed tests because of the possible drawbacks of the one-tailed test.

The second subsection contains six articles, which center around the debate on hypothesis testing and the logic of statistical inference. The first article, written by W. W. Rozeboom in 1960, sets the tone of the debate with its title, "The Fallacy of the Null Hypothesis Significance Test." Rozeboom presents an argument that ". . . traditional null-hypothesis procedure has already been superseded in modern statistical theory by a variety of more satisfactory inferential techniques." As an alternative he offers the "Bayesian approach" to hypothesis testing with its use of inverse probability.

The term "Bayesian" comes from a theorem named after an eighteenth century English clergyman, T. Bayes, who did work in probability and decision theory. The Bayesian approach to hypothesis testing differs from the null hypothesis procedure in that the prior probability distribution (before the experiment) of the alternative outcomes is based upon our knowledge up to that point. Under the null hypothesis procedure we assume all alternatives to be equally likely. This is the same as saying that we have no idea how our experiment will turn out, so we are giving each alternative the same probability of occurrence. If we were using the Bayesian approach we would establish our prior probability distribution of alternatives based on our knowhow or previous information. For example, if you have been selling shoes for years and wish to test the likelihood that given styles of shoes will sell, we can use your expertise to establish a prior distribution. We can run an observational study and obtain information that we use to modify the prior probability distribution to obtain the posterior distribution. The posterior distribution comes after the observation or experiment. The way we modify the prior probability distribution to obtain the posterior distribution has been specified by Bayes' theorem. Rozeboom covers this approach in an entertaining way.

Following Rozeboom's article things were quiet for a few years until D. A. Grant published his article on the strategy and tactics of investigating theoretical models in 1962. In this article he relates the use of the null hypothesis to the specific theoretical problem under investigation. Grant points out that two strategies can be used. The first is to couple the theory under test with the null hypothesis and cite support for the theory if one fails to reject the null hypothesis. The second strategy is to couple the theory under examination with the alternative hypothesis; if one rejects the null hypothesis, one then cites support for the theoretical notion under study.

A. Binder takes issue with Grant's proposal for testing theoretical models and suggests that Grant's argument is "only a special case of the argument against bad experimentation." Binder also extends Grant's arguments and suggests a few more considerations in the use of the null hypothesis procedure when testing a theoretical position.

The eleventh article, by W. Edwards, continues the debate between Grant and Binder and re-raises the issue of null hypothesis testing. Edwards argues that the null hypothesis procedure can be made to be biased in favor of rejecting the null hypothesis or with sloppy experimentation to be biased in favor of failing to reject the null hypothesis. Since conflicting conclusions are easily reached depending upon one's research strategy, Edwards suggests dropping the null procedures for the Bayesian approach.

Of course things didn't end there. W. Wilson, H. Miller and J. Lower present an argument against Edwards' assertion that the classical null approach is violently biased against the null hypothesis. At the same time they rework Edwards' example to demonstrate that it is not as bad as Edwards would have one believe. They also imply that a lot of the trouble with the current use of the classical statistical testing procedures is with the experimenter and not with the statistical test.

D. Bakan's article, the last in this second subsection, reviews the use of the test of significance in psychological research. Bakan comes to the conclusion that "we need to get on with the business of generating *psychological* hypotheses . . . and make inferences which bear on them. . . ." He also suggests that social scientists stop looking for some "automatic inference" procedure. The statistical procedures can only help test the conclusions on estimates of populations; we must then make inferences about the psychological or social phenomena under investigation.

The last subsection contains only two articles, but these provide a fine grounding in the proper use of the *t* and *F* statistics. The first, by C. A Boneau, is concerned with Student's *t* test and what happens to the statistical conclusions one reaches if in employing this statistic one violates the assumptions underlying it. The findings provide evidence that violations of the assumptions are not as serious as one might think.

The last article, by W. G. Cochran, presents the same type of review of violations of assumptions for the analysis of variance that Boneau discusses for *t*. The conclusions of Cochran are similar to those of Boneau, and both point to the simple fact that there is no substitute for careful, controlled experimentation and proper use of statistical techniques to enhance the validity of one's scientific conclusions.

4.1

Two kinds of experiments distinguished in terms of statistical operations

Melvin R. Marks

The work which a scientist does when he is "finding out" something has been variously called research, investigation, study, inquiry, examination, etc. Underwood (1949) dichotomizes experimental problems into the "I wonder-what-would-happen," and the "I-bet-this-would-happen" types. Mann[1] selects for distinction the terms "inquiry" and "research": The former proceeds without theory, whereas the latter is generated by theory and is reflective, *i.e.*, the results of research modify the constructs of the basic theoretical matrix. Both Underwood and Mann point out that basically there is a difference between just "finding-out," and "finding-out-if-a-hypothesis-is-consistent-with-fact." From this standard a distinction might be made on the basis of necessity for control. In "finding-out" control might be dispensed with, since control must always be with respect to something, and in "finding-out" that something is what is sought. On the contrary, in "finding-out-if- . . . ," control is necessary to properly evaluate results.

The writer believes that some sort of distinction is necessary, that it is best made in terms of statistical operations, and that it has important implications for the validity of conclusions drawn from results. What is needed first are two terms relatively free of connotations bearing on possible distinctions. While "inquiry" has been little used, "research" has been overused—and in a wide variety of meanings. To avoid ambiguity, the terms "experimentation$_I$" and "experimentation$_{II}$" have been selected.

A brief review of the statistical concepts relevant to the distinction

Melvin R. Marks, "Two Kinds of Experiments Distinguished in Terms of Statistical Operations," *Psychological Review*, **58**, 1951, 179–184. Copyright (1951) by the American Psychological Association, and reproduced by permission.

[1] From unpublished seminar discussions.

will be helpful. Most current texts discuss them in part, but it is believed that the synthesis proposed here is new. At any rate such synthesis should be of value to those psychologists engaged in "experimentation" who have not found time to peruse thoroughly the introductory material of the more advanced statistical tests.

STATISTICAL HYPOTHESIS

This is any testable assumption about parametric values. Typical parameters are population mean, population variance, difference between two population means, etc. As examples, we might make several statistical hypotheses about the populations from which two sample groups, say A and B, were drawn:

"There is no difference between the means of the populations from which A and B were drawn"—algebraically,

$$(m_A - m_B) - 0 = 0. \tag{1}$$

"The population means of A and B differ by 5 points."

$$(m_A - m_B) - 5 = 0, \tag{2}$$

or, more generally,

$$(m_A - m_B) - K = 0 \qquad (K = \text{any constant}). \tag{3}$$

Note that in each of the three equations, the quantity $(m_A - m_B)$ is given a positive sign. In effect, a definite prediction is being made about the direction of the difference in magnitudes. A completely generalized equation,

$$(|m_A - m_B|) - K = 0, \tag{4}$$

states, "the absolute difference between the means of the populations from which A and B were drawn does not differ from some constant K."

It is assumed frequently that (4) represents the classical "null hypothesis." Actually, (1), (2), and (3) are null hypotheses as well. In each case the hypothesis is that the parametric difference between some assumed difference and a given constant is zero. In each instance the right side of the equation is the null which is to be tested.

LEVEL OF CONFIDENCE, STATISTICAL SIGNIFICANCE, ETC.

When the test of a statistical hypothesis yields an estimate of the probability that the null is tenable, it is customary to say that the results are "significant" if the associated probability, p, is less than some value chosen arbitrarily in advance of the test. For the purpose of this paper and the statement, "significant at the 5 per cent level," means that the hypothesis tested was of the form (1), (2), (3), or (4), and that the associated $p < .05$.

TYPE I ERROR

Suppose that A and B are each groups of 11 children for whom I.Q.s have been determined. Let the mean difference in I.Q. be 4.172, and the standard error of that difference be 2. With these data, $t = 2.086$, $d.f. = 20$, and $p = .05$. An investigator using the 5 per cent level would conclude that there is a significant difference. Two possibilities exist: either there is a difference in the population means, or there is not. If, actually, there is a difference, his conclusion would be correct; if, actually, there is no difference, *i.e.*, both groups come from the same population, he has erroneously falsified the null. The false rejection of the null hypothesis is called a Type I Error. Note that the probability of occurrence of a Type I Error is determined absolutely by the level of confidence adopted. In the example the 5 per cent level was chosen; hence, by definition, there are 5 chances in 100 of a random fluctuation which would lead to a t significant at that level.

TYPE II ERROR

This may be illustrated by recasting our example. Suppose that the two groups of children came from populations whose means differed by 2 I.Q. points; that the difference between the means of the samples was also 2 points (*i.e.*, "perfect" sampling), and that the standard error estimated from the sample difference was again 2 points. In this case, $t = 1$, $d.f. = 20$, $p > .30$. The same investigator would conclude erroneously that the null was tenable. The failure to reject the false null is called a Type II Error.

The *power* of a statistical test is defined as the probability that a Type II Error will *not* occur. Unlike the case of the Type I Error, the probability of a Type II Error cannot be fixed by the investigator, since it is a function of three variables, only two of which are under the investigator's control. The variables are: (1) the sample size; as N increases probability of Type II Error decreases (power increase), *i.e.*, a parametric difference is more likely to be found if the sample is large; (2) the probability of Type I Error (per cent level of confidence chosen by the investigator); as the level increases (in the sense that a 5 per cent level is an increase over the 1 per cent level), probability of Type II Error decreases (power increases), *i.e.*, it is more likely that a parametric difference will be detected with the 5 per cent level than with the 1 per cent level; (3) the relative magnitude of the parametric difference; as the magnitude (in standard error units) increases, probability of Type II Error decreases (power increases), *i.e.*, if the population difference is large it is likely that such difference will be detected in the sampling.

It is obvious that (3) above is unknown to the investigator. It should be noticed that, unless the hypothesis tested includes a definitive statement of the relative magnitude of the parametric difference, the power of the test is indeterminate. For example, given the hypothesis that $m_A > m_B$, without qualification as to the magnitude of the difference, the power cannot be computed; but, if m_A and m_B are assigned definite values, the power may be computed. Neyman and Tokarska (1936) have tabulated the power of the t-test for various levels of confidence (Type I Error), sample sizes, and parametric differences. For example, with 20 d.f. at the 5 per cent level, if the parametric difference is 4.12 standard error units, power $= .99$, *i.e.*, probability of Type II Error $= 0.1$; with 1.70 parametric difference, power decreases to .50; with .39 units of parametric difference, power decreases to .10. Increase in power with increase in N is slight. For infinite d.f. at the 5 per cent level, 3.97 units of parametric difference give power of .99; 1.64 give power of .50, and .36 give power of .10. Johnson (1949, p. 68) cites several papers where power functions of other tests are tabulated.

TWO-TAILED TESTS

In the two examples concerning the difference in group I.Q., the *sign* of that difference was not stipulated. Those examples were tests of the null expressed as in equation (4). The sign of t was not at issue; m_A could have exceeded m_B or vice versa. Tests made on this basis are called two-tailed test, because both tails of the symmetrical distribution of t or x/σ are considered in determining the associated probability. The tables of t and x/σ are misleading in that only positive values of the statistic are tabulated, but the associated probabilities are for two-tailed tests. Although the distribution of Chi^2 is not symmetrical, here too tabular probabilities are for two-tailed tests, *i.e.*, the probability of frequency discordance in *either* direction from the hypothetical; note, however, that Chi^2, being a sum of squares, is necessarily always positive. Similarly, the F-ratio is always positive, but here too, tabled values refer to probabilities of chance *plus or minus* differences among the means of the treatments—thus the F-test as usually used is a two-tailed test.

ONE-TAILED TESTS

Suppose that, in the example concerning the children's I.Q.s the investigator had predicted *before* the measurements were taken that $m_A > m_B$. The experimental hypothesis would fail obviously if $m_A < m_B$, by no matter how slight an amount. The investigator in making his test should exclude exactly half of the values which t might have, since he is interested only in the occurrence of t's which bear a positive sign. This is quite legitimate since, in effect, he has agreed that all negative t's will indicate an insignificant difference. His question is, given that m_A does exceed m_B, is the difference too great to be attributed to chance variation at the level of confidence selected? The two-tailed test does not apply here, since only positive values of t are considered. The hypothesis is properly evaluated by using only one tail of the sampling distribution. The t sought is that which might occur 10 per cent of the time if the null were true, since 5 per cent, or half of the possible values, would occur in the positive direction. This $t = \pm 1.725$, since $+1.725$ would occur by chance 5 per cent of the time. Recall that the 5 per cent value for the two-tailed test is 2.086, and it will be seen that the investigator is much more likely to detect the parametric difference if he uses the one-tailed test.

The one-tailed test leads to a similar increase in power with other statistical tests as well. For the sampling distribution of x/σ, ± 1.96 and ± 2.58 are the values required for the 5 per cent and 1 per cent levels, respectively, when a two-tailed test is employed; for the one-tailed test the corresponding values are 1.65 and 2.33. When using the Chi^2 and F-tables with a one-tailed test, the value of the statistic which has an associated probability double that of the level of confidence selected is the value which must be reached for falsification of the null. That value is always less than would have been required if a two-tailed test were used.

We are now prepared to examine, or better to define, the distinction between experimentation$_I$ and experimentation$_{II}$. *Experimentation$_I$ is that form of experiment to which two-tailed tests of the data are appropriate; experimentation$_{II}$ is that form of experiment to which one-tailed tests of the data are appropriate.*

It follows that (a) preëxperimental predictions and (b) use of one-tailed tests are severally necessary and together sufficient for experimentation$_{II}$. The *intention* of the experimenter is *not* the controlling factor. Thus, if he decides to do experimentation$_{II}$ but either fails to make a prediction or employs a two-tailed test he is actually engaged in experimentation$_I$. Conversely, if he decides on experimentation$_I$ but predicts the direction of his results and employs a one-tailed test, he is actually engaged in experimentation$_{II}$.

The experimental distinctions are not mere verbal quibbles. We may exemplify their importance first with a fictitious problem, then with illustrations chosen from psychological literature.

Suppose that the investigator of group differences in I.Q. found that his sample means differed by 3.9 I.Q. points, and that the standard error of the mean differences was again 2.0. On these data $t = 1.95$. If we suppose further that group B had been given special coaching designed to "increase" I.Q., while group A had been maintained as a control, it would appear that the investigator thought he was engaged in experimentation$_{II}$. However, unaware of the distinction made here, he consults the t-table and finds that the obtained value of 1.95 falls short of the 2.086 required for the 5 per cent level. He concludes that the special treatment given to group B was ineffective in increasing I.Q. Actually, by using the right test (one-tailed), he would fulfill the requirements for experimentation$_{II}$, and his results would be significant at the indicated level.

It is not difficult to find instances where investigators have done experimentation$_I$, under the illusion that they were doing experimentation$_{II}$. Tinker (1948) had Ss read two forms of his Speed of Reading Test, the first under "ordinary" conditions, the second while the reading table was vibrated mechanically. He predicted that vibration would cause fatigue which would in turn lead to a decrement in reading speed. His results bore out the prediction. He also used a control group who read both forms under "ordinary" conditions. Here there was a critical ratio of 2.52 in favor of the first reading. He employed a two-tailed test at the 1 per cent level and found the difference could be attributed to chance (2.58 was needed). He might have predicted that there would be some fatigue in the second reading, even though the table was not being vibrated. Such predicted decrement with a one-tailed test would have been significant at the 1 per cent level. Tinker's results with the control group amount to experimentation$_I$, not experimentation$_{II}$.

Kircheimer, Axelrod and Hickerson (1949) studied college students who: (1) changed their majors and (a) were counseled, (b) were not counseled; and (2) did not change their majors and (a) were counseled, (b) were not counseled. They predicted that, of the groups who changed majors, those who were also counseled would show the greatest increase in grade point average. The critical ratio (in the direction of the prediction) was 1.84. The writers said that this was significant at the 7 per cent level, and had they employed a 5 per cent level criterion, the result would not have been significant statistically. Had they employed a one-tailed test as they should, the result would clearly have been significant at the 5 per cent level. These writers too have failed to draw the distinction between experimentation$_I$ and experimentation$_{II}$ and their results have appeared less critical thereby.

Gagné (1950) predicted that when highly similar items were grouped together, learning of serial lists would be accomplished in a shorter time than when such items were separated. The results were in the predicted direction with a critical ratio of 1.69. Gagné stated that the probability of occurrence of a deviation in the expected direction was 4.6 per cent. He correctly used a one-tailed test. Had he used the two-tailed test and adopted the 5 per cent level, he would not have found significant differences, and the prediction would appear to have failed. He rightly distinguished between experimentation$_I$ and experimentation$_{II}$.

The example concerned with Tinker's results was predicated on the basis that the direction of the decrement in reading decrement *could* have been predicted *before* the data were in. It must be emphasized that the one-tailed test is not justified unless the prediction is made prior to the data. If an investigator begins a study without preconceptions of results, and on studying those results generates a theory which will account for them, he cannot accept such afterthoughts as predictions, *i.e.*, switch—on the spot—from a two-tailed to a one-tailed test and pride himself on doing experimentation$_{II}$. He can *use* the results to predict *future* data. If he makes the switch after collecting the data, he has, in effect, increased the probability of committing a Type I Error without realizing it. If he uses the 5 per cent level on a one-tailed test, this is equivalent to the 10 per cent level with a two-tailed test, and he has unwittingly doubled his level of confidence: a dangerous departure from scientific conservatism. Burke (1950) and Cronbach (1949) have pointed out the inflation of level of confidence which occurs with either adding more subjects on the basis of already gathered data, or "shopping around" in compiled data for comparisons which are likely to show significant differences.

The experimenter$_{II}$ is in possession of a statistical tool not at the disposal of the experimenter$_I$. He is not limited to the prediction of results in a particular direction, but he may frame his hypothesis in such a manner that the magnitude of the parametric difference is included as well. For example, again referring to the comparison of group I.Q.s mentioned previously, the research worker might use an hypothesis of the form of equation (3), where the stipulated difference attributable to the "coaching" may be arbitrarily small. In this case, if the null remains tenable in light of the data, the theory is not rejected, whereas in experimentation$_I$, the null is of the form of equation (1) and, if there were theory, that theory would appear to have failed. This tool should be used with caution. No theory is particularly valuable if it predicts differences which are impracticably small. For instance, if the same coaching were expected to increase I.Q. by .5 points, the theory might remain tenable but the question "So what?" suggests itself. In this vein, Edwards (1950) has distinguished "practical" from "statistical" significance. The experimenter$_{II}$ should ever attempt to increase the power of his tests. If he is at liberty to set up experimental situations he should endeavor to maximize parametric differences since these contribute to test power. It follows that, in new areas experimentation$_{II}$ might wisely be limited to the comparison of extreme instances, rather than include the attempted distinction of subtle differences.

In summary, experimentation$_{II}$—as distinguished from experimentation$_I$—is characterized by predictions and one-tailed tests of significance. It has inherently greater statistical power and increases the productivity of investigation.

References

Burke, C. J. The effect of postponed decisions on statistical tests. Paper read at 22nd annual meeting of the Midwestern Psychological Association. May 5, 6, 1950, Detroit, Mich.

Cronbach, L. J. Statistical methods applied to Rorschach scores: A review. *Psychological Bulletin,* 1949, **46**, 393–429.

Edwards, A. L. *Experimental design in psychological research.* New York: Holt, Rinehart & Winston, Inc., 1950.

Gagné, R. M. The effect of sequence of presentation of similar items on the learning of paired associates. *Journal of Experimental Psychology,* 1950, **40**, 61–73.

Johnson, P. O. *Statistical methods in research.* Englewood Cliffs, N. J.: Prentice-Hall, 1949.

Kircheimer, B. A., Axelrod, D. W., & Hickerson, G. X., Jr. An objective evaluation of counseling. *Journal of Applied Psychology,* 1949, **33**, 249–257.

Neyman, J., & Tokarska, B. Errors of the second kind in testing Student's hypothesis. *Journal of the American Statistical Association,* 1936, **31**, 318–326.

Tinker, M. A. Effect of vibration on reading. *American Journal of Psychology,* 1948, **51**, 386–390.

Underwood, B. J. *Experimental psychology.* New York: Appleton-Century-Crofts, 1949.

4.2

Tests of hypotheses:
one-sided vs.
two-sided alternatives

Lyle V. Jones[1]

Psychological literature abounds with experimental studies which utilize statistical tests of the significance of differences between two groups of subjects. Most of these studies present tests based upon either the distribution of Student's t or upon the distribution of χ^2. Since the comparison of an experimental group with a control group of subjects is so fundamental to the experimental method, and since statistical tests of significance are appropriate for testing hypotheses regarding differences between two groups of subjects, it would seem important to correct a common misconception concerning the application of these tests of hypotheses.

One model for a test of significance of mean difference, the more familiar model, is that in which we test the null hypothesis, H_0, against a set of two-sided alternatives, H_1. We might formalize this test,

$$H_0 : \mu_1 - \mu_2 = 0$$
$$H_1 : \mu_1 - \mu_2 \neq 0,$$

where μ_1 is the mean of the population represented by one sample and μ_2 is the mean of the population represented by a second sample. Assuming scores X_1, from the first population, and scores X_2, from the second, both to be distributed normally, and assuming the population standard deviations to be equal, we may find

Lyle V. Jones, "Tests of Hypotheses: One-Sided vs. Two-Sided Alternatives," *Psychological Bulletin,* **46,** 1949, 43–46. Copyright (1949) by the American Psychological Association, and reproduced by permission.

[1] This article was prepared while the writer was a National Research Council Fellow.

$$t = \frac{\bar{X}_1 - \bar{X}_2}{s\sqrt{\dfrac{1}{N_1} + \dfrac{1}{N_2}}},$$

where

$$s = \sqrt{\frac{\displaystyle\sum_{i=1}^{N_1} (X_{1i} - \bar{X}_1)^2 + \sum_{j=1}^{N_2} (X_{2j} - \bar{X}_2)^2}{N_1 + N_2 - 2}}$$

and N_1 and N_2 are the numbers of individuals in the samples from the first and second populations.[2] Having stipulated a desired confidence level, α, we may enter the t table (Fisher & Yates, 1938) with $N_1 + N_2 - 2$ degrees of freedom and a p-value equal to α to find a critical value of t, t_c. If the absolute value of the observed t exceeds t_c, we reject H_0 in favor of H_1; otherwise, we accept H_0. In Figure 1 appears a distribution of t showing, graphically, the nature of this decision. This distribution corresponds to the sampling distribution of mean differences, under the null hypothesis. For any value of t to the right of t_c or to the left of $-t_c$, we reject H_0. The two shaded tails of the distribution, taken together, make up α per cent of the total area under the curve.

The model above, the test of the null hypothesis against two-sided alternatives, is the one used most often by investigators in psychology. Yet in many cases, probably in most cases, it is not the test most appropriate for their experimental problems. More often than not, in psychological research, our hypotheses have a *directional* character. We are interested in whether or not a given diet *improves* maze performance in the rat. We hypothesize that the showing of a particular motion picture to a group of individuals would lead to a *more tolerant* attitude toward certain racial minorities. We wish to test whether or not anxious subjects will respond *more actively* than normal subjects to environmental changes which might be perceived as threatening. In each case, theoretical

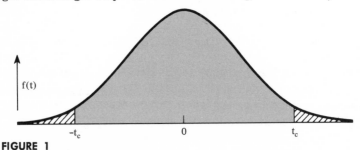

FIGURE 1
The two-tailed test model.

[2] Of course, if the two samples are not independently selected, we should make use of the correlation between them in the determination of t.

considerations allow the postulation of the direction of experimental effects. The appropriate experimental test is one which takes this into account, a test of the null hypothesis against a one-sided alternative.

In the one-sided case we test H_0 against H_1, where

$$H_0:\mu_1 - \mu_2 = 0$$
$$H_1:\mu_1 - \mu_2 > 0.$$

Under the identical assumptions of the two-sided model we may calculate t as before. Again a confidence level, α, is stipulated. The distinction between the one-sided test and the two-sided test arises in the determination of the critical value, t_c. In the present case this critical value is found by entering the t table with $N_1 + N_2 - 2$ degrees of freedom, as before, but with a p-value equal to 2α. If our observed t is greater than this t_c we reject H_0 in favor of H_1; if t is less than t_c we accept H_0. The t distribution in Figure 2 exemplifies this procedure. A value of t to the right of t_c leads to the rejection of H_0, the acceptance of H_1. While the shaded area under the curve once again represents α per cent of the total area, the shaded portion is restricted, in this case, to one tail of the distribution.

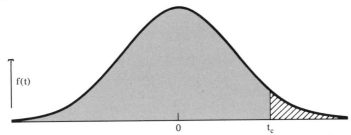

FIGURE 2
The one-tailed test model.

It might be noted that with this formulation of the one-tailed test there is no allowance for the possibility that the true difference, $\mu_1 - \mu_2$, is negative. In the type of problem for which the one-tailed test is suited, such a negative mean difference is no more interesting than a zero difference. In fact, the hypotheses for the one-sided case might be

$$H_0:\mu_1 - \mu_2 \leqq 0$$
$$H_1:\mu_1 - \mu_2 > 0.$$

In order to determine a sampling distribution under H_0 we should consider the "worst" of the infinite alternatives under H_0, i.e., that alternative which would make the decision between H_0 and H_1 a most difficult one. Clearly, the decision would be more difficult if the true mean difference were zero than if the true difference were any negative value.

Hence we would proceed exactly as in the preceding one-tailed case, utilizing, for our test, the distribution of t based upon the same sampling distribution of mean differences as before. The confidence level should be doubled to provide the p-value for entering a table to find a critical t_c, or, if it is desired to ascertain the p-value corresponding to an observed t, the correct value is one-half that given in the typical table of t.

While the one-tailed test has been exemplified here as a test of mean difference, based upon the t distribution, it is limited in application neither to mean difference problems nor to the t statistic. Indeed, whereever an alternative to the null hypothesis is stated in terms of the direction of expected results, the one-tailed test is applicable.

The failure, among psychologists, to utilize the one-tailed statistical test, where it is appropriate, very likely is due to the propagation of the two-tailed model by writers of text books in psychological statistics. It is typical, in such texts, to find little or no attention given to one-tailed tests.[3] Since the test of the null hypothesis against a one-sided alternative is the most powerful test for all directional hypotheses, it is strongly recommended that the one-tailed model be adopted wherever its use is appropriate.

Reference

Fisher, R. A., & Yates, F. *Statistical tables for biological, medical, and agricultural research*. Edinburgh: Oliver and Boyd, 1938.

[3] One notable exception occurs in A. L. Edwards' *Experimental Design in Psychological Research*, where the two cases are clearly differentiated.

4.3

A note on one-tailed
and two-tailed tests

W. E. Hick

The following remarks were provoked by the recent paper by Marks[1] on this subject. Although the impracticability, in most cases, of deciding just what level of significance is "significant" makes it superfluous to fuss about minor inaccuracies, the difference between one-tailed and two-tailed tests can be serious. Moreover, the problem of which to use in a given case does give an appreciable amount of trouble to the research worker. It appears to me that the interpretation given by Marks errs on the unsafe side.

In the first place, it should be clearly understood—though Marks, perhaps inadvertently, implies the opposite—that it makes no difference *when* a theory or hypothesis is conceived, as far as its logical content and implications are concerned. Logic is timeless; it does not, and cannot, matter whether a theory was conceived before, during, or after the experiment; it may be suggested by the data, or it may be revealed in a dream. An hypothesis derived from the data may be expected to fit the data, more or less; but the statistical calculations are intended to show how well or how badly it does fit, and, in the vast majority of cases, to show this on the assumption that personal opinions and collateral evidence are set on one side.

This is quite separate from the subtraction of degrees of freedom when parameters of the chance hypothesis are calculated from the data. When we do this, we are not depending on the data to provide the chance hypothesis, then we can *a fortiori* exclude all others, since they are less will act as the most stringent test. Ideally, we should choose the hypothesis which best fits the data, with the exception that it must have no

W. E. Hick, "A Note on One-Tailed and Two-Tailed Tests," *Psychological Review*, 59, 1952, 316–318. Copyright (1952) by the American Psychological Association, and reproduced by permission.
[1] M. R. Marks, Two kinds of experiment distinguished in terms of statistical operations. *Psychol. Rev.*, 1951, 58, 179–184.

preferential tendency to produce the effect in question—e.g., a deviation of the mean, or something of that kind. If we can exclude this chance hypothesis, then we can *a fortiori* exclude all others, since they are less likely. This is the step which requires us to deduct degrees of freedom, because it is only this step which gives us any appreciable amount of information from the data. To be able to reject one chance hypothesis tells us nothing, since there is an infinity of others; but to be able to reject a whole class really means something, and the data must pay.

However, though robbing the data in this way affects the statistical significance, it does not appear to have anything to do with the problem of one-tailed and two-tailed tests. To deal with this, we must first consider the statistical question asked—namely, "Could my results reasonably be due to chance, or must I assume that they were due to some cause or systematic bias?" Or, since the experimenter is not necessarily required to make up his mind on this point, he may go no further than the question, "What is the probability, on the chance hypothesis, of my results occurring?"

Now this, unfortunately, is just what the ordinary statistical methods fail to answer explicitly. As is well known, the probability of the particular sample depends, not only on the shape of the distribution, but also on the number of possible alternative samples. On the other hand the P of the ordinary significance test is, strictly speaking, a probability irrelevant to the question of whether the chance hypothesis should be rejected on the grounds of unlikelihood. The best we can say of it is that it happens, in the ordinary cases, to be vaguely related to what we want to know.

If the reader doubts this, he has only to consider what "significance" he would attach to the occurrence of an observation in, say, a 5 per cent slice at the end of a rectangular distribution. Obviously such an event would not give us the least grounds for rejecting the hypothesis, and, in fact, there is no significance test, in the ordinary sense, for this distribution, except for the trivial case of an observation right outside it. This is almost a *reductio ad absurdum*, but it serves to emphasize that P, though it is a probability, is not the one which, by itself, makes us view the chance hypothesis with suspicion.

In the Neyman and Pearson theory of critical regions P has a definite place. But some statisticians think this theory approaches the problem from the wrong end. It is certainly difficult, and often impossible, to apply. If then we are compelled in practice to use some rough and ready method, which is the least rough and the most ready? It is clear that the actual sample probability is what matters, and it is possible to make some sense of it by dividing it by the probability of the most probable sample. This relative probability answers the question, "How much less likely are my results than what I should have expected on the chance hypothesis?" However, this also leads to absurdity in special cases, and we must admit that there is no easy method that is quite foolproof.

The essence of the ordinary P, correctly used, is that it is the class probability of a critical region composed solely of all the samples whose individual probabilities are below a certain critical level. The distribution may go up and down like a mountain range, but wherever the curve falls below this level, there is a piece of the critical region in that place. This is not the case, incidentally, for the tabular probabilities of Chi^2, which are only the probabilities of *exceeding* a given value. Marks states that these probabilities are for two-tailed tests, but this is only true in the rather different sense that they refer indiscriminately to positive or negative discrepancies. But the Chi^2 distribution has a short tail as well as a long one, and both should properly be included in the critical region. As it is, we just have to remember that small values are "suspicious."

In attempting to justify the one-tailed test, Marks appears to regard as equivalent such states of mind as *knowing* that, say, a negative deviation cannot occur except by chance, not being interested if it does, guessing that a positive deviation will occur, and so on. What the experimenter is interested in has nothing to do with the case. What he guesses is relevant, if he can show that it is a rational guess and can assess it in terms of probability. What he knows for certain is still more important. If he knows that a negative bias is impossible, it means that no negative deviation, however great, will shake his faith in "pure chance" as the cause of it. If he is in as strong a position as that, of course he is entitled to put the whole critical region in the positive tail and call it a one-tailed test. What he is doing is to accept a more lenient degree of significance, though it is difficult to see why he should bother with this form of test at all; if the bias cannot be negative, there is little risk of it being exactly zero.

But suppose he is only moderately sure that the bias is positive; is he to do a one-and-a-half-tailed test, or something like that? By all means, if he can assess his prior knowledge in terms of fractions of a tail, except that it then follows that there are not just two kinds of experimentation, as Marks suggests, but one for every degree of prior expectancy.

It is far better to carry out the statistical work in the ordinary way, without tampering with the critical region, and let prior probability (or any other relevant consideration) determine what level of significance you will regard as decisive. You can still believe in a positive deviation more readily than a negative one, if you like; but you do not confuse the issue for others who may have different ideas.

4.4

One- and
two-tailed tests

Melvin R. Marks

Hick, in "A Note on One-Tailed and Two-Tailed Tests,"[1] purports to vitiate the recommendations made by me in a recent paper.[2] I think a reply is indicated.

1. My paper was not intended to precipitate a controversy over the philosophical foundations of mathematical statistics. It *was* intended to show that, by taking advantage of available statistical techniques, the experimenter could increase the precision of his investigation by recognizing and minimizing the incidence of Type I and Type II errors. The paper was expository rather than argumentative, a restatement rather than a presentation *de novo*. All statements made therein relative to the testing of hypotheses have the acceptance of statisticians generally.

2. The experimenter's decision to adopt a particular level of confidence is not made with reference to statistical considerations. Such a decision depends entirely on the assessment of the likelihood of Type I and Type II errors and the practical importance to be attached to their occurrence. Thus, the propriety of the one- or two-tailed test depends only upon the nature of the hypothesis to be tested—not on the level of confidence adopted.

3. Although logic is timeless, the "personal equation" of the experimenter is not. The legitimate use of particular data for the test of a particular hypothesis *does* depend on why (the basis for), if not when, the hypothesis was formulated. Hick says, ". . . it does not and cannot matter whether a theory (read hypothesis) was conceived before, during,

Melvin R. Marks, "One- and Two-Tailed Tests," *Psychological Review*, 60, 1953, 207–208. Copyright (1953) by the American Psychological Association, and reproduced by permission.

[1] W. E. Hick, A note on one-tailed and two-tailed tests, *Psychol. Rev.*, 1952, 59, 316–318.

[2] M. R. Marks, Two kinds of experiments distinguished in terms of statistical operations. *Psychol. Rev.*, 1951, 58, 179–184.

or after the experiment; it may be suggested by the data, or it may be revealed in a dream." Now this statement is true in *all* respects *if* conclusions about the tenability of the hypothesis are to be restricted to the data at hand. But if these conclusions are to be extrapolated, we must strike the phrase, "it may be suggested by the data." For consider, with reference to any particular data, if the hypothesis is to proceed from the data, the best hypothesis is that the data are as they are. In such case, we might solemnly aver that they are as they are by definition, and never bother to test any hypothesis. We shall always be right, if only we select the hypothesis carefully enough—until the next time!

4. Hick challenges, at least implicitly, the Neyman and Pearson theory as it applies to critical regions. The pros and cons of such a controversy have no place in this discussion. However, it may be stated that, when the critical region selected is appropriate to the hypothesis being tested, then the level of confidence (by definition) is known exactly, and the probability that a Type II error will occur (false acceptance of the null hypothesis) is minimized.

5. Hick's remark on my treatment of the chi-square test has merit. Technically, the term "two-tailed," as applied to the chi-square distribution, refers to the actual tails of that distribution—i.e., the regions of exceptionally large and exceptionally small values of chi square. However, it is still legitimate to cut the level of confidence in half when we predict direction of frequency discordance. This is so because when we eliminate exactly half of the possible values of chi square (that half corresponding to frequency discordances in the unwanted direction), we also eliminate half of the Type I errors.

6. The same remarks apply, with somewhat less force, to the indicated use of the F test. Bendig[3] has pointed out that only one tail of the F distribution is tabled—i.e., values of F equal to or greater than unity. When I discussed the increased precision which could be achieved by predicting the hierarchy of magnitude of the means, I was referring to the fact that, in the case of simple classification analysis of variance with only two columns (when $t^2 = F$), the number of Type I errors is cut in half when the negative values of t are eliminated beforehand. The number of Type I errors would be reduced still further in the case of more than two columns (variables) if the ordering of the means was predicted beforehand, although in such case power would suffer—i.e., the chances of committing a Type II error would increase since a slight inversion from the predicted order would necessitate rejection of the experimental hypothesis.

[3] Personal communication.

4.5

A brief note on
one-tailed tests

C. J. Burke

Concurrent with the recent discussions of one-tailed and two-tailed tests by Hick (1952), Jones (1952), and Marks (1951) there has been a disturbing increase in the use of one-tailed tests in student experimental reports as well as in published and not-yet-published manuscripts. While the popularity of one-tailed tests is undoubtedly attributable in part to the overwillingness of psychologists as a group to make use of the statistical recommendations they have most recently read, there seems to be a certain residual of bad logic, so far as both statistics and psychology are concerned, which merits examination. The writer takes the position already taken by Hick (1952) in all important essentials but the argument to be presented differs, at least in emphasis, from that of Hick. It should be noted that some tests, χ^2 and F, for example, are naturally single-ended. Nothing said here should be construed so as to apply to them.

Both Jones (1952) and Marks (1951) seem to the writer to confuse somewhat two quite different notions—that an experimental hypothesis is often directional and that an experimenter may be willing to accept a deviation of any size in the unexpected direction as consonant with the null hypothesis. We shall consider two quotations from Jones.

> The model above, the test of the null hypothesis against two-sided alternatives, is the one used most often by investigators in psychology. Yet in many cases . . . it is not the test most appropriate for their experimental problems. More often than not, in psychological research, our hypotheses have a *directional* character . . . theoretical considerations allow the postulation of the direction of experimental effects. The

appropriate experimental test is one which takes this into account, a test of the null hypothesis against a one-sided alternative (1952, p. 44).

It is a fact that many hypotheses in psychological research, experimentally conceived, are directional for the investigator conducting the experiment, but it does not follow from this that one-sided tests should be used in experimental reports.

To amplify these considerations we point out that there are, in many experiments, two statistical decisions to be made and two different levels of confidence may be involved. The first is the decision made by the individual experimenter who frequently plans one experiment from his evaluation of a previous one. We concede that here a one-tailed test is often proper. The second is the decision which determines the place of his findings in the literature of psychology. Here the one-tailed test seems inadmissible. It is the second type of decision with which we are concerned. Marks (1951) has in essence repeated from statistical sources a discussion of the Type I and Type II errors which shows that the decisions made in any statistical interpretation depend only upon the underlying populations and the rule of procedure used. Any comparison of alternative rules of procedure must take into account errors of both types, the error of rejecting a hypothesis when it is true and the error of failing to reject it when it is false, but the underlying statistical considerations do not provide automatically a criterion for the selection of one rule over another. Such a criterion is to be sought in the number and kinds of errors the experimenter will tolerate. Roughly, an acceptable criterion is to make the over-all number of errors as small as possible and at the same time to render large and serious errors relatively impossible. Within the class of hypotheses which are considered to be directional it is likely that a one-tailed test might yield a smaller over-all number of errors than a two-tailed test, but there is, under the single-tailed rule, no safeguard whatsoever against occasional large and serious errors when the difference is in the unexpected direction. If one is less willing to commit a large error than to commit a small one, it does not follow from the theory of testing statistical hypotheses that the experimenter's expectation of a given direction for the result necessarily makes the one-tailed test desirable.

To advance this point in our case against the use of the one-tailed test in the public report, we next take up the second quotation from Jones.

It might be noted that with this formulation of the one-tailed test there is no allowance for the possibility that the true difference . . . is negative. In the type of problem for which the one-tailed test is suited, such a negative mean difference is no more interesting than a zero difference (1952, p. 45).

This statement is perfectly correct.[1] If we consider it carefully we discover its import to be that the investigator should use a one-tailed test when he is willing to accept a difference in the unexpected direction, *no matter how large*, as consonant with the hypothesis of zero difference. This is quite a different matter from using a one-tailed test whenever the direction of the difference is predicted, on some grounds or other, in advance. It is to be doubted whether experimental psychology, in its present state, can afford such lofty indifference toward experimental surprises.

The questions raised by the one-tailed test are to be answered finally by considering the effect of general use of this procedure on the content of psychological literature. The writer cannot agree with Hick (1952) that its use makes little difference since there is no practical rule for deciding what "significance" really is. In some super-scientific world this point might be well taken, but there is evidence that in our workaday world (where we sometimes read only the concluding sections of reports) it does make a difference whether the investigator has stated that his results were significant. The controversy over the Blodgett effect is a case in point (Blodgett, 1929, 1946; Kendler, 1952; Maltzman, 1952; Meehl & MacCorquodale, 1951; Reynolds, 1945; Thistlethwaite, 1951, 1952).

Remembering that the problem of testing a statistical hypothesis is a statistical problem in which each individual experiment is viewed only as a member of a class of similar experiments and recalling that the properties of any statistical test are determined solely by the procedure followed and by the populations underlying the class, it is pertinent to inquire into the effects of widespread adoption of one-tailed tests upon the literature. The writer believes the following statements to be reasonable forecasts.

1. The discovery of new psychological phenomena will be hindered. Our literature abounds with instances in which the outcome of a given experiment has differed reliably and sharply from expectation. These experiments are usually of great interest—new psychological concepts arise from them. Our science is not yet so mature that these can be expected to occur infrequently. The most recent instance, known to the writer, of conflicting results from experiments thought to be highly similar was reported by Underwood (1952) at the 1952 meetings of the American Psychological Association. From any careful examination of contemporary psychological literature we must conclude that nowhere in the field can we have sufficient a priori confidence in the outcome of any genuinely new experiment to justify the neglect of differences in the unexpected direction.

---

[1] In his subsequent discussion, Jones spoils the force of this point by confusing hypotheses to be tested with classes of hypotheses to be guarded against as alternatives.

2. There will be an increase in barren controversy. Fruitless controversies arise from unreliable results. Conclusions at low levels of confidence tend to be unreliable, and the adoption of one-tailed tests is equivalent to a general lowering of levels of confidence. At a time of severe journal overload this is especially pernicious. There is no substitute in statistical methodology for the carefully designed and controlled experiment in which any important difference between groups will show up at a high enough level of confidence to insure a certain reliability in the conclusion.

3. Abuses will be rampant. It is no criticism of the position held on statistical grounds by Jones and Marks to point out that the considerations involved in the choice of a one-tailed test are really rather delicate. A nice instance of what can happen is seen in an experimental report by Gwinn (1951). Gwinn reports two experiments which are not markedly different from each other. They turn out in opposite directions, and, by appropriate selection of the position of his "critical tail," Gwinn establishes significance and near significance (1 per cent and 8 per cent levels, approximately) for his results on the basis of one-tailed tests.

The moral can be pointed with advice. We counsel anyone who contemplates a one-tailed test to ask of himself (before the data are gathered): "If my results are in the wrong direction and significant at the one-billionth of 1 per cent level, can I publicly defend the proposition that this is evidence of no difference?" If the answer is affirmative we shall not impugn his accuracy in choosing a one-tailed test. We may, however, question his scientific wisdom.

References

Blodgett, H. C. The effect of the introduction of a reward upon maze performance of rats. *University of California Publications in Psychology*, 1929, **4**, 113–134.

Blodgett, H. C. Reynolds' repetition of Blodgett's experiment on "latent" learning. *Journal of Experimental Psychology*, 1946, **36**, 184–186.

Gwinn, G. T. Resistance to extinction of learned fear drives. *Journal of Experimental Psychology*, 1951, **42**, 6–12.

Hick, W. E. A note on one-tailed and two-tailed tests. *Psychological Review*, 1952, **59**, 316–318.

Jones, L. V. Tests of hypotheses: one-sided vs. two-sided alternatives. *Psychological Bulletin*, 1952, **49**, 43–46.

Kendler, H. H. Some comments on Thistlethwaite's perception of latent learning. *Psychological Bulletin*, 1952, **49**, 47–51.

Maltzman, I. The Blodgett and Haney types of latent learning experiment: reply to Thistlethwaite. *Psychological Bulletin*, 1952, **49**, 52–60.

Marks, M. R. Two kinds of experiment distinguished in terms of statistical operations. *Psychological Review*, 1951, **58**, 179–184.

Meehl, P. E., & MacCorquodale, K. A. Failure to find the Blodgett effect and
some secondary observations on drive conditioning. *Journal of
Comparative Physiological Psychology*, 1951, **44**, 178–183.

Reynolds, B. A repetition of the Blodgett experiment on "latent" learning.
Journal of Experimental Psychology, 1945, **35**, 504–516.

Thistlethwaite, D. A critical review of latent learning and related experiments.
Psychological Bulletin, 1951, **48**, 97–129.

Thistlethwaite, D. Reply to Kendler and Maltzman. *Psychological Bulletin*, 1952,
49, 61–71.

Underwood, B. J. The learning and retention of serial nonsense lists as a function
of distributed practice and intralist similarity. Paper read at the
American Psychological Association, Washington, D. C., September,
1952.

4.6

A rejoinder on
one-tailed tests

Lyle V. Jones

In a recent issue of this journal, Burke (1953) criticizes
earlier discussions of one-tailed and two-tailed tests (Hick, 1952; Jones,
1952; Marks, 1951) and suggests need for caution in the application of
one-tailed statistical tests to psychological research designs. The writer
is in accord with several implications of Burke's note. As is true for most
statistical designs, abuses would be reduced markedly were the test
model completely specified and justified in terms of the purpose of
investigation, before data are viewed by the investigator. To be guided
by the data in the specification of hypotheses and statistical tests is a
grave breach of the rules of experimental verification (Marks, 1953).

The argument presented by Burke, however, is more than a plea for
careful consideration of the choice of test models for given experimental
problems. It is stated that the selection of a one-tailed test model requires

Lyle V. Jones, "A Rejoinder on One-Tailed Tests," *Psychological Bulletin*, **51**, 1954,
585–586. Copyright (1954) by the American Psychological Association, and reproduced by permission.

that an investigator be willing to "publicly defend the proposition . . . of no difference" if results actually show a large difference in the direction opposite to that predicted (Burke, 1953, p. 387). The proposition appears indefensible, since it demands arguing that a particular observed difference, no matter how large, is only a sampling departure from zero. If accepted, Burke's argument should lead to universal avoidance of one-tailed tests.

Consider the following experimental problems, selected for simplicity as single variable designs: (a) On the basis of a certain behavioral theory, we might predict that an experimental condition imposed upon subjects in the population under study would raise the mean level of performance on a given task. The theory provides a prediction (an alternative hypothesis) that the mean for the experimental group population will exceed the mean for the control group population. The hypothesis under test is that the mean for the experimental population is the same as or less than that for the control population. We should like a statistical test that will yield a decision: either the data are consistent with the hypothesis under test, or we reject the hypothesis in favor of the alternative. (b) In a field of applied psychology a new diagnostic technique is developed and is to be adopted if, and only if, we are confident that it is better than the current technique which would be replaced. Assuming the availability of a suitable criterion, the two techniques are applied to comparable samples or to the same sample; interest resides in the extent to which the parametric proportion of successful predictions using the new technique exceeds that using the old. A statistical test is to supply a decision: either the new technique is no more adequate than the old, or the new technique is more adequate.

For the class of problems illustrated by these two examples, the hypothesis under test is not simply one of no difference. We wish to test the hypothesis that the algebraic difference between parametric mean performance under experimental and control conditions is zero *or negative* against the alternative hypothesis that the difference is positive. It is meant to stress this formulation of the one-tailed statistical test with the greatest possible emphasis. With this formulation, it is apparent that acceptance of the hypothesis under test does not demand defense of a proposition of no difference; the observed difference, whether negative, zero, or slightly positive, simply does not allow acceptance of the alternative hypothesis at the level of stringency (values of α and β) chosen for the test.

In a footnote, Burke (1953, p. 385) criticizes this formulation as it appeared earlier (1952, p. 45) on the grounds that it confuses hypotheses to be tested with hypotheses to be guarded against as alternatives. To the contrary, this statement of the problem clarifies the nature of the hypothesis under test. The hypothesis to be tested is[1]

$$H_0: \mu_e - \mu_c \leqq 0,$$

where μ_0 is the population mean for the experimental condition, μ_0 is the population mean for the control condition, and the experimental prediction yields an alternative hypothesis,

$$H_1: \mu_e - \mu_c > 0.$$

Burke's primary argument seems to rest upon the contention that "there is, under the single-tailed rule, no safeguard whatsoever against occasional large and serious errors when the difference is in the unexpected direction" (Burke, 1953, p. 385). Our formulation of the hypothesis under test completely resolves this difficulty, for no error is committed when that hypothesis is accepted on the basis of a large observed difference in the unexpected direction. The event is one of a class of events consistent with the hypothesis tested.

If one were to retain the alternative hypothesis, H_1, above, and to adopt a two-tailed statistical test, accepting H_1 when there were observed large differences between the means *in either direction*, his position would be unenviable. For, following the rules of his test, he would have to reject the hypothesis under test in favor of the alternative, even though an observed difference, $\mu_e - \mu_c$, was a substantial negative value.

The remaining discussion by Burke consists of pragmatic arguments against the adoption of one-tailed tests. The arguments appear valid only under the assumption that every application of the one-tailed test is an abuse of experimental methodology.

Certainly, if (*a*) the test model is specified completely (including specification of the confidence level to be adopted) before the data are gathered, and if (*b*) the purpose of the test is only to determine whether a particular directional prediction is supported by the data, then the one-tailed test not only is appropriate, but it is in error to use a two-tailed test model.

References

Burke, C. J. A brief note on one-tailed tests. *Psychological Bulletin*, 1953, **50**, 384–387.

Dixon, W. J., & Massey, F. J., Jr. *Introduction to statistical analysis*. New York: McGraw-Hill, 1951.

Hick, W. E. A note on one-tailed and two-tailed tests. *Psychological Review*, 1952, **59**, 316–318.

[1] An equivalent formulation of the one-tailed test model, of which the writer was unaware at the time of his earlier note, is that proposed by Dixon and Massey (1951, pp. 100–104).

Jones, L. V. Tests of hypotheses: one-sided vs. two-sided alternatives. *Psychological Bulletin,* 1952, **49**, 43–46.

Marks, M. R. Two kinds of experiment distinguished in terms of statistical operations. *Psychological Review,* 1951, **58**, 179–184.

Marks, M. R. One- and two-tailed tests. *Psychological Review,* 1953, **60**, 207–208.

4.7

Further remarks on one-tailed tests

C. J. Burke[1]

In the discussion of one-tailed tests there are a number of issues which have not been clearly separated. An effort will here be made to separate them.

TWO TYPES OF MODELS

Psychologists have used two different types of models in interpreting their data: statistical models and specifically psychological models. The two types have been employed in different ways and for entirely different purposes. Statistical models have been used to determine whether data are intrinsically interesting or, in other words, whether the data indicate relationships between psychological variables. Psychological models have been utilized in attempts to organize the knowledge attained in various areas of the field.

Early uses of statistical methods mixed the psychological and statistical models. The substantive biases of the experimenter led him to seek out, with each set of data, those statistical techniques which supported his views and to neglect the techniques which did not. Gradually, it was recognized that the function of the statistical model was prior to,

C. J. Burke, "Further Remarks on One-Tailed Tests," *Psychological Bulletin,* **51**, 1954, 587–590. Copyright (1954) by the American Psychological Association, and reproduced by permission.

[1] The writer thanks his colleagues, A. M. Binder and W. K. Estes, for their suggestions after reading an earlier draft of this reply.

and separate from, the function of the substantive model. Cognizant of the fallibility of their data and the greater fallibility of their theories, psychologists developed the view that data should be interpreted without the intrusion of the biases, however well founded, of the experimenter and accordingly accepted certain statistical procedures as conventions. The advocates of one-tailed tests would take us a large step backward, for they openly favor the mixing of the statistical and substantive models.

Any psychological theory worth bothering with will generate a number of predictions which can be directly checked against experimental data. Such direct determination of the consistency of data with the model is the only proper test of a psychological theory. In a check of this kind, there is a point by point verification of the theory, and strictly speaking, no statistical model is necessary. This does not mean that there should be no statistical interpretation of the data; it is usually wise to make the ordinary statistical interpretation prior to the theoretical check—to determine whether the data show enough to warrant a theoretical analysis. If, in any given experiment, a theory provides only a directional prediction, there seems to be nothing wrong with our traditional procedure of first establishing the presence of a difference between groups and subsequently noting its direction. A statistical model should enter directly into the verification of a substantive theory only when it is based on a null hypothesis with respect to the residuals of the data from theoretical predictions.

A statistical model can be viewed as a technique for assessing the reliability of an experiment without repeating it. The question answered by a statistical test is whether or not the sample obtained is a member of a certain class of samples. This class is usually defined by the null hypothesis to be tested and by the conditions under which the experiment would be repeated, if repetition were undertaken; frequently certain supplementary parametric assumptions must be made. Since an experiment can be repeated without any appeal to a body of substantive theory, no such appeal need ever be made for purposes of statistical analysis. Experiments are often designed from theoretical considerations, and the conditions of repetition may, therefore, be dictated by a substantive model. In some analysis of variance designs there may be freedom in the choice of an error term. This freedom indicates that the single experiment which has been performed, even in conjunction with a null hypothesis, does not define a unique class of repetitions. When this is the case, appeal to substantive theory may lead to the choice of a proper error term, and this appeal may seem to involve a direct intrusion of substantive theory upon the statistical interpretation. As we have put the matter, however, it is readily seen that in this instance the theory does no more than define the conditions under which the experiment would be repeated.

THE ONE-TAILED TEST IN THEORETICAL STATISTICS

Properly described, the concept of a one-tailed test is clear and free from any objection on mathematical grounds. There is no disagreement on this point. Disagreement on what constitutes a proper description of a one-tailed test centers specifically on the nature of the hypothesis that is tested. Jones has emphasized his conviction that the hypothesis of a nonpositive difference is tested against the alternative of a positive difference. I have stated that a null hypothesis is tested with safeguards against only a subset of the possible alternatives.

As it now stands, the mathematical theory of testing statistical hypotheses requires that the hypothesis under test lead to the calculation of a level of confidence. (Until the onslaught of the one-tailers, books in psychological statistics cautioned us to test only "exact" hypotheses.) If we ask Jones to exhibit his calculation of the level of confidence in a one-tailed test, we shall probably discover that he bases the calculation on the distribution obtained under the null hypothesis. Then, according to the usual statistical logic, he is in fact testing the null hypothesis against selected alternatives. The comments of my earlier note were made within this framework.

An easy modification of statistical logic can make the one-tailed test of a nonpositive difference tenable. Since the probability that he rejects the hypothesis of a nonpositive difference when it is true can, in his procedure, never exceed the level of confidence calculated under the null hypothesis, Jones can declare that he has a bound on his level of confidence even though he does not know its precise value. Utilizing this bound, he can set up consistent rules for one-tailed interpretation of the hypothesis he has stated.

The two paragraphs above present what seem to be the only two positions open to Jones. Whichever he accepts, I am in complete accord with him so far as mathematical statistics goes, for both are clear and statistically defensible. Yet his succinct assertion that I advocate universal avoidance of one-tailed tests is correct. The locus of disagreement is to be found in practice, not in theory.

APPLIED PROBLEMS

In applied problems like Jones's example of the drugs, it is often true that certain practices or routines will be changed if, and only if, an experiment yields a reliable difference *in a specified direction*—if the new is reliably better than the old. For problems like this, Jones maintains that the two-tailed test is inappropriate. Verbally his argument proceeds well, but before we accept it, let us note that problems of this kind have never given any difficulty—we have always understood what we did and why. Let Jones use his one-tailed test at the 5 per cent confidence level to analyze data from a large number of such experiments. We shall analyze the same data using a two-tailed test of the null hypothesis but permitting ourselves a luxuriously loose 10 percent confidence level. Being sensible men, we shall not advocate a change to the new routine if the old one proves reliably superior. Unless we wish to differentiate a null from a negative difference (we might, for example, consider a replication necessary unless the null hypothesis were rejected), the two procedures will obviously lead to the same recommendations with every set of data. Thus, they have identical consequences for the immediate practical decision.

For myself, I do not much care whether one uses a one-tailed or two-tailed model in such applications. Since the choice of a level of confidence is arbitrary and since identical decisions will always be made, we have complete equivalence between the one-tailed model and a doubled level of confidence in our traditional procedure. It would be wrong for either Jones or me to pretend that compelling reasons exist for preferring one of these ways of talking about applied experiments over the other. Any appeal to the risks of the two kinds of errors is little more than an uninteresting parlor game. Since no gain results, there seems to be insufficient reason for changing the traditional two-tailed interpretation, but this is not a very strong objection. In my earlier note, I conceded the admissibility of the one-tailed test for the experimenter's own laboratory planning; I am quite willing to add the applied experiment to the list.

THE PUBLIC LITERATURE

There do seem to be compelling reasons why the use of one-tailed tests in our permanent reports must be rejected. These reasons are of two kinds: one kind residing in the nature of the scientific enterprise and the functions of the experimental literature, and the other associated with the debasing effect upon the literature that is almost certain to result from adopting one-tailed tests.

Experimental scientists must have for data a permanent respect that transcends their passing interest in the stories they make up about their data. Science is a public enterprise. These are statements to which most experimental scientists would subscribe. One-tailed tests violate both the spirit and the letter of these statements. Interest in whether a relationship is found between two variables in a given experiment is seldom confined to those who share the theoretical preconceptions of the experimenter, and the right to discuss experimental data in relation to a particular theory does not remove the obligation to interpret the experiment according to rules that can be accepted by the reader who rejects your views as well as by the reader who shares them. Our experience has shown that experimental results can be viable for years after they are first reported even when the experiments are designed on the basis of moribund theoretical considerations. Let us consider our obligation to future psychologists who might not be able to understand why we were so naive in 1954 as to predict *that* direction for our experimental results. Why should we put needless difficulties in the way of any reader, present or future?

Why, particularly, when the change over to one-tailed test is almost certain to gain nothing and lose much? Jones asserts that my dire forecasts on the effects of one-tailed tests on our literature are predicated on the assumption that every one-tailed test is a misuse. Here he has missed the fundamental point of my argument. As in applied experiments, there is pragmatic equivalence between using the one-tailed test and doubling the level of confidence in the two-tailed test, for if the discrepancy is in the expected direction, the power of these two tests is identical, and experiments should be, and usually are, set up with power as a foremost consideration.

Most psychologists would resist the general adoption of the 10 per cent level for rejecting hypotheses. Such a low level of aspiration would lead to a decreased number of subjects per experiment and thus accomplish a gradual attrition in the soundness of our reported conclusions. This is precisely what will happen if there is general, and editorial, acceptance of the one-tailed test.

CONCLUDING REMARKS

A diminution in the quality of our psychological literature may be of little moment to the theoretical statistician but must be of grave concern to the experimental psychologist. Once they have the assurance of the mathematical statistician that the procedure is logically defensible, experimenters seem so eager to employ one-tailed tests that, sociologically speaking, I have lost the argument almost before it has been joined. Of course, the procedure is mathematically sound, but psychologists are mistaken if they believe the mathematical statistician can speak with special authority in the matter. The decision is for us alone; it is for us to ponder whether, in the interpretation of our data, the one-tailed test is wise. We are drifting into an unfortunate decision, and I, for one, wish to enter a plea for a little less methodology and a little more wisdom. We expound a distinction between experimental hypotheses and statistical hypotheses to our beginning students; let us maintain the distinction for ourselves.

4.8

The fallacy of the null-hypothesis significance test

William W. Rozeboom

The theory of probability and statistical inference is various things to various people. To the mathematician, it is an intricate formal calculus, to be explored and developed with little professional concern for any empirical significance that might attach to the terms and propositions involved. To the philosopher, it is an embarrassing mystery whose justification and conceptual clarification have remained stubbornly refractory to philosophical insight. (A famous philosophical epigram has it that induction [a special case of statistical inference] is the glory of science and the scandal of philosophy.) To the experimental scientist, however, statistical inference is a research instrument, a processing device by which unwieldy masses of raw data may be refined into a product more suitable for assimilation into the corpus of science, and in this lies both strength and weakness. It is strength in that, as an ultimate *consumer* of statistical methods, the experimentalist is in position to demand that the techniques made available to him conform to his actual needs. But it is also weakness in that, in his need for the tools constructed by a highly technical formal discipline, the experimentalist, who has specialized along other lines, seldom feels competent to extend criticisms or even comments; he is much more likely to make unquestioning application of procedures learned more or less by rote from persons assumed to be more knowledgeable of statistics than he. There is, of course, nothing surprising or reprehensible about this—one need not understand the principles of a complicated tool in order to make effective use of it, and the research scientist can no more be expected to have sophistication in the theory of

William W. Rozeboom, "The Fallacy of the Null-Hypothesis Significance Test," *Psychological Bulletin,* **57,** 1960, 416–428. Copyright (1960) by the American Psychological Association, and reproduced by permission.

statistical inference than he can be held responsible for the principles of the computers, signal generators, timers, and other complex modern instruments to which he may have recourse during an experiment. Nonetheless, this leaves him particularly vulnerable to misinterpretation of his aims by those who build his instruments, not to mention the ever present dangers of selecting an inappropriate or outmoded tool for the job at hand, misusing the proper tool, or improvising a tool of unknown adequacy to meet a problem not conforming to the simple theoretical situations in terms of which existent instruments have been analyzed. Further, since behaviors once exercised tend to crystallize into habits and eventually traditions, it should come as no surprise to find that the tribal rituals for data-processing passed along in graduate courses in experimental method should contain elements justified more by custom than by reason.

In this paper, I wish to examine a dogma of inferential procedure which, for psychologists at least, has attained the status of a religious conviction. The dogma to be scrutinized is the "null-hypothesis significance test" orthodoxy that passing statistical judgment on a scientific hypothesis by means of experimental observation is a decision procedure wherein one rejects or accepts a null hypothesis according to whether or not the value of a sample statistic yielded by an experiment falls within a certain predetermined "rejection region" of its possible values. The thesis to be advanced is that despite the awesome pre-eminence this method has attained in our experimental journals and textbooks of applied statistics, it is based upon a fundamental misunderstanding of the nature of rational inference, and is seldom if ever appropriate to the aims of scientific research. This is not a particularly original view—traditional null-hypothesis procedure has already been superceded in modern statistical theory by a variety of more satisfactory inferential techniques. But the perceptual defenses of psychologists are particularly efficient when dealing with matters of methodology, and so the statistical folkways of a more primitive past continue to dominate the local scene.

To examine the method in question in greater detail, and expose some of the discomfitures to which it gives rise, let us begin with a hypothetical case study.

A CASE STUDY IN NULL-HYPOTHESIS PROCEDURE; OR, A QUORUM OF EMBARRASSMENTS

Suppose that according to the theory of behavior, T_0, held by most right-minded, respectable behaviorists, the extent to which a certain behavioral manipulation M facilitates learning in a certain complex learning situation C should be null. That is, if "ϕ" designates the degree to which manipulation M facilitates the acquisition of habit H under circumstances C, it follows from the orthodox theory T_0 that $\phi = 0$. Also suppose, however, that a few radicals have persistently advocated an alternative theory T_1 which entails, among other things, that the facilitation of H by M in circumstances C should be appreciably greater than zero, the precise extent being dependent upon the values of certain parameters in C. Finally, suppose that Igor Hopewell, graduate student in psychology, has staked his dissertation hopes on an experimental test of T_0 against T_1 on the basis of their differential predictions about the value of ϕ.

Now, if Hopewell is to carry out his assessment of the comparative merits of T_0 and T_1 in this way, there is nothing for him to do but submit a number of Ss to manipulation M under circumstances C and compare their efficiency at acquiring habit H with that of comparable Ss who, under circumstances C, have *not* been exposed to manipulation M. The difference, d, between experimental and control Ss in average learning efficiency may then be taken as an operational measure of the degree, ϕ, to which M influence acquisition of H in circumstances C. Unfortunately, however, as any experienced researcher knows to his sorrow, the interpretation of such an observed statistic is not quite so simple as that. For the observed dependent variable d, which is actually a performance measure, is a function not only of the extent to which M influences acquisition of H, but of many additional major and minor factors as well. Some of these, such as deprivations, species, age, laboratory conditions, etc., can be removed from consideration by holding them essentially constant. Others, however, are not so easily controlled, especially those customarily subsumed under the headings of "individual differences" and "errors of measurement." To curtail a long mathematical story, it turns out that with suitable (possibly justified) assumptions about the distributions of values for these uncontrolled variables, the manner in which they influence the dependent variable, and the way in which experimental and control Ss were selected and manipulated, the observed sample statistic d may be regarded as the value of a normally distributed random variate whose average value is ϕ and whose variance, which is independent of ϕ, is unbiasedly estimated by the square of another sample statistic, s, computed from the data of the experiment.[1]

The import of these statistical considerations for Hopewell's dissertation, of course, is that he will not be permitted to reason in any simple way from the observed d to a conclusion about the comparative merits of T_0 and T_1. To conclude that T_0, rather than T_1, is correct, he must argue that $\phi = 0$, rather than $\phi > 0$. But the observed d, what-

---

[1] s is here the estimate of the standard error of the difference in means, not the estimate of the individual SD.

ever its value, is logically compatible both with the hypothesis that $\phi = 0$ and the hypothesis that $\phi > 0$. How, then can Hopewell use his data to make a comparison of T_0 and T_1? As a well-trained student, what he *does*, of course, is to divide d by s to obtain what, under H_0 is a t statistic, consult a table of the t distribution under the appropriate degrees-of-freedom, and announce his experiment as disconfirming or supporting T_0, respectively, according to whether or not the discrepancy between d and the zero value expected under T_0 is "statistically significant"—i.e., whether or not the observed value of d/s falls outside of the interval between two extreme percentiles (usually the 2.5th and 97.5th) of the t distribution with that df. If asked by his dissertation committee to justify this behavior, Hopewell would rationalize something like the following (the more honest reply, that this is what he has been taught to do, not being considered appropriate to such occasions) : "In deciding whether or not T_0 is correct I can make two types of mistakes: I can reject T_0 when it is in fact correct [Type I error], or I can accept T_0 when in fact it is false [Type II error]. As a scientist, I have a professional obligation to be cautious, but a 5% chance of error is not unduly risky. Now if all my statistical background assumptions are correct, then, if it is really true that $\phi = 0$ as T_0 says, there is only one chance in 20 that my observed statistic d/s will be smaller than $t_{.025}$ or larger than $t_{.975}$, where by the latter I mean, respectively, the 2.5th and 97.5 percentiles of the t distribution with the same degrees-of-freedom as in my experiment. Therefore, if I reject T_0 when d/s is smaller than $t_{.025}$ or larger than $t_{.975}$, and accept T_0 otherwise, there is only a 5% chance that I will reject T_0 incorrectly. If asked about this Type II error, and why he did not choose some other rejection region, say between $t_{.475}$ and $t_{.525}$, which would yield the same probability of Type I error, Hopewell should reply that although he has no way to compute his probability of Type II error under the assumptions traditionally authorized by null-hypothesis procedure, it is presumably minimized by taking the rejection region at the extremes of the t distribution.

Let us suppose that for Hopewell's data, $d = 8.50$, $s = 5.00$, and $df = 20$. Then $t_{.975} = 2.09$ and the acceptance region for the null hypothesis $\phi = 0$ is $-2.09 < d/s < 2.09$, or $-10.45 < d < 10.45$. Since d does fall within this region, standard null-hypothesis decision procedure, which I shall henceforth abbreviate "NHD," dictates that the experiment is to be reported as supporting theory T_0. (Although many persons would like to conceive NHD testing to authorize only rejection of the hypothesis, not in addition, its acceptance when the test statistic fails to fall in the rejection region, if failure to reject were not taken as grounds for acceptance, then NHD procedure would involve no Type II error, and no justification would be given for taking the rejection at the extremes of the distribution, rather than in its middle.) But even as Hopewell reaffirms T_0 in his dissertation, he begins to feel uneasy. In fact, several disquieting thoughts occur to him:

1. Although his test statistic falls within the orthodox acceptance region, a value this divergent from the expected zero should nonetheless be encountered less than once in 10. To argue in favor of a hypothesis on the basis of data ascribed a p value no greater than .10 (i.e., 10%) by that hypothesis certainly does not seem to be one of the more impressive displays of scientific caution.

2. After some belated reflection on the details of theory T_1, Hopewell observes that T_1 not only predicts that $\phi > 0$, but with a few simplifying assumptions no more questionable than is par for this sort of course, the value that ϕ should have can actually be computed. Suppose the value derived from T_1 in this way is $\phi = 10.0$. Then, rather than taking $\phi = 0$ as the null hypothesis, one might just as well take $\phi = 10.0$; for under the latter, $(d - 10.0)/s$ is a 20 df t statistic, giving a two-tailed, 95% significance, acceptance region for $(d - 10.0)/s$ between $-.209$ and 2.09. That is, if one lets T_1 provide the null hypothesis, it is accepted or rejected according to whether or not $-.45 < d < 20.45$, and by this latter test, therefore, Hopewell's data must be taken to support T_1—in fact, the likelihood under T_1 of obtaining a test statistic this divergent from the expected 10.0 is a most satisfactory three chances in four. Thus it occurs to Hopewell that had he chosen to cast his professional lot with the T_1-ists by selecting $\phi = 10.0$ as his null hypothesis, he could have made a strong argument in favor of T_1 by precisely the same line of statistical reasoning he has used to support T_0 under $\phi = 0$ as the null hypothesis. That is, he could have made an argument that persons partial to T_1 would regard as strong. For behaviorists who are already convinced that T_0 is correct would howl that since T_0 is the dominant theory, only $\phi = 0$ is a legitimate null hypothesis. (And is it not strange that what constitutes a valid statistical argument should be dependent upon the majority opinion about behavior theory?)

3. According to the NHD test of a hypothesis, only two possible final outcomes of the experiment are recognized—either the hypothesis is rejected or it is accepted. In Hopewell's experiment, all possible values of d/s between -2.09 and 2.09 have the same interpretive significance, namely, indicating that $\phi = 0$, while conversely, all possible values of d/s greater than 2.09 are equally taken to signify that $\phi \neq 0$. But Hopewell finds this disturbing, for all of the various possible values that d/s might have had, the significance of $d/s = 1.70$ for the "comparative merits of T_0 and T_1 should surely be more similar to that of, say, $d/s = 2.10$ than to that of, say, $d/s = 1.70$.

4. In somewhat similar vein, it also occurs to Hopewell that had he opted for a somewhat riskier confidence level, say a Type I error of 10% rather than 5%, d/s would have fallen outside the region of accept-

ance and T_0 would have been rejected. Now surely the degree to which a datum corroborates or impugns a proposition should be independent of the datum-assessor's personal temerity. Yet according to orthodox significance-test procedure, whether or not a given experimental outcome supports or disconfirms the hypothesis in question depends crucially upon the assessor's tolerance for Type I risk.

Despite his inexperience, Igor Hopewell is a sound experimentalist at heart, and the more he reflects on these statistics, the more dissatisfied with his conclusions be comes. So while the exigencies of graduate circumstances and publication requirements urge that his dissertation be written as a confirmation to T_0, he nonetheless resolves to keep an open mind on the issue, even carrying out further research if opportunity permits. And reading his experimental report, so of course would we—has any responsible scientist ever made up his mind about such a matter on the basis of a single experiment? Yet in this obvious way we reveal how little our actual inferential behavior corresponds to the statistical procedure to which we pay lip-service. For if we did, in fact, accept or reject the null hypothesis according to whether the sample statistic falls in the acceptance or in the rejection region, then there would be no replications of experimental designs, no multiplicity of experimental approaches to an important hypothesis—a single experiment would, by definition of the method, make up our mind about the hypothesis in question. And the fact that in actual practice, a single finding seldom even tempts us to such closure of judgment reveals how little the conventional model of hypothesis testing fits our actual evaluative behavior.

DECISIONS VS. DEGREES OF BELIEF

By now, it should be obvious that something is radically amiss with the traditional NHD assessment of an experiment's theoretical import. Actually, one does not have to look far in order to find the trouble—it is simply a basic misconception about the purpose of a scientific experiment. The null-hypothesis significance test treats acceptance or rejection of a hypothesis as though these were *decisions* one makes on the basis of the experimental data—i.e., that we elect to adopt one belief, rather than another, as a result of an experimental outcome. *But the primary aim of a scientific experiment is not to precipitate decisions, but to make an appropriate adjustment in the degree to which one accepts, or believes, the hypothesis or hypotheses being tested.* And even if the purpose of the experiment *were* to reach a decision, it could not be a decision to accept or reject the hypothesis, for decisions are voluntary commitments to action—i.e., are *motor* sets—whereas acceptance or rejection of a hypothesis is a *cognitive* state which may provide the basis for rational decisions, but is not itself arrived at by such a decision (except perhaps indirectly in that a decision may initiate further experiences which influence the belief.)

The situation, in other words, is as follows: As scientists, it is our professional obligation to reason from available data to explanations and generalities—i.e., beliefs—which are supported by these data. But belief in (i.e., acceptance of) a proposition is not an all-or-none affair; rather, it is a matter of degree, and the extent to which a person believes or accepts a proposition translates pragmatically into the extent to which he is willing to commit himself to the behavioral adjustments prescribed for him by the meaning of that proposition. For example, if that inveterate gambler, Unfortunate Q. Smith, has complete confidence that War Biscuit will win, then odds are irrelevant—it is simply a matter of arranging to collect some winnings after the race. On the other hand, the more that Smith has doubts about War Biscuit's prospects, the higher the odds he will demand before betting. That is, the *extent* to which Smith accepts or rejects the hypothesis that War Biscuit will win the fifth at Belmont is an important determinant of his betting decisions for that race.

Now, although a scientist's data supply *evidence* for the conclusions he draws from them, only in the unlikely case where the conclusions are logically deducible from or logically incompatible with the data do the data warrant that the conclusions be entirely accepted or rejected. Thus, e.g., the fact that War Biscuit has won all 16 of his previous starts is strong evidence in favor of his winning the fifth at Belmont, but by no means warrants the unreserved acceptance of this hypothesis. More generally, the data available confer upon the conclusions a certain *appropriate degree of belief*, and it is the inferential task of the scientist to pass from the data of his experiment to whatever *extent* of belief these and other available information justify in the hypothesis under investigation. In particular, the proper inferential procedure is *not* (except in the deductive case) a matter of deciding to accept (without qualification) or reject (without qualification) the hypothesis: even if adoption of a belief were a matter of voluntary action—which it is not— neither such extremes of belief or disbelief are appropriate to the data at hand. As an example of the disastrous consequences of an inferential procedure which yields only two judgment values, acceptance and rejection, consider how sad the plight of Smith would be if, whenever weighing the prospects for a given race, he always worked himself into either supreme confidence or utter disbelief that a certain horse will win. Smith would rapidly impoverish himself by accepting excessively low odds on horses he is certain will win, and failing to accept highly favorable odds on horses he is sure will lose. In fact, Smith's two judgment values need not be *extreme* acceptance and rejection in order for his inferential procedure to be maladaptive. All that is required is that the degree of belief arrived at be in general inappropriate to the likelihood conferred on the hypothesis by the data.

Now, the notion of "degree of belief appropriate to the data at hand" has an unpleasantly vague, subjective feel about it which makes it unpalatable for inclusion in a formalized theory of inference. Fortunately, a little reflection about this phrase reveals it to be intimately connected with another concept relating conclusion to evidence which, though likewise in serious need of conceptual clarification, has the virtues both of intellectual respectability and statistical familiarity. I refer, of course, to the *likelihood*, or *probability*, conferred upon a hypothesis by available evidence. Why should not Smith *feel* certain, in view of the data available, that War Biscuit will win the fifth at Belmont? Because it *is* not certain that War Biscuit will win. More generally, what determines how strongly we should accept or reject a proposition is the probability given to this hypothesis by the information at hand. For while our voluntary actions (i.e., decisions) are determined by our intensities of belief in the relevant propositions, not by their actual probabilities, expected utility is maximized when the cognitive weights given to potential but not yet known-for-certain pay-off events are represented in the decision procedure by the probabilities of these events. We may thus reliquish the concept of "appropriate degree of belief" in favor of "probability of the hypothesis," and our earlier contention about the nature of data-processing may be rephrased to say that the proper inferential task of the experimental scientist is not a simple acceptance or rejection of the tested hypothesis, but determination of the probability conferred upon it by the experimental outcome. This likelihood of the hypothesis relative to whatever data are available at the moment will be an important determinant for decisions which must currently be made, but is not itself such a decision and is entirely subject to revision in the light of additional information.

In brief, what is being argued is that the scientist, whose task is not to prescribe actions but to establish rational beliefs upon which to base them, is fundamentally and inescapably committed to an explicit concern with the problem of inverse probability. What he wants to know is how plausible are his hypotheses, and he is interested in the probability ascribed by a hypothesis to an observed experimental outcome only to the extent he is able to reason backwards to the likelihood of the hypothesis, given this outcome. Put crudely, no matter how improbable an observation may be under the hypothesis (and when there are an infinite number of possible outcomes, the probability of any particular one of these is, usually, infinitely small—the familiar p value for an observed statistic under a hypothesis H is not actually the probability of that outcome under H, but a partial integral of the probability-density function of possible outcomes under H), it is still confirmatory (or at least nondisconfirmatory, if one argues from the data to rejection of the background assumptions) so long as the likelihood of the observation is even smaller under the alternative hypotheses. To be sure, the theory of hypothesis-likelihood and inverse probability is as yet far from the level of development at which it can furnish the research scientist with inferential tools he can apply mechanically to obtain a definite likelihood estimate. But to the extent a statistical method does not at least move in the *direction* of computing the probability of the hypothesis, given the observation, that method is not truly a method of *inference*, and is unsuited for the scientist's cognitive ends.

THE METHODOLOGICAL STATUS OF THE NULL-HYPOTHESIS SIGNIFICANCE TEST

The preceding arguments have, in one form or another, raised doubts about the appropriateness of conventional significance-test decision procedure for the aims it is supposed to achieve. It is now time to bring these charges together in an explicit bill of indictment.

1. The null-hypothesis significance test treats "acceptance" or "rejection" of a hypothesis as though these were decisions one makes. But a hypothesis is not something, like a piece of pie offered for dessert, which can be accepted or rejected by a voluntary physical action. Acceptance or rejection of a hypothesis is a cognitive process, a *degree* of believing or disbelieving which, if rational, is not a matter of choice, but determined solely by how likely it is, given the evidence, that the hypothesis is true.

2. It might be argued that the NHD test may nonetheless be regarded as a legitimate decision procedure if we translate "acceptance (rejection) of the hypothesis" as meaning "acting as though the hypothesis were true (false)." And to be sure, there are many occasions on which one must base a course of action on the credibility of a scientific hypothesis. (Should these data be published? Should I devote my research resources too and become identified professionally with this theory? Can we test this new Z bomb without exterminating all life on earth?) But such a move to salvage the traditional procedure only raises two further objections. (a) While the scientist—i.e., the person—must indeed make decisions, his *science* is a systematized body of (probable) *knowledge*, not an accumulation of decisions. The end product of a scientific investigation is a degree of confidence in some set of propositions, which then consitutes a *basis* for decisions. (b) Decision theory shows the NHD test to be woefully inadequate as a decision procedure. In order to decide most effectively when or when not to act as though a hypothesis

is correct, one must know both the probability of the hypothesis under the data available and the utilities of the various decision outcomes (i.e., the values of accepting the hypothesis when it is true, of accepting it when it is false, of rejecting it when it is true, and of rejecting it when it is false). But traditional NHD procedure pays no attention to utilities at all, and considers the probability of the hypothesis, given the data— i.e., the inverse probability—only in the most rudimentary way (by taking the rejection region at the extremes of the distribution rather than in its middle). Failure of the traditional significance test to deal with inverse probabilities invalidates it not only as a method of rational inference, but *also* as a useful decision procedure.

3. The traditional NHD test unrealistically limits the significance of an experimental outcome to a mere two alternatives, confirmation or disconfirmation of the null hypothesis. Moreover, the transition from confirmation to disconfirmation as a function of the data is discontinuous—an arbitrarily small difference in the value of the test statistic can change its significance from confirmatory to disconfirmatory. Finally, the point at which this transition occurs is entirely gratuitous. There is absolutely no reason (at least provided by the method) why the point of statistical "significance" should be set at the 95% level, rather than, say the 94% or 96% level. Nor does the fact that we sometimes select a 99% level of significance, rather than the usual 95% level, mitigate this objection—one is as arbitrary as the other.

4. The null-hypothesis significance test introduces a strong bias in favor of one out of what may be a large number of reasonable alternatives. When sampling a distribution of unknown mean μ, different assumptions about the value of μ furnish an infinite number of alternate null hypotheses by which we might assess the sample mean, and whichever hypothesis is selected is thereby given an enormous, in some cases almost insurmountable, advantage over its competitors. That is, NHD procedure involves an inferential double standard—the favored hypothesis is held innocent unless proved guilty, while any alternative is held guilty until no choice remains but to judge it innocent. What is objectionable here is not that some hypotheses are held more resistant to experimental extinction than others, but that the differential weighing is an all-or-none side effect of a personal choice, and especially, that the method *necessitates* one hypothesis being favored over all the others. In the classical theory of inverse probability, on the other hand, all hypotheses are treated on a par, each receiving a weight (i.e., its "a priori" probability) which reflects the credibility of that hypothesis on grounds other than the data being assessed.

5. Finally, if anything can reveal the practical irrelevance of the conventional significance test, it should be its failure to see genuine application to the inferential behavior of the research scientist. Who has

ever given up a hypothesis just because one experiment yielded a test statistic in the rejection region? And what scientist in his right mind would ever feel there to be an appreciable difference between the interpretive significance of data, say, for which one-tailed $p = .04$ and that of data for which $p = .06$, even though the point of "significance" has been set at $p = .05$? In fact, the reader may well feel undisturbed by the charges raised here against traditional NHD procedure precisely because, without perhaps realizing it, he has never taken the method seriously anyway. Paradoxically, it is often the most firmly institutionalized tenet of faith that is most susceptible to untroubled disregard— in our culture, one must early learn to live with sacrosanct verbal formulas whose import for practical behavior is seldom heeded. I suspect that the primary reasons why null-hypothesis significance testing has attained its current ritualistic status are (a) the surcease of methodological insecurity afforded by having an inferential algorithm on the books, and (b) the fact that a by-product of the algorithm is so useful, and its end product so obviously inappropriate, that the latter can be ignored without even noticing that this has, in fact, been done. What has given the traditional method its spurious feel of usefulness is that the *first*, and by far most laborious, step in the procedure, namely, estimating the probability of the experimental outcome under the assumption that a certain hypothesis is correct, is also a crucial first step toward what one is genuinely concerned with, namely, an idea of the likelihood of that hypothesis, given this experimental outcome. Having obtained this most valuable statistical information under pretext of carrying through a conventional significance test, it is then tempting, though of course quite inappropriate, to heap honor and gratitude upon the method while overlooking that its actual *result*, namely, a decision to accept or reject, is not used at all.

TOWARD A MORE REALISTIC APPRAISAL OF EXPERIMENTAL DATA

So far, my arguments have tended to be aggressively critical—one can hardly avoid polemics when butchering sacred cows. But my purpose is not just to be contentious, but to help clear the way for more realistic techniques of data assessment, and the time has now arrived for some constructive suggestions. Little of what follows pretends to any originality; I merely urge that ongoing developments along these lines should receive maximal encouragement.

For the statistical theoretician, the following problems would seem to be eminently worthy of research:

1. Of supreme importance for the theory of probability is analysis

of what we mean by a proposition's "probability," relative to the evidence provided. Most serious students of the philosophical foundations of probability and statistics agree (cf. Braithwaite, pp. 119f.) that probability of a proposition (e.g., the probability that the General Theory of Relativity is correct) does not, prima facie, seem to be the same sort of thing as the probability of an event-class (e.g., the probability of getting a head when this coin is tossed). Do the statistical concepts and formulas which have been developed for probabilities of the latter kind also apply to hypothesis likelihoods? In particular, are the probabilities of hypotheses quantifiable at all, and for the theory of inverse probability, do Bayes' theorem and its probability-density refinements apply to hypothesis probabilities? These and similar questions are urgently in need of clarification.

2. If we are willing to assume that Bayes' theorem, or something like it, holds for hypothesis probabilities, there is much that can be done to develop the classical theory of inverse probability. While computation of inverse probabilities turns essentially upon the parametric a priori probability function, which states the probability of each alternative hypothesis in the set under consideration prior to the outcome of the experiment, it should be possible to develop theorems which are invariant over important subclasses of a priori probability functions. In particular, the difference between the a priori probability function and the "a posteriori" probability function (i.e., the probabilities of the alternative hypotheses after the experiment), perhaps analyzed as a difference in "information," should be a potentially fruitful source of concepts with which to explore such matters as the "power" or "efficiency" of various statistics, the acquisition of inductive knowledge through repeated experimentation, etc. Another problem which seems to me to have considerable import, though not one about which I am sanguine, is whether inverse-probability theory can significantly be extended to hypothesis-probabilities, given knowledge which is only probabilistic. That is, can a theory of sentences of form "The probability of hypothesis H, given that E is the case, is p," be generalized to a theory of sentences of form "The probability of hypothesis H, given that the probability of E is q, is p"? Such a theory would seem to be necessary, e.g., if we are to cope adequately with the uncertainty attached to the background assumptions which always accompany a statistical analysis.

My suggestions for applied statistical analysis turn on the fact that while what is desired is the a posteriori probabilities of the various alternative hypotheses under consideration, computation of these by classical theory necessitates the corresponding a priori probability distribution, and in the more immediate future, at least, information about this will exist only as a subjective feel, differing from one person to the next, about the credibilities of the various hypotheses.

3. Whenever possible, the basic statistical report should be in the form of a *confidence interval*. Briefly, a confidence interval is a subset of the alternative hypotheses computed from the experimental data in such a way that for a selected confidence level α, the probability that the true hypothesis is included in a set so obtained is α. Typically, an α-level confidence interval consists of those hypotheses under which the p value for the experimental outcome is larger than $1 - \alpha$ (a feature of confidence intervals which is sometimes confused with their definition), in which case the confidence-interval report is similar to a simultaneous null-hypothesis significance test of each hypothesis in the total set of alternative. Confidence intervals are the closest we can at present come to quantitative assessment of hypothesis-probabilities (see technical note below), and are currently our most effective way to eliminate hypotheses from practical consideration—if we choose to act as though none of the hypotheses not included in a 95% confidence interval are correct, we stand only a 5% chance of error. (Note, moreover, that this probability of error pertains to the incorrect simultaneous "rejection" of a major part of the total set of alternative hypotheses, not just to the incorrect rejection of one as in the NHD method, and is a *total* likelihood of error, not just of Type I error.) The confidence interval is also a simple and effective way to convey that all-important statistical datum, the conditional probability (or probability density) function—i.e., the probability (probability density) of the observed outcome under each alternative hypothesis—since for a given kind of observed statistic and method of confidence-interval determination, there will be a fixed relation between the parameters of the confidence interval and those of the conditional probability (probability density) function, with the end-points of the confidence interval typically marking the points at which the conditional probability (probability density) function sinks below a certain small value related to the parameter α. The confidence-interval report is not biased toward some favored hypothesis, as is the null-hypothesis significance test but makes an impartial simultaneous evaluation of all the alternatives under consideration. Nor does the confidence interval involve an arbitrary decision as does the NHD test. Although one person may prefer to report, say, 95% confidence intervals while another favors 99% confidence intervals, there is no conflict here, for these are simply two ways to convey the same information. An experimental report can, with complete consistency and some benefit, simultaneously present several confidence intervals for the parameter being estimated. On the other hand, different choices of significance level in the NHD method is a clash of incompatible decisions, as attested by the fact that NHD analysis which simultaneously presented two different significance levels would yield a logically inconsistent conclusion when the observed statistic has a value in the

acceptance region of one significance level and the rejection region of the other.

Technical note: One of the more important problems now confronting theoretical statistics is exploration and clarification of the relationships among inverse probabilities derived from confidence-interval theory, fiducial-probability theory (a special case of the former in which the estimator is a sufficient statistic), and classical (i.e., Bayes') inverse-probability theory. While the interpretation of confidence intervals is tricky, it would be a mistake to conclude, as the cautionary remarks usually accompanying discussions of confidence intervals sometimes seem to imply, that the confidence-level α of a given confidence interval I should not really be construed as a probability that the true hypothesis, H, belongs to the set I. Nonetheless, if I is an α-level confidence interval, the probability that H belongs to I as computed by Bayes' theorem given an a priori probability distribution will, in general, not be equal to α, nor is the difference necessarily a small one—it is easy to construct examples where the a posteriori probability that H belongs to I is either 0 or 1. Obviously, when different techniques for computing the probability that H belongs to I yield such different answers, a reconciliation is demanded. In this instance, however, the apparent disagreement is largely if not entirely spurious, resulting from differences in the evidence relative to which the probability that H belongs to I is computed. And if this is, in fact, the correct explanation, then fiducial probability furnishes a partial solution to an outstanding difficulty in the Bayes approach. A major weakness of the latter has always been the problem of what to assume for the a priori distribution when no pre-experimental information is available other than that supporting the background assumptions which delimit the set of hypotheses under consideration. The traditional assumption (made hesitantly by Bayes, less hesitantly by his successors) has been the "principle of insufficient reason," namely, that given no knowledge at all, all alternatives are equally likely. But not only is it difficult to give a convincing argument for this assumption, it does not even yield a unique a priori probability distribution over a continuum of alternative hypotheses, since there are many ways to express such a continuous set, and what is an equilikelihood a priori distribution under one of these does not necessarily transform into the same under another. Now, a fiducial probability distribution determined over a set of alternative hypotheses by an experimental observation is a measure of the likelihoods of these hypotheses relative to all the information contained in the experimental data, but based on no pre-experimental information beyond the background assumptions restricting the possibilities to this particular set of hypotheses. Therefore, it seems reasonable to postulate that the no-knowledge a priori distribution in classical inverse probability theory should be that distribution which, when experimental data capable of yielding a fiducial argument are now given, results in an a posteriori distribution identical with the corresponding fiducial distribution.

4. While a confidence-interval analysis treats all the alternative hypotheses with glacial impartiality, it nonetheless frequently occurs that our interest is focused on a certain selection from the set of possibilities. In such case, the statistical analysis should also report, when computable, the precise p value of the experimental outcome, or better, though less familiarly, the probability density at that outcome, under each of the major hypotheses; for these figures will permit an immediate judgment as to which of the hypotheses is most favored by the data. In fact, an even more interesting assessment of the postexperimental credibilities of the hypotheses is then possible through use of "likelihood ratios" if one is willing to put his pre-experimental feelings about their relative likelihoods into a quantitative estimate. For let Pr (H,d), Pr (d,H), and Pr (H) be, respectively, the probability of a hypothesis H in light of the experimental data d (added to the information already available), the probability of data d under hypothesis H, and the pre-experimental (i.e., a priori) probability of H. Then for two alternative hypotheses H_0 and H_1, it follows by classical theory that

$$\frac{\text{Pr } (H_0, d)}{\text{Pr } (H_1, d)} = \frac{\text{Pr } (H_0)}{\text{Pr } (H_1)} \times \frac{\text{Pr } (d, H_0)}{\text{Pr } (d, H_1)} \qquad (1)^2$$

Therefore, if the experimental report includes the probability (or probability density) of the data under H_0 and H_1, respectively, and its reader can quantify his feelings about the relative pre-experimental merits of H_0 and H_1 (i.e., Pr (H_0)/Pr (H_1)), he can then determine the judgment he should make about the relative merits of H_0 and H_1 in light of these new data.

5. Finally, experimental journals should allow the researcher much more latitude in publishing his statistics in whichever form seems most insightful, especially those forms developed by the modern theory of estimates. In particular, the stranglehold that conventional null-hypothesis significance testing has clamped on publication standards must be broken. Currently justifiable inferential algorithm carries us only through computation of conditional probabilities: from there, it is for every man's clinical judgment and methodological conscience to see him through to a final appraisal. Insistence that published data must have the biases

---

[2] When the numbers of alternative hypotheses and possible experimental outcomes are transfinite Pr (ddH) = Pr (H,d) = Pr (H) = 0 in most cases. If so, the probability ratios in Formula 1 are replaced with the corresponding probability-density ratios. It should be mentioned that this formula rather idealistically presupposes there to be no doubt about the correctness of the background statistical assumptions.

of the NHD method built into the report, thus seducing the unwary reader into a perhaps highly inappropriate interpretation of the data, is a professional disservice of the first magnitude.

SUMMARY

The traditional null-hypothesis significance-test method, more appropriately called "null-hypothesis decision (NHD) procedure," of statistical analysis is here vigorously excoriated for its inappropriateness as a method of inference. While a number of serious objections to the method are raised, its most basic error lies in mistaking the aim of scientific investigation to be a decision, rather than a cognitive evaluation of propositions. It is further argued that the proper application of statistics to scientific inference is irrevocably committed to extensive consideration of inverse probabilities, and to further this end, certain suggestions are offered, both for the development of statistical theory and for more illuminating application of statistical analysis to empirical data.

Reference

Braithwaite, T. B. Scientific explanation. Cambridge, England: Cambridge University Press, 1953.

4.9

Testing the null hypothesis and the strategy and tactics of investigating theoretical models

David A. Grant[1]

Testing the null hypothesis, H_0, against alternatives, H_1, is well established and has a proper place in scientific research. However, this testing procedure, when it is routinely applied to comparing experimental outcomes with outcomes that are quantitatively predicted from a theoretical model, can have unintended results and bizarre implications. This paper first outlines three situations in which testing H_0 has conventionally been done by psychologists. In terms of the probable intentions or strategy of the experimenter testing H_0 turns out to be an appropriate tactic in the first situation, but it is inadequate in the second situation, and it is self-defeating with curious implications in the last situation. Alternatives to this conventional procedure are then presented along with the considerations which make the alternative preferable to testing the usual H_0.

THREE APPLICATIONS OF H_0 TESTING

Probably the most common application of the tactic of testing H_0 arises when the independent variable has produced a sample difference or set of differences in the magnitude of the dependent variable. Quantita-

David A. Grant, "Testing the Null Hypothesis and the Strategy and Tactics of Investigating Theoretical Models," *Psychological Review*, **69**, 1962, 54–61. Copyright (1962) by the American Psychological Association, and reproduced by permission.

[1] The author is indebted to Arnold M. Binder of Indiana University whose arguments inspired him to make explicit some of the issues involved in using conventional analysis of variance procedures in testing the adequacy of a theoretical model. As this paper went through various revisions over a period of time the writer is correspondingly indebted to a number of supporting agencies: the Graduate Research Committee and the College of Letters and Science of the University of

tive predictions of the size of the difference or differences are not available. The experimenter wishes to know whether or not differences of the size obtained could have occurred by virtues of the operation of the innumerable nonexperimental factors conventionally designated as random. He sets up H_0 that the differences are zero; chooses a significance level, α; determines the set of hypotheses alternative to H_0 that he is willing to entertain, H_1; selects an appropriate test statistic, t, F, χ^2, U, T, or the like; and proceeds with the test. Rejection of H_0 permits him to assert, with a precisely defined risk of being wrong, that the obtained differences were not the product of chance variation. Failure of the test to permit rejection of H_0, which, unfortunately, is commonly termed "accepting" H_0, means that the obtained differences or greater ones would occur by chance with a probability greater than α. This situation is straightforward. The experimenter has limited aims. He has asked a simple question, and he has received a simple answer, subject only to those ambiguities which attend all experimental and inductive inferences. His tactics are admirably suited to his strategic objective.

Another common but less satisfactory instance of testing H_0 arises when the results of pre-experimental matching or pretesting are to be evaluated. Here the experimenter has measured the dependent variable or some related variable before operation of the independent variable, and he devoutly hopes that the experimental and control groups are alike except for random differences. He is now relieved or chagrined, depending upon whether H_0 is "accepted" or "rejected" as a consequence of his test. Even if H_0 is accepted his relief is tempered by some uneasiness. He knows that he has not proved, and indeed cannot prove, that H_0 is "true." His tactics in testing H_0 seem to be appropriate to the impossible strategic aim of proving the truth of H_0. Certainly, if he had a more reasonable aim he has adopted inappropriate tactics. Utilizing these tactics, the best he can do is to beat a strategic retreat, and if H_0 is accepted he can perhaps point out that he has used a very powerful test and that if there were real differences they were most likely very small. Although psychologists have never to my knowledge done so, he might be able to go one step further and point out that his testing procedure would reject H_0 a given percentage of the time, say, 90% if the "true" difference had been as little as, say, one-tenth of an SD. This sort of statement of the power of a test is a commonplace in acceptance inspection (Grant, 1952, Ch. 13).

With the advent of more detailed mathematical models in psy-

Wisconsin, the National Science Foundation, and finally to the Department of Psychology of the University of California, Berkeley, his host during final preparation of the manuscript.

chology (e.g., Bush, Abelson, & Hyman, 1956; Bush & Estes, 1959; Goldberg, 1958; Kemeny, Snell, & Thompson, 1957) a new statistical testing situation is arising more and more frequently. The specificity of the predictions and perhaps the whole philosophy behind model construction pose a different kind of statistical problem than those faced by most psychological investigators in the past. It seems obvious that as the use of models become more widespread a greater number of investigators will face the problem of evaluating the correspondence between empirical data points and precise numerical predictions of these points. Unfortunately most of the procedures used to date in testing the adequacy of such theoretical predictions set rather bad examples. Probably the least adequate of these procedures has been that in which an H_0 of exact correspondence between theoretical and empirical points is tested against H_1 covering any discrepancy between predictions and experimental results.

Most models predict a considerable number of different aspects of the data, and some of these aspects are predicted with greater success than others (Bush & Estes, 1959, Chs. 14, 15, 17, 18). We shall restrict our discussion to the prediction of values along a curve which might be a learning curve. An idealized version of such a typical situation is presented in Figure 1. Here, the dependent variable, Y, is plotted on the

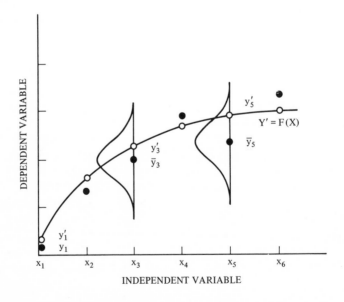

FIGURE 1

Idealized situation involving the test of a theoretical function, $Y' = f(X)$. (Theoretical points, Y'_i, are represented by open circles; obtained means, Y_i, are represented by solid circles.)

vertical axis against the independent variable, X, on the horizontal axis. The theoretical model has led to an expression, $Y' = f(X)$, giving a set of k theoretical predictions, Y'_1, Y'_2, $\cdots$, Y'_k. The experiment has produced k empirical data points, a set of mean values, $\bar{Y}_1, \bar{Y}_2, \cdots, \bar{Y}_k$ corresponding to the values of the independent variable that were investigated, namely, X_1, X_2, $\cdots$, X_k. Individual observations tend to form normal distributions about each of the $\bar{Y}_i$, and these normal distributions tend to have equal σ's for all data points. In further discussion we shall assume that inaccuracies in the manipulation of the independent variable, X, can be ignored. The problem now is to investigate the goodness of fit of the Y'_i to the $\bar{Y}_i$ or the correspondence between the Y'_i and the $\bar{Y}_i$.

The tactics oriented toward accepting H_0 as corroborating the theory involve breaking down the jth individual observation from the general mean of all of the observations, as follows:

$$Y_{ij} - \bar{Y} = (Y_{ij} - \bar{Y}_i) + (\bar{Y}_i - Y'_i) + (Y'_i - \bar{Y}) \tag{1}$$

where Y_{ij} is the jth observation in the ith normal distribution, and $\bar{Y}$ is the general mean of all observations.

The total sum of squares may then be partitioned as follows:

$$SS_{\text{Tot}} = SS_{\text{Dev Est}} + SS_{\text{Dev Theory}} + SS_{\text{Theory}} \tag{2}$$

where $SS_{\text{Dev Est}}$ is the sum of squares associated with the variation of individual measures from their means, $SS_{\text{Dev Theory}}$ is the sum of squares associated with the systematic departures of empirical data points from the theoretical points, and SS_{Theory} is the sum of squares associated with departures of the theoretical points from the general mean of the whole experiment.

If we suppose that the linear model for the analysis of variance holds, then:

$$Y_{ij} = \mu + T_i + D_i + e_{ij} \tag{3}$$

where μ is the population mean for all Y_{ij} over the specific values of the independent variable, X_i; T_i is the departure of the "true" theoretical value of Y'_i from μ; D_i is the discrepancy of the "true" value of $\bar{Y}_i$ from the true value of Y'_i; and e_{ij} is a random element from a normal distribution with a mean of zero and variance, σ_e^2, for all i.

For a fixed set of X_i the Ts and Ds may be defined so that $\Sigma T_i = \Sigma D_i = 0$. Under H_0 each $D_i = 0$. Under H_1 some $D_i \neq 0$, and the variance of the D_i, $\sigma_D^2 \neq 0$. This last variance may be termed the true variance of the discrepancies from the theory over the particular set of X_i that was investigated.

The foregoing is a conventional analysis of variance model, and the F ratio of the $MS_{\text{Dev Theory}}$ divided by the $MS_{\text{Dev Est}}$ provides an excellent and powerful test of H_0 against H_1. The number of degrees of freedom for SS_{DevEst} will be $k(n-1)$ where n is the number of observations per data point, and the degrees of freedom for $SS_{DevTheory}$ will be $k - n_T$, where n_T is the number of degrees of freedom lost in the process of fitting the model to the data. If this F is significant, we reject H_0, concluding that the discrepancies between the $\bar{Y}_i$ and the Y'_i are too great to be accounted for by the observed random variation in the experiment. In this conclusion we accept the 5% or 1% risk implied by our choice of α.

Logical difficulties arise when F fails of significance. H_0 remains tenable but is not proved to be correct. A tenable H_0 provides some support for the theory but only in the negative sense of failing to provide evidence that the theory is faulty. To assert that accepting the H_0 proves that the model provides a satisfactory fit to the data is an inaccurate and misleading statement. We may mean that we are satisfied, but others, especially proponents of other theories, will tend to regard our test as too lenient.

Failure to reject H_0, instead of producing closure, leaves certain annoying ambiguities, but the tactics of testing this particular H_0 imply a strategy that suffers from more serious defects that are readily apparent when the whole conception of testing a theory is carefully considered. To begin with, in view of our present psychological knowledge and the degree of refinement of available theoretical models it seems certain that even the best and most useful theories are not perfect. This means, in terms of the analysis of variance model, there will be some nonzero D_i's. H_0, then, is never really "true." Its "acceptance," rather than "proving" the theory, merely indicates that in this instance the D_i's were too small to be demonstrated by the sensitivity of the experiment in question. The tactics of accepting H_0 as proof and rejecting H_0 as disproof of a theory lead to the anomalous results that a small-scale, insensitive experiment will most often be interpreted as favoring a theory, whereas a large-scale, sensitive experiment will usually yield results opposed to the theory!

Curiously enough, even rejection of H_0 by means of a very stringent experimental test may be quite misleading as far as casting light on the adequacy of the theory is concerned. If the D_i's are very small indeed the theoretical model may be a great improvement over anything else that is available and satisfactory for many purposes even though an extremely sensitive experiment were to reveal the nonzero D_i's. If our task, as scientists, were to test and accept or reject theories as they came off some assembly line the tactics of testing H_0 could be made in a satisfactory manner simply by requiring that the test be "sufficiently" stringent. In fact, our task and our intentions are usually different from testing products; what we are really up to resembles *quality control* rather than *acceptance inspection,* and statistical procedures suitable for the latter are rarely optimal for the former (Grant, 1952, Chs. 1, 13).

HYPOTHESIS TESTING VERSUS STATISTICAL ESTIMATION

An analogy will make clear the relation between testing tactics and the intention of the tester. Suppose that I wish to test a parachute; how should I go about it? How I should test depends upon my general intentions. If I want to sell the parachute and am testing it only to be able to claim that it has been tested and I do not care what happens to the purchaser, then I should give the parachute a most lenient, nonanalytic test. If, however, I am testing the parachute to be sure of it for my own use, then I should subject it to a very stringent, nonanalytic test. But if I am in the competitive business of manufacturing and selling parachutes, then I should subject it to a searching, analytic test, designed to tell me as much as possible about the locus and cause of any failure in order that I may improve my product and gain a larger share of the parachute market. My contention is that the last situation is the one that is most analogous to that facing the theoretical scientist. He is not accepting or rejecting a finished theory; he is in the long-term business of constructing better versions of the theory. Progress depends upon improvement or providing superior alternatives, and improvement will ordinarily depend upon knowing just how good the model is and exactly where it seems to need alteration. The large D_i's designate the next point of attack in the continuing project of refining the existing model. Therefore attention should be focused upon the various discrepancies between prediction and outcome instead of on the over-all adequacy of the model.

In view of our long-term strategy of improving our theories, our statistical tactics can be greatly improved by shifting emphasis away from over-all hypothesis testing in the direction of statistical estimation. This always holds true when we are concerned with the actual size of one or more differences rather than simply in the existence of differences. For example, in the second instance of hypothesis testing cited at the beginning of this paper, where the investigator tests a pre-experimental difference, he would do better to obtain 95% or 99% confidence interval for the pre-experimental difference. If the interval is small and includes zero, he (and any other moderately sophisticated person) knows immediately that he is on fairly safe ground; but if the interval is large, even though it includes zero, it is immediately apparent that the situation is more uncertain. In both instances H_0 would have been accepted.

TESTING A REVISED H_0

Before turning to estimation procedures that are useful in examining the correspondence between experimental outcomes and predictions from a mathematical model, I shall digress briefly to outline a statistical testing method which can legitimately be used in appraising the fit of a model to data as shown in Figure 1. Basically the statistical argument in the proper test is reoriented so that rejection of H_0 constitutes evidence favoring the theory. The new H_0 is that the correlation between the predicted values, Y'_i, and the obtained values $\bar{Y}_i$, is zero, after all correlation due to the fitting process has been eliminated. The alternative, H_1, against which H_0 is tested is that there is a correlation greater than zero between theoretical and empirical points. The four simple steps required to obtain the necessary F test are as follows:

    1. Calculate $t_i = Y'_i - \bar{Y}$ for all i. $\Sigma t_i = 0$.[2]

    2. Calculate $SS_{\text{Correspondence}} = n(\Sigma t_i \bar{Y}_i)^2/\Sigma t_i^2$, where n is the number of observations upon which each $\bar{Y}_i$ is based. Negative values of $(\Sigma t_i \bar{Y}_i)$ are treated as zero.

    3. Obtain $MS_{\text{Correspondence}} = SS_{\text{Correspondence}}/n_T$, where n_T, the number of degrees of freedom involved in fitting the theoretical points to the empirical data, will ordinarily be the number of linearly independent fitting constants in the mathematical expression of the model.

    4. Divide $MS_{\text{Correspondence}}$ by MS_{DevEst} to give $F_{\text{Correspondence}}$ which has n_T degrees of freedom for its numerator and $k(n-1)$ degrees of freedom for its denominator, k being the number of $\bar{Y}_i$. The test is one-tailed in the sense that negative values of $\Sigma t_i \bar{Y}_i$ are treated as zero values, so that the probability values of the F distribution must be halved, an unusual procedure with F tests in analysis of variance.

Following the above procedure, rejection of H_0 now means that there is more than random positive covariation between predicted and obtained values of the dependent variable.

    This test is admirable in that it puts the burden of proof on the investigator, because a small-scale, insensitive experiment is unlikely to produce evidence favoring the model. Furthermore, if the model has any merit, the more sensitive the experiment, the more likely it is that a significant F, favoring the theory, will be obtained. Actually, the test is extremely sensitive to virtue in the theory, and therefore in the case of a moderately successful model and a moderately sensitive experiment both this F and the one testing the significance of systematic deviations from the model ($F = MS_{\text{DevTheory}}/MS_{\text{DevEst}}$) will tend to be significant. This outcome is no anomaly; it merely indicates that the model predicts some

---

[2] In the unusual event where the general mean of the observations, $\bar{Y}$, is not used as a fitting constant for $Y' = f(X)$, the t_i must be computed as deviations from the mean of all the Y'_i, $\bar{Y}'$. The test will then be insensitive to discrepancies between Y' and $\bar{Y}$, and the interpretation will be somewhat equivocal. A separate test of H_0 that $\bar{Y}_{\text{population}}$ equals $\bar{Y}$, is feasible, but here the experimenter is forced into the illicit posture of seeking to embrace H_0.

but not all of the systematic variation in the data. In short, progress is being made, but improvement is possible. The fact that simultaneous significance of both Fs, indicating general success and specific failures of a model, should be commonplace points up the necessity of turning to methods of statistical estimation for a more adequate examination of the workings of a theoretical model.

PRACTICAL ESTIMATION METHODS FOR INVESTIGATION OF MODELS

As is true of statistical tests, each method of statistical estimation has its advantages and limitations. In the investigation of the adequacy of theoretical curves in psychology there are reasons to believe that the simpler estimation methods have practical advantages over some of the more elegant procedures. To give a fairly complete view of the situation, methods of point and interval estimation of $\sigma_D{}^2$ and of the individual D_i will be described, and a brief evaluation of each method will be given.

Estimating $\sigma_D{}^2$

The variance of the discrepancies between the Y'_i and the $\bar{Y}_i$ condenses into a single number the adequacy of fit of the theoretical model. As such it is an excellent index for the evaluation of the model. The smaller the variance, $\sigma_D{}^2$, the better the model, and vice versa. As an estimate of the size of the discrepancies one might expect in future similar applications of the model, $\sigma_D{}^2$ is far more informative than any F test. Furthermore $\sigma_D{}^2$ is readily estimated in the case of homogeneity of the error variance, $\sigma_e{}^2$. The expected values of the relevant mean squares are as follows:

$$\mathrm{Exp}\,(MS_{\mathrm{Dev\,Theory}}) = \sigma_e^2 + n\sigma_D^2 \tag{4}$$

$$\mathrm{Exp}\,(MS_{\mathrm{Dev\,Est}}) = \sigma_e^2 \tag{5}$$

A maximum likelihood estimate of the variance of the discrepancies, $\hat{\sigma}_D{}^2$ is then:

$$\hat{\sigma}_D^2 = (MS_{\mathrm{Dev\,Theory}} - MS_{\mathrm{Dev\,Est}})/n \tag{6}$$

The accuracy of this estimator depends upon the number of degrees of freedom associated with $SS_{\mathrm{Dev\,Theory}}$ and $SS_{\mathrm{Dev\,Est}}$. The latter rarely poses any practical problem, but the former, in view of the predilection of psychologists for minimizing the number of data points, is quite critical. This is readily apparent when interval estimation of $\sigma_D{}^2$ is attempted.

Bross (1950) gives a convenient method for accurate approximation of the fiducial interval for $\sigma_D{}^2$, and in this case the fiducial and confidence intervals are essentially equal. The method will be outlined below for the 5% interval.

1. Obtain $\hat{\sigma}_D{}^2$ from Equation 6, above. (If the estimate is negative or zero, meaningful limits cannot be obtained.)
2. Find:

$$L = \frac{\dfrac{F}{F_{025(k-n_T,\,k[n-1])}} - 1}{\dfrac{F \cdot F_{025(k-n_T,\,\infty)}}{F_{025(k=n_T,\,k[n=1)]}} - 1}$$

where:

$$F = MS_{\text{Dev Theory}} / MS_{\text{Dev Est}}$$

$F_{0.25\,(k-nT,\,k[n-1])}$ is the entry in the 2.5% F table (Pearson & Hartley, 1954) for $n_1 = k - n_T$ and $n_2 = k[n-1]$; and $F_{025(k-nT,\,\infty)}$ is the entry for $n_1 = k - n_T$ and $n_2 = \infty$.

3. Find:

$$L = \frac{F \cdot F_{025(k[n-1],\,k-n_T)} - 1}{\dfrac{F \cdot F_{025(k[n-1],\,k-n_T)}}{F_{025(\infty,\,k-n_S)}} - 1}$$

where $F_{025(k(n-1),\,k-nT)}$ is the entry in the 2.5% F table for $n_1 = k[n-1]$, and $n_2 = k - n_T$; $F_{025(\infty,\,k-nT)}$ is the entry for $n_1 = \infty$, and $n_2 = k - n_T$.

4. The upper and lower limits are then $\bar{L}\sigma_D{}^2$ and $L\sigma_D{}^2$, respectively. With less than 15–20 data points these limits will be found to be uncomfortably wide, a fact to bear in mind when designing an experimental test of a theoretical model. For example, in Figure 1, with 6 data points and two degrees of freedom for curve fitting, the limits might plausibly be 0–40, whereas with 14 data points the limits might be 0–12.

Aside from the considerable variability in the estimate of $\sigma_D{}^2$ which can be reduced by increasing the number of data points, there are two other important limitations to the use of estimates of the variance of the discrepancies in evaluating a model. First of all, the population value of $\sigma_D{}^2$ is completely dependent upon the particular values of the independent variable, X, which are chosen for the test of the model. Choice of two different sets of Xs could well lead to two entirely different values of $\sigma_D{}^2$, and both of these values could be perfectly accurate. Secondly, although $\sigma_D{}^2$ gives an over-all index of the adequacy of the model being tested, it condenses so much information into one measure that it does not permit pinpointing the especially large D_i's so that they can be given proper attention in considering revision of the model.

Estimating the D_i

The individual D_i may be estimated as points, or intervals may be established for the D_i, collectively or individually. As before, each method has its good points and its limitations.

Point estimation of the individual D_i's consists simply in comparing the individual data points, the $\bar{Y}_i$, with the fitted curve. It is a crude method, but it has served well in the past and represents the beginning of wisdom. For example, in Figure 1, the model builder might well note that the first three data points lie below the curve and ask himself if there is some special reason for this. He would also note that the greatest discrepancy occurs at $\bar{Y}_5$, where the neighboring discrepancies are in the other direction. The weakness of this simple method lies in the absence of a criterion which will assist the investigator in deciding which discrepancies should be singled out for further attention and which may be disregarded because they are within the range of expected random variation. This defect is remedied by the interval estimation techniques.

Probably the ideal method of interval estimation is that in which intervals are established for the whole curve in one operation by finding the 95% confidence band. The method takes the theoretical curve as a point of departure, and the result is a pair of curves above and below the theoretical curve, which will tend in the case of random variation to contain between them 95% of the data points. Points lying outside the band are immediately suspect; they are the most promising candidates for attention in the next version of the model. There are two practical difficulties with this method. First, homogeneity of the error variance, σ_e^2, over all the X_i is required. And secondly, because errors in estimation of each fitting parameter must be taken into account, for all but the simplest curves (Cornell, 1956, pp. 184–186) the bands may be difficult[3] to obtain. Although the method is elegant, in practice it will rarely represent sufficient improvement over the final method, given below, to justify its use.

The last method seems to me to be the most useful and most robust and most flexible method. It can be widely applied, and the relative ease of application, coupled with its ability to discriminate between significant and random discrepancies make it superior to the other estimation methods. It also possesses the homely virtue of being readily understood. In contrast to the preceding method, this one takes as its point of departure the empirical means, and consists, simply, in computing the 95% confidence limits for each of the $\bar{Y}_i$. If there is homogeneity of variance, the error variance of each mean is taken simply as σ_e^2/n; in cases of suspected heterogeneity, each mean must have its own estimate of error

[3] A sufficient estimate of the error variance of each parameter must be available and independent of the estimates of all other parameters or else the covariances of all parametric estimates must be found and the theoretical function must have continuous first partial derivatives with respect to the parameters in order that the confidence bands may be found in the asymptotic case (Rao, 1952, pp. 207–208). Where an asymptote is involved in the fitting of the theoretical function, satisfactory independent estimators can rarely be obtained.

variance. This will, of course, be the variance of the distribution of Y_{ij} for each i, divided by n. When these limits have been obtained, attention is directed to instances where the theoretical curve lies outside the limits. In some cases, the investigator might choose to establish the 80% or 90% limits in order to direct his attention to less drastic departures of the experimental results from the model. Choice of an optimum level for the limits is hard to establish on a general a priori basis, but it is likely that limits narrower than the traditional 95% will be found more useful than the broader limits. Simple as this method is, it is hard to improve upon in actual practice. Instead of giving an almost meaningless over-all acceptance or rejection of a model, it directs attention to specific defects, its functioning improves as the precision of the experimental test is improved, and the investigator can set the confidence coefficient so as to increase its sensitivity to defect at a cost of a fairly well-specified percentage of false positives or wild goose chases. A final and often crucial advantage is that the confidence intervals, based as they are upon the experimental means, can be obtained in cases where the form of the theoretical function does not permit satisfactory estimation of its parameters, and the analysis of variance and confidence bands methods cannot properly be applied.

SUMMARY AND CONCLUSIONS

In this paper I have attempted to show that the traditional procedure of testing a null hypothesis (H_0) of a zero difference or set of zero differences is quite appropriate to the experimenter's intentions or scientific strategy when he is unable to predict differences of a specified size. When theory or other circumstances permit the prediction of differences of specified size, using these predictions as the values in H_0 is tactically inappropriate, frustrating and self-defeating. This is particularly true when a theoretical curve has been predicted, and H_0 is framed in terms of zero discrepancies from the curve. If rejection of H_0 is interpreted as evidence against the theory, and "acceptance" of H_0 is interpreted as evidence favoring the theory, we find that the larger and more sensitive the experiment is, the more likely it will lead to results opposed to the theory; whereas the smaller and less sensitive the experiment, the more likely the results will favor the theory. Aside from this anomaly, which can be corrected by recasting H_0 in terms of a zero covariance between theoretical prediction and experimental outcome, hypothesis testing as a statistical tactic in this case implies an acceptance-inspection strategy. Acceptance-inspection properly involves examination of finished products with a view to accepting them if they are good enough and rejecting them if they are shoddy enough. The theoretician is not a purchaser but rather he is a producer of goods in a competitive market so that his examination of his theory should be from the standpoint of quality control. His idealized intentions are to detect and correct defects, if possible, so that he can produce a more adequate, more general theoretical model. Because his ideal strategy is not to prove or disprove a theory but rather to seek a better theory, his appropriate statistical tactics should be those involving estimation rather than hypothesis testing.

Examination of alternative techniques available for point or interval estimation of discrepancies between theoretical predictions and experimental outcomes or the over-all variance of these discrepancies suggests strongly that estimation of the confidence intervals for the means found along a theoretical curve is the most practical and most widely applicable general procedure. Other writers have recently emphasized the values of various estimation as opposed to hypothesis testing techniques (e.g., Bolles & Messick, 1958; Gaito, 1958; Savage, 1957) and it is hoped that considerations pointed out by them and points raised in this paper will be helpful to investigators who are in the process of examining theoretical models which lead to specific numerical predictions of experimental outcomes.

References

Bolles, R., & Messick, S. Statistical utility in experimental inference. *Psychological Reports*, 1958, **4**, 223–227.

Bross, I. Fiducial intervals for variance components. *Biometrics*, 1950, **6**, 136–144.

Bush, R. R., Abelson, R. P., & Hyman, R. *Mathematics for psychologists: Examples and problems.* New York: Social Science Research Council, 1956.

Bush, R. R., & Estes, W. K. (Eds.) *Studies in mathematical learning theory.* Stanford, Calif.: Stanford University Press, 1959.

Cornell, F. G. *The essentials of educational statistics.* New York: Wiley, 1956.

Gaito, J. The Bolles-Messick coefficient of utility. *Psychological Reports*, 1958, **4**, 595–598.

Goldberg, S. *Introduction to difference equations.* New York: Wiley, 1958.

Grant, E. L. *Statistical quality control.* New York: McGraw-Hill, 1952.

Kemeny, J. G., Snell, J. L., & Thompson, G. L. *Introduction to finite mathematics.* Englewood Cliffs, N. J.: Prentice-Hall, 1957.

Pearson, E. S., & Hartley, H. O. (Eds.) *Biometrika tables for statisticians.* Cambridge, England: Cambridge University Press, 1954.

Rao, C. R. *Advanced statistical methods in biometric research.* New York: Wiley, 1952.

Savage, I. R. Nonparametric statistics. *Journal of the American Statistical Association*, 1957, **52**, 331–344.

4.10

Further considerations on testing the null hypothesis and the strategy and tactics of investigating theoretical models

David A. Grant has argued that it is inappropriate to design experiments such that support for a theory comes from acceptance of the null hypothesis. The present article points out that while this position could be defended in Fisher's approach to testing statistical hypotheses, it could not in the Neyman-Pearson approach or on more general scientific grounds. It is emphasized that one optimally designs experiments with enough sensitivity for rejecting poor theories and accepting useful theories, whether acceptance or rejection of the null hypothesis leads to empirical support. The argument that, in the procedure to which Grant objects, an insensitive experiment is more likely to lead to support for a theory is shown to be only a special case of the argument against bad experimentation.

The arguments in a recent article by Grant (1962) are directed against experimental designs oriented toward acceptance of the null hypothesis, that is, where support for an empirical hypothesis depends upon acceptance of the null hypothesis. Atkinson and Suppes (1958) provide an excellent example of the type of experimental logic to which Grant objects. These investigators postulated a one-stage Markov model for a zero-sum, two-person game. On the basis of the model they predicted, first, the mean proportion of various responses over asymptotic trials and, second, that the probability of State k given States i and j on

Arnold Binder, "Further Considerations on Testing the Null Hypothesis and the Strategy and Tactics of Investigating Theoretical Models," *Psychological Review*, **70**, 1963, 107–115. Copyright (1963) by the American Psychological Association, and reproduced by permission.

the two previous trials is equal to the probability of State k given only State j on the immediately preceding trial (i.e., that a one-stage Markov model accounts for the data). The predictions were then compared with the obtained results by means of a series of t tests, in the former case, and a χ^2 test, in the latter. One of the t tests, for example, involved a comparison of the predicted proportion of .600 against the observed mean proportion of .605, while another a comparison of a predicted value of .667 and an observed value of .670. Support for the one-stage Markov model was then inferred by the failure of the t tests and the χ^2 to reach the .05 level of significance. That is, support for the empirical model came from acceptance of the null hypotheses. Other examples may be found in Binder and Feldman, 1960; Bower, 1962; Brody, 1958; Bush and Mosteller, 1955; Grant and Norris, 1946; Harrow and Friedman, 1958; Weinstock, 1958; and Witte, 1959.

To facilitate future discussion it is convenient to refer to the procedure where acceptance of the null hypothesis leads to support for an empirical hypothesis as acceptance-support $(a\text{-}s)$, and to the procedure where empirical support comes from rejection of the null hypothesis as rejection-support $(r\text{-}s)$.

In addition to the objections to $a\text{-}s$, Grant argues that the method of testing statistical hypothesis may not be a very good idea in any case He thus argues it is wise to shift away from the current emphasis in psychological research on hypothesis testing in the direction of statistical estimation.

STATISTICAL LOGIC

There have been two principal schools of thought in regard to the logical and procedural ramifications of statistical inference. The older of these stems from the writings of Yule, Karl Pearson, and Fisher, while the other comes from the early work of Neyman and Pearson and the more recent developments of Wald. The respective influences of each of these schools on experimental statistics is abundantly evident, but a difficulty in separating these influences is that the actual recommendations for tests and interval estimates in a field like psychology are similar for both.

In the Fisher school one starts the testing process with a hypothesis, called the "null hypothesis," which states that the sample at issue comes from a hypothetical population with a sampling distribution in a certain known class. Using this distribution one rejects the null hypothesis whenever the discrepancy between the statistic and the relevant parameter of the distribution of interest is so large that the probability of obtaining that discrepancy or a larger one is less than the quantity designated α (the significance level). No clear statement is provided for the manner in which the null hypothesis is chosen, but the tests with which Fisher (1949) has been associated are in the form where the null hypothesis is equated with the statement "the phenomenon to be demonstrated is in fact absent" (p. 13).

The concept "rejection of the null hypothesis" is therefore unambiguous in the context of Fisher's viewpoint, but what about "acceptance of the null hypothesis?" Fisher (1949) provides the following statement "the null hypothesis is never proved or established, but is possibly disproved, in the course of experimentation. Every experiment may be said to exist only in order to give the facts a chance of disproving the null hypothesis" (p. 16). This is not very edifying since one does not expect to prove any hypothesis by the methods of probabilistic inference. Hogben (1957) has interpreted these and similar statements of the Yule-Fisher group to mean that a test of significance can lead to one of two decisions: the null hypothesis is rejected at the α level or judgment is reserved in the absence of sufficient basis for rejecting the null hypothesis.

Papers by Neyman and Pearson (1928a, 1928b) pointed out that the choice of a statistical test must involve consideration of alternative hypotheses as well as the hypothesis of central concern. They introduced the distinction between the error of falsely rejecting the null hypothesis and the error of falsely accepting it (rejecting its alternative). Neyman and Pearson's (1933) general theory of hypothesis testing, based on the concepts Type I error, Type II error, power, and critical region, was presented later.

The possible parameters for the distribution of the random variable or variables in a given investigation are conceptually represented by a set of points in what is called a parameter space. This space is considered to be divided into two or more subsets, but we shall restrict our present discussion to the classical case in which there are exactly two subsets of points.

The statistical hypothesis specifies that the parameter point lies in a particular one of these two subsets while the alternative hypothesis specifies the other subset for the point. A statistical test is a procedure for deciding, on the basis of a set of observations, whether to accept or reject the hypothesis. Acceptance of the hypothesis is precisely the same as deciding that the parameter point lies in the set encompassed by the hypothesis, while rejection of the hypothesis is deciding that the point lies in the other subset. A typical test procedure assigns to each possible value of the random variable (statistic) one of the two possible decisions.

Sets of distributions (or their associated parameters), in this mathematical model, may be considered to correspond to the explanations in the empirical world which may account for the possible outcomes of a given experiment. Empirical hypotheses, which specify values or relationships in the scientific world, are translatable on this basis into statistical hypotheses. But the distinction between empirical and statistical hypotheses is quite important: The former refer to scientific results and relationships, the latter to subsets of points in a parameter space; they are related by a set of correspondences between scientific events and parameter sets.

The term "null hypothesis" does not occur in the writings of many of the advocates of the Neyman-Pearson view. Except for one pejorative footnote I was unable to find the term used by Neyman (1942), for example, in any of an extensive array of his publications. In general, these people prefer the term "statistical hypothesis" or simply "hypothesis" in designating the subset of central concern and alternative hypothesis for the other subset. However, null hypothesis has taken on meaning over the years in the context of the Neyman-Pearson tradition among many writers of statistics, particularly those with expository proclivities. In the *Dictionary of Statistical Terms* (Kendall & Buckland, 1957) we find the following definition for null hypothesis: "In general, this term relates to a particular hypothesis under test, as distinct from the alternative hypotheses which are under consideration. It is therefore the hypothesis which determines the Type 1 Error" (p. 202).

AN EVALUATION[1]

Grant's position in regard to *a-s* is certainly not new or novel since it has been implicit in the writings of Fisher for the past 25 years. Moreover, it has been part of the folklore of statistical advising in psychology

---

[1] There is a third viewpoint, represented in the psychological literature by Rozeboom's (1960) article, from which Grant's position could be evaluated. This viewpoint emphasizes the importance of the a posteriori probabilities of alternative explanations, in the Bayes sense, rather than the decision aspects of experimentation. However, the philosophical and practical problems of this approach remain enormous as is evident in the debates on this and related topics over the years. See, for example, Jeffreys (1957), Neyman (1952), Hogben (1957), Savage (1954), Chernoff and Moses (1959), von Mises (1942, 1957), and particularly Parzen (1960) who discusses the dangers of using Bayesian inverse probability in applied problems. It is typically not the case in basic research that one can assume that an unknown parameter is a random variable with some specified a priori distribution, and in such cases this approach does not presently provide any adequate answers to the problems of hypothesis evaluation.

While of a markedly different philosophical persuasion than the present writer, Rozeboom (1960) is equally unsympathetic with the inferential bias represented by Grant. He cuts into an essential component of this bias in the following succinct and effective manner,

> Although many persons would like to conceive NHD [the null hypothesis decision procedure] testing to authorize only rejection of the hypothesis, not, in addition, its acceptance when the test statistic fails to fall in the rejection region,

at least as far back as my initial exposure to psychological statistics (see Footnote 2). And, in fact, if Grant wishes to argue that his position holds only in the very narrowest interpretation of the Yule-Karl Pearson-Fisher structure, I see no grounds for contesting it. If there are only two possible decisions—reject the null hypothesis or reserve judgment—one would surely not wish to equate the null hypothesis with the empirical hypothesis designating a specific value. Using this logic an investigator could just as well discard as retain a theory when it has led to perfect predictions over a wide range.

In this context I would like to point out that there are many logical difficulties connected with the Fisher formulations which have been brought out dramatically in years of debate (Fisher, 1935, 1950, 1955, 1959, 1960; Neyman, 1942, 1952, 1956, 1961). Moreover there are some people who, while generally sympathetic with the Fisher viewpoint, are quite willing to accept the null hypothesis and conclude that this provides support for an empirical hypothesis (Mather, 1943; Snedecor, 1956).

In the pursuit of evaluating Grant's position from the Neyman-Pearson theory we must remember that the null hypothesis is a statistical hypothesis which designates a particular subset of parameter points. Moreover, the null hypothesis and the alternative hypothesis (the other subset) are mutually exhaustive so that rejection of the one implies acceptance of the other; acceptance of a hypothesis being the belief, at a certain probability level, that the subset specified by the hypothesis includes the parameter point. There can be no question about the legitimacy or acceptability of acceptance of the null hypothesis within this purely mathematical scheme since acceptance and rejection are perfectly complementary.

Consequently any interpretive difficulties which result from accepting the null hypothesis must be in the rules for or manner of relating empirical and statistical (null) hypotheses. The null hypothesis is of course that hypothesis for which the probability of erroneous rejection is fixed at α (or set at a maximum of α); the test (critical region) is chosen so as to maximize power for the given α and the alternative hypothesis. Since therein lies the only feature of the process that differentiates the null hypothesis from the other subset, the relating of empirical and statistical hypotheses must be based upon it.

While there are no firm rules for deciding with which of the two subsets a given empirical hypothesis should be associated, there have been certain practices or conventions used by different writers. Neyman (1942), for example, suggested a most reasonable convention for relating empiri-

if failure to reject were not taken as grounds for acceptance, then NHD procedure would involve no Type II error, and no justification would be given for taking the rejection region at the extremes of the distribution, rather than in its middle (p. 419).

cal and statistical hypotheses which is to equate with the null hypothesis that empirical hypothesis for which the error of erroneous rejection is more serious than the error of erroneous acceptance so that the more important error is under the direct control of the experimenter. There are a few other conventions based upon the derivational advantages of fixing α for a simple (rather than a composite) hypothesis, but it is quite clear that Grant has not merely restated any of these. In fact, Grant's (1962) strong statement that "using these predictions as the values in H_0 [the null hypothesis] is tactically inappropriate, frustrating, and self-defeating," (p. 61) indicates that his position is much more than a convention of convenience.

The position which I will develop over the remainder of this paper is not that a-s is preferable to r-s, but that there are no sound foundations for damning a-s. In this process let me initially point out that one can be led astray unless he recognizes that when one tests a point prediction he usually knows before the first sample element is drawn that his empirical hypothesis is not precisely true. Consider testing the hypothesis that two groups differ in means by some specified amount. We might test the hypothesis that the difference in means is 0, or perhaps 12, or perhaps even 122.5. But in each case we are certain that the difference is not precisely 0.0000 . . . ad inf., or 12.0000 . . . , or 122.50000 . . . ad inf.

Recognition of this state of affairs leads to thinking in terms of differences or deviations that are or are not of importance for a given stage of theory construction or of application. Some express this in terms of differences which do and do not have practical importance, but I prefer the term zone of indifference which is used with important implications in sequential analysis. That is, if, for example, the difference in mean performance between two groups is less than, say, ϵ the two means may be considered equivalent for the given stage of theoretical development. In the case of a prediction of one-third for the proportion of right turns of rats in a maze, one would expect the same courses of action to be followed if the figure were actually .334 or .335. Thus, although we may specify a point null hypothesis for the purpose of our statistical test, we do recognize a more or less broad indifference zone about the null hypothesis consisting of values which are essentially equivalent to the null hypothesis for our present theory or practice.

While the formal procedures for testing statistical hypotheses are based upon the assumption that the sample size (n) is fixed prior to consideration of alternative test procedures, the user of statistical techniques is faced with the problem of choosing n and does so with regard for the magnitude of the discriminations which are or are not important for his particular application or level of theory development. In the typical case we choose the conditions of experimentation, including sample size, such that we will reject the null hypothesis with a given probability when the parameter difference is a certain magnitude. This is frequently done very formally in fields like agriculture, although rather informally in psychology. For example, in Cochran and Cox (1957) there is an extended discussion of the procedures for choosing the number of replications for an experiment on the basis of the practical importance of true differences. Thus, in one of their examples, a difference of 20% of the mean of two values is considered sufficiently important to warrant a sensitive enough experiment to have an .80 probability of detecting it; that is, if the difference is 20% a large enough n is desired to insure that the power of the test is .80. Although it may happen that the required sample size is a function of an unknown distribution and not determinable in advance, it can usually be approximated with the tests used most frequently by psychologists.

The choice of sample size is but one feature in the overall planning to obtain an experiment of the desired precision with due consideration for the level of theory development (including alternate theories), the zones of indifference, and the related consequences of dcision. However, such other features as the standard error per unit observation and the design efficiency do not have the flexibility of sample size, and, moreover, are usually chosen to maximize precision for reasons of economy. The choice of optimum sample size applies to all experimental strategies, including the non-objectionable (to Grant) and more usual r-s. It is surely apparent that anyone who wants to obtain a significant difference badly enough can obtain one—if his only consideration is obtaining that significant difference. Accepting that the means, for example, of two groups are never perfectly equal, the difference between them is some value ϵ. It is obviously an easy matter to choose a sample size large enough, for the ϵ, such that we will reject the null hypothesis with a given probability. But the difference may be so slight as to have no practical or theoretical consequences for the given stage of measurement and theory construction. As McNemar (1960) has recently pointed out, in his objections to the use of extreme groups, significant differences may be obtained even when the underlying correlation is as low as .10 which implies a proportion of predicted variance equal to .01.

After arguing against a-s on the basis of the dangers of tests that tend toward leniency, Grant points out that the procedure may be equally objectionable when the test is too stringent. He illustrates the latter by an example of a theory which is useful though far from perfect in its predictions. This particular point is perfectly in accord with my arguments since it demonstrates the parallelism of a-s and r-s. First, one does not usually want an experiment that is too stringent: in a-s because it may not be desirable to reject a useful, though inaccurate theory; in r-s because one may accept an extremely poor and practically useless theory. Second, one does not want an experiment that is too lenient or insensitive; in a-s because one may accept an extremely poor and practically useless theory; in r-s because it may not be desirable to reject a useful, though inaccurate theory (that is, to accept the null hypothesis which implies rejection of its alternative). The identical terms were chosen in the preceding sentences to dramatize the parallel implications for a-s and r-s of the general desirability of a test that is neither too stringent nor too insensitive. Whether or not the experiment is precise enough is, then, a function of theoretical and practical consequences, and not of whether acceptance or rejection of the null hypothesis leads to support for an empirical theory.

But, one may argue, while there is logical equivalence as stated above, there is not motivational equivalence. That is, while it is agreed that ideally investigators design their experiments (including their choice of sample sizes) in order to be reasonably certain of detecting only differences which are of practical or theoretical importance, in actual practice they are neither so wise nor so pure as to be influenced by these factors to the exclusion of social motivations. And it is indeed much easier to do insensitive rather than precise experimentation. This phenomenon is of course what Grant (1962) referred to in his statement,

> The tactics of accepting H_0 as proof and rejecting H_0 as disproof of a theory lead to the anomalous results that a small-scale, insensitive experiment will most often be interpreted as favoring a theory, whereas a large-scale, sensitive experiment will usually yield results opposed to the theory! (p. 56).

Perhaps that reflects the essential point of Grant's presentation—merely to caution imprudent experimenters that the combination of personal desire to establish one's hypothesis and the ease of performing insensitive experimentation produce a particularly troublesome interaction.

Before proceeding it should be remembered that scientific considerations may be made secondary to personal desires to establish a theory whether the procedure be a-s or r-s in a perfectly analogous fashion. The only difference involves such practical considerations as the fact that it is usually easier to run 5 or 10 subjects than 100 or 500.

If Grant (1962) merely intended his article to convey this obvious warning, I cannot understand the discussions which involve such statements as the following:

> Unfortunately most of the procedures used to date in testing the adequacy of such theoretical predictions [from mathematical models] set rather bad examples. Probably the least adequate of these procedures has been that in which an H_0 of exact correspondence between theoretical and empirical points is tested against H_1 covering any discrepancy between predictions and experimental results (p. 55).

If one is pointing out the dangers of using insensitive a-s tests (rather than condemning a-s on logical grounds), one would be expected to object to a particular or general use of *a-s* only if the use involved insensitive tests. Thus, it might be argued that experimenter WKE obtained support for his quantitative prediction by the use of *a-s* with a test so insensitive that it could not reasonably detect important discrepancies between predictions and observations. Or, as another fictional example, it might be stated that RRB always used *a-s* and always found support for his linear models, but his *n* was uniformly less than 5. But, unless there were almost uniform use of insensitive tests with *a-s*, this cautionary position could not reasonably lead to a condemnation of *a-s*.

As I see it, moreover, the argument against insensitive *a-s* tests is nothing but a particular form of the more general argument against bad experimentation. It is unquestionably the case than an *a-s* experiment that is too small and insensitive is poor, but the poorness is a property of the insensitivity and not of the *a-s* procedure. An *r-s* experiment that is too small and insensitive is equally poor. Due to the interaction between personal achievement desires and the ease of sloppy experimentation, as referred to previously, it may be necessary to be particularly alert to the usual scientific safe guards when using *a-s*, but that is a trivial matter and hardly worthy of an article.

In summary, it would be perfectly justifiable to argue that *n* is too small (or even too large) for a particular degree of sensitivity required at a given level of scientific development, but that is far from a proscription of designs where acceptance of the null hypothesis is in some way to the experimenter's social or personal advantage.

GRANT'S POSITION FROM THE VIEWPOINT OF SCIENTIFIC DEVELOPMENT

In the process of concluding this discussion I would like to emphasize and expand on certain factors which seem most critical in the process of evaluating scientific theories, as well as to indicate that my objections to Grant's apparent position are justifiable beyond the confines of the Neyman-Pearson theory.

It is surely clear that at various phases in the development of a scientific field one is faced with the problem of deciding about the suitability of different theories. When a discipline is at an early stage of development, knowledge of empirical relationships is crude so that broad isolation of explanatory constructs may be the most that is obtainable. At this stage one might consider as a significant accomplishment the ruling out of the hypothesis that observed differences are chance phenomena. The empirical hypothesis of central concern would be that there is some relationship of unknown magnitude, while its alternative would be the chance or noise explanation.

With increasing sophistication in the discipline the alternative hypotheses may represent different, but more or less equally well-developed theories. One does not choose between theory and chance, but between theory and theory or between theory and theories. Another aspect of increased sophistication is frequently the greater precision in the prediction of empirical results for the various theories.

The decision as to which of the theories is admissible on the basis of the available data may be accomplished directly within the Neyman-Pearson framework, but that is not necessarily the case. Sometimes the choice among theories depends upon a succession of tests of hypotheses or possibly even upon quite informal considerations; as an example of the latter, one theory may lead to a prediction which is perfectly in accord (within rounding errors) with the observations while the other theory is off by quite a margin—a statistical test would be considered foolish indeed. In disciplines that have markedly smaller observational variability than psychology the most common procedure consists of a subjective comparison between predictions and observations. Moreover, the point that one chooses among alternative hypotheses at various stages of scientific development (whether by statistical methods or otherwise) most certainly does not imply that his efforts stop once he has accepted or rejected a given hypothesis as Grant implies; if the accepted theory, for example, is of any interest he proceeds to make finer analyses and comparisons which may range from orthogonal subcomparisons in the analysis of variance to intuitive rumination. This provides a basis for objecting to Grant's arguments to the effect that hypothesis testing should be replaced (not supplemented) by estimation. The point is that both are usable, but at different phases of investigation.

I will again refer to the Atkinson and Suppes (1958) experiment to illustrate the relative roles of hypothesis testing and subsequent analysis in scientific advancement. Their first strategy was to decide which of two theories—game theory or the Markov model—was most adequate in the given experimental context. This clearly was a problem of testing hypotheses; a choice had to be made and the procedures of estimation could at best provide a substage on the way to the decision. The Markov model was accepted and game theory rejected, as noted above, but this certainly did not lead to a cessation of activity. Instead the investigators initially compared theoretical and observed transition matrices (and found them distinctly different). They then tested the more specific hypothesis of a one-stage Markov model against the alternative of a two-stage model, and finally investigated the stationarity of the Markov process.

During its early phases, Einstein's general theory of relativity was equivalent to Newtonian theory in the success of explaining various common phenomena, and a choice between them could not be made. But the Einstein theory led to certain predictions differing from Newtonian and these in turn led to a series of "crucial" tests. Among these were the exact predictions as to the magnitude of the bending of a ligth ray from a star by the gravitational field of the sun and the shift of wavelength of light emitted from atoms at the surface of stars. The general theory of relativity, thus, led to predictions which differed from the predictions of the alternative theory (Newton's), and the utimate correspondence between these predictions and empirical observations (acceptance of no difference between predicted and obtained results) led to support for general relativity. While agreements between theory and observational results have been close, they certainly have not been perfect—even physicists have problems of measurement precision and intricacy of mathematical derivation. But to the best judgment of the scientists the closeness of the fit between predictions and observations warrants the conclusion that the data provide support for the theory. Surely, however, despite its tremendous power, physicists do not claim that Einstein's general theory has been proved, nor are they convinced that it will not be ultimately replaced by a better theory.

It does not seem reasonable to argue that this method of scientific procedure is not suitable for psychology—just because our measurement precision happens to be lower than in physics and we use statistical tests rather than purely observational comparison.

References

Atkinson, R. C., & Suppes, P. An analysis of two-person game situations in terms of statistical learning theory. *Journal of Experimental Psychology,* 1958, **55**, 369–378.

Binder, A., & Feldman, S. E. The effects of experimentally controlled experience upon recognition responses. *Psychological Monographs,* 1960, **74** (9, Whole No. 496).

Bower, G. H. An association model for response and training variables in paired-associate learning. *Psychological Review,* 1962, **69**, 34–53.

Brody, A. L. Independence in the learning of two consecutive responses per trial. *Journal of Experimental Psychology,* 1958, **56**, 16–20.

Bush, R. R., & Mosteller, F. *Stochastic models for learning.* New York: Wiley, 1955.

Chernoff, H., & Moses, L. E. *Elementary decision theory.* New York: Wiley, 1959.

Cochran, W. G., & Cox, Gertrude M. *Experimental designs.* (2nd ed.) New York: Wiley, 1957.

Fisher, R. A. The fiducial argument in statistical inference. *Annals of Eugenics,* 1935, **6**, 391–398.

Fisher, R. A. *Statistical methods for research workers.* (10th ed.) Edinburgh: Oliver & Boyd, 1948.

Fisher, R. A. *The design of experiments*. (5th ed.) Edinburgh: Oliver & Boyd, 1949.

Fisher, R. A. The comparison of samples with possibly unequal variances. In *Contributions to mathematical statistics*. New York: Wiley, 1950.

Fisher, R. A. Statistical methods and scientific induction. *Journal of the Royal Statistical Society, Series B*, 1955, **17**, 69–78.

Fisher, R. A. *Statistical methods and scientific inference*. (2nd ed.) Edinburgh: Oliver & Boyd, 1959.

Fisher, R. A. Scientific thought and the refinement of human reasoning. *Journal Operations Research Society, Japan*, 1960, **3**, 1–10.

Grant, D. A. Testing the null hypothesis and the strategy and tactics of investigating theoretical models. *Psychological Review*, 1962, **69**, 54–61.

Grant, D. A., & Norris, Eugenia B. Dark adaptation as a factor in the sensitization of the beta response of the eyelid to light. *Journal of Experimental Psychology*, 1946, **36**, 390–397.

Harrow, M., & Friedman, G. B. Comparing reversal and nonreversal shifts in concept formation with partial reinforcement controlled. *Journal of Experimental Psychology*, 1958, **55**, 592–598.

Hogben, L. *Statistical theory*. London: Allen & Unwin, 1957.

Jeffreys, H. *Scientific inference*. (2nd ed.) Cambridge: Cambridge University Press, 1957.

Kendall, M. G., & Buckland, W. R. *A dictionary of statistical terms*. Edinburgh: Oliver & Boyd, 1957.

Mather, K. *Statistical analysis in biology*. New York: Interscience, 1943.

McNemar, Q. At random: Sense and nonsense. *American Psychologist*, 1960, **15**, 295–300.

Neyman, J. Basic ideas and theory of testing statistical hypothesis. *Journal of the Royal Statistical Society*, 1942, **105**, 292–327.

Neyman, J. *Lectures and conferences on mathematical statistics and probability*. (2nd ed.) Washington: United States Department of Agriculture, Graduate School, 1952.

Neyman, J. Note on article by Sir Ronald Fisher. *Journal of the Royal Statistical Society*, 1956, **18**, 288–294.

Neyman, J. Silver jubilee of my dispute with Fisher. *Journal of Operations Research Society, Japan*, 1961, **3**, 145–154.

Neyman, J., & Pearson, E. S. On the use and interpretation of certain test criteria for purposes of statistical inference. Part I. *Biometrika*, 1928, **20A**, 175–240. (a)

Neyman, J., & Pearson, E. S. On the use and interpretation of certain test criteria for purposes of statistical inference. Part II. *Biometrika*, 1928, **20A**, 263–294. (b)

Neyman, J., & Pearson, E. S. On the problem of the most efficient tests of statistical hypotheses. *Philosophic Transactions of Royal Society, Series A*, 1933, **231**, 289–337.

Parzen, E. *Modern probability theory and its applications*. New York: Wiley, 1960.

Rozeboom, W. W. The fallacy of the null-hypothesis significance test. *Psychological Bulletin*, 1960, **57**, 416–428.

Savage, L. J. *The foundations of statistics*. New York: Wiley, 1954.

Snedecor, G. W. *Statistical methods*. (5th ed.) Ames: Iowa State College Press, 1956.

von Mises, R. On the correct use of Bayes' formula. *Annals of Mathematical Statistics*, 1942, **13**, 156–165.

von Mises, R. *Probability, statistics and truth*. (2nd ed.) London: Allen & Unwin, 1957.

Weinstock, S. Acquisition and extinction of a partially reinforced running response at a 24-hour intertrial interval. *Journal of Experimental Psychology*, 1958, **56**, 151–158.

Witte, R. S. A stimulus-trace hypothesis for statistical learning theory. *Journal of Experimental Psychology*, 1959, **57**, 273–283.

4.11

Tactical note on the relation between scientific and statistical hypotheses

Ward Edwards[1]

Grant, Binder, and others have debated what should be the appropriate relationship between the scientific hypotheses that a scientist is interested in and the customary procedures of classical statistical inference. Classical significance tests are violently biased against the null hypothesis. A conservative theorist will therefore associate his theory with the null hypothesis, while an enthusiast will not—and they may often reach conflicting conclusions, whether or not the theory is correct. No procedure can satisfactorily test the goodness of fit of a single model to data. The remedy is to compare the fit of several models to the same data. Such procedures do not compare null with alternative hypotheses, and so are in this respect unbiased.

Ward Edwards, "Tactical Note on the Relation Between Scientific and Statistical Hypotheses," *Psychological Bulletin*, **63**, 1965, 400–402. Copyright (1965) by the American Psychological Association, and reproduced by permission.

[1] This research was supported by the United States Air Force under Contract AF 19 (628)-2823 monitored by the Electronics Systems Division, Air Force Systems Command. I am grateful to L. J. Savage, D. A. Grant, and W. R. Wilson for helpful criticisms of an earlier draft.

Grant (1962), Binder (1963), and Wilson and Miller (1964) have been debating the question of what should be the appropriate relationship between the scientific hypotheses or theories that a scientist is interested in and the statistical hypotheses, null and alternative, that classical statistics invites him to use in significance tests. Grant rightly notes that using the value predicted by a theory as a null hypothesis puts a premium on sloppy experimentation, since small numbers of observations and large variances favor acceptance of the null hypothesis and "confirmation" of the theory, while sufficiently precise experimentation is likely to reject any null hypothesis and so the theory associated with it, even when that theory is very nearly true. Grant's major recommendation for coping with the problem is to use confidence intervals around observed values; if the theoretical values do not lie within these limits, the theory is suspect. With this technique also, sloppy experimentation will favor acceptance of the theory—but at least the width of the intervals will display sloppiness. Grant also suggests testing the hypothesis that the correlation between predicted and observed values is zero (in cases in which a function rather than a point is being predicted), but notes that an experiment of reasonable precision will nearly always reject this hypothesis for theories of even very modest resemblance to the truth. Binder, defending the more classical view, argues that the inference from outcome of a statistical procedure to a scientific conclusion must be a matter of judgment, and should certainly take the precision of the experiment into account, but that there is no reason why the null hypothesis should not, given an experiment of reasonable precision, be identified with the scientific hypothesis of interest. Wilson and Miller point out that the argument concerns not only statistical procedures but also choice of theoretical prediction to be tested, since some predictions are of differences and some of no difference. Their point seems to apply primarily to loosely formulated theories, since precise theories will make specific numerical predictions of the sizes of differences and it would be natural to treat these as null hypothesis values.

Edwards, Lindman, and Savage (1963), in an expository paper on Bayesian statistical inference, have pointed out that from a Bayesian point of view, classical procedures for statistical inference are always violently biased against the null hypothesis, so much so that evidence that is actually in favor of the null hypothesis may lead to its rejection by a properly applied classical test. This fact implies that, other things being equal, a theory is likely to look better in the light of experimental data if its prediction is associated with the alternative hypothesis than if it is associated with the null hypothesis.

For a detailed mathematical exposition of the bias of classical significance tests, see Edwards, Lindman, and Savage (1963) and Lindley (1957). Lindley has proven a theorem frequently illustrated in Edwards,

Lindman, and Savage (1963) that amounts to the following. An appropriate measure of the impact of evidence on one hypothesis as against another is a statistical quantity called the likelihood ratio. Name any likelihood ratio in favor of the null hypothesis, no matter how large, and any significance level, no matter how small. Data can always be invented that will simultaneously favor the null hypothesis by at least that likelihood ratio and lead to rejection of that hypothesis at at least that significance level. In other words, data can always be invented that highly favor the null hypothesis, but lead to its rejection by an appropriate classical test at any specified significance level. That theorem establishes the generality and ubiquity of the bias. Edwards, Lindman, and Savage (1963) show that data like those found in psychological experiments leading to .05 or .01 level rejections of null hypotheses are seldom if ever strong evidence against null hypotheses, and other actually favor them.

The following example gives the flavor of the argument though it is extremely crude and makes no use of such tools as likelihood ratios. The boiling point of statistic acid is known to be exactly 150°C. You, an organic chemist, have attempted to synthesize statistic acid: in front of you is a beaker full of foul-smelling glop, and you would like to know whether or not it is indeed statistic acid. If it is not, it may be any of a large number of related compounds with boiling points diffusely (for the example, that means uniformly) distributed over the region from 130°C to 170°C. By one of those happy accidents so common in statistical examples, your thermometer is known to be unbiased and to produce normally distributed errors with a standard deviation of 1°. So you measure the boiling point of the glop, once.

The example, of course, justifies the use of the classical critical ratio test with a standard deviation of 1°. Suppose that the glop really is statistic acid. What is the probability that the reading will be 151.96° or higher? Since 1.96 is the .05 level on a two-tailed critical ratio test, but we are here considering only the upper tail, that probability is .025. Similarly, the probability that the reading will be 152.58° or greater is .005. So the probability that the reading will fall between 151.96° and 152.58°, if the glop is really statistic acid, is .025–.005 = .02.

What is the probability that the reading will fall in that interval if the glop is not statistic acid? The size of the interval is .62°. If the glop is not statistic acid, the boiling points of the other compounds that it might be instead are uniformly disributed over a 40° region. So the probability of any interval within that region is simply the width of the interval divided by the width of the region, .62/40 = .0155. So if the compound is statistic acid, the probability of a reading between 151.96° and 152.58° is .02, while if it is not statistic acid that probability is only .0155. Clearly the occurrence of a reading in the region, especially a reading near its lower end, would favor the null hypothesis, since a reading in that region is more likely if the null hypothesis is true than if it is false. And yet, any such reading would lead to a rejection of the null hypothesis at the .05 level by the critical ratio test.

Obviously the assumption made about the alternative hypothesis was crucial to the calculation. (Such special features as normality, the literal uniformity of the distribution under the alternative hypothesis, and the particular regions and significance levels chosen are not at all important; they affect only the numerical details, not the basic phenomenon.) The narrower the distribution under the alternative hypothesis, the less striking is the paradox; the wider that distribution, the more striking. That distribution is narrowest if it is a single point, and favors the alternative hypothesis most if that point happens to coincide with the datum. And yet Edwards, Lindman, and Savage (1963) show that even a single-point alternative hypothesis located exactly where the data fall cannot bias the likelihood ratio against the null hypothesis as severely as classical significance tests are biased.

This violent bias of classical procedures is not an unmitigated disaster. Many null hypotheses tested by classical procedures are scientifically preposterous, not worthy of a moment's credence even as approximations. If a hypothesis is preposterous to start with, no amount of the bias against it can be too great. On the other hand, if it is preposterous to start with, why test it?

The implication of this bias of classical procedures against null hypotheses seems clear. If classical procedures are to be used, a theory identified with a null hypothesis will have several strikes against it just because of that identification, whether or not the theory is true. And the more thorough the experiment; the larger that bias becomes. The scientific conservative, eager to make sure that error is scotched at any cost, will therefore prefer to test his theories as null hypotheses—to their detriment. The scientific enthusiast, eager to make sure that his good new ideas do not die premature or unnecessary deaths, will if possible test his theories as alternative hypotheses—to their advantage. Often these men of different temperament will reach different conclusions.

The subjectivity of this conclusion is distressing, though realistic. There should be a better, less subjective approach—and there is. The trouble is that in classical statistics the alternative hypothesis is essentially undefined, and so provides no standard by means of which to judge the congruence between datum and null hypothesis; hence the arbitrariness of the .05, .01, and .001 levels, and their lack of agreement with less arbitrary measures of congruence. A man from Mars, asked whether or not your suit fits you, would have trouble answering. He could notice the discrepancies between its measurements and yours, and might answer no; he could notice that you did not trip over it, and might answer yes. But give him two suits and ask him which fits you better, and his task starts to make sense, though it still has its difficulties. I believe that the argument between Grant and Binder is essentially unresolvable; no procedure can test the goodness of fit of a single model to data in any satisfactory way. But procedures for comparing the goodness of fit of two or more models to the same data are easy to come by, entirely appropriate, and free of the difficulties Binder and Grant have been arguing about. (They do have difficulties. Most important, either these models must specify to some extent the error characteristics of the data-generating process, or else a special model of the data-generating process, such as the normality assumption concerning the thermometer in the statistic acid example, must also be supplied. But of course this difficulty is common to all of statistics, and is fully as much a difficulty for the approaches I am rejecting as for those I am espousing.) The likelihood-ratio procedures I advocate do not make any use of classical null-hypothesis testing, and so the question of which model to associate with the null hypothesis does not arise. While there is nothing essentially Bayesian about such procedures, I naturally prefer their Bayesian to their non-Bayesian version, and so refer you to Savage (1962), Raiffa and Schlaifer (1961), Schlaifer (1959, 1961), and Edwards, Lindman, and Savage (1963) as appropriate introductions to them. Unfortunately, I cannot refer you to literature telling you how to invent not just one but several plausible models that might account for you data.

References

Binder, A. Further considerations on testing the null hypothesis and the strategy and tactics of investigating theoretical models. *Psychological Review,* 1963, **70**, 107–115.

Edwards, W., Lindman, H., & Savage, L. J. Bayesian statistical inference for psychological research. *Psychological Review,* 1963, **70**, 193–242.

Grant, D. A. Testing the null hypothesis and the strategy and tactics of investigating theoretical models. *Psychological Review,* 1962, **69**, 54–61.

Lindley, D. V. A statistical paradox. *Biometrika,* 1957, **44**, 187–192.

Raiffa, H., & Schlaifer, R. *Applied statistical decision theory.* Boston: Harvard University, Graduate School of Business Administration, Division of Research, 1961.

Savage, L. J., et al. *The foundations of statistical inference: A discussion.* New York: Wiley, 1962.

Schlaifer, R. *Probability and statistics for business decisions.* New York: McGraw-Hill, 1959.

Schlaifer, R. *Introduction to statistics for business decisions.* New York: McGraw-Hill, 1961.

Wilson, W. R., & Miller, H. A note on the inconclusiveness of accepting the null hypothesis. *Psychological Review,* 1964, **71**, 238–242.

4.12

Much ado about the null hypothesis

Warner Wilson
Howard L. Miller
Jerold S. Lower[1]

Edwards has charged that classical statistics, in contrast to Bayesian statistics, is always violently biased against the null hypothesis. Edwards has advised the conservative classical investigator that he should, therefore, always identify his theory with the null hypothesis, so as minimize specious claims for theoretical support. This paper reinterprets the so-called bias in terms of differential assumptions about the nature of the alternatives which must be considered; its main purpose, however, is to point out that insensitive experiments, in contrast to sensitive ones, are always biased for, rather than against, the null hypothesis. It is this 2nd bias (which exists independently of the 1st) that prompts the conservative investigator not to identify his theory with the null hypothesis.

Grant (1962) and Binder (1963) have clarified the fact that two strategies can be used in theory testing. First, one can identify the theory test with the null hypothesis and claim support for the theory if the null hypothesis is accepted. The second approach, presumably more orthodox and traditional, is to identify the theory under test with the alternative hypothesis and claim support for the theory if the null hypothesis is rejected. Binder has referred to these two approaches as

Warner Wilson, Howard L. Miller, and Jerold S. Lower, "Much Ado About the Null Hypothesis," *Psychological Bulletin,* **67,** 1967, 188–196. Copyright (1967) by the American Psychological Association, and reproduced by permission.

[1] The authors wish to express their thanks to Ward Edwards for his personal communications with them and for the patient forbearance he has demonstrated in attempting to help clarify the ideas that are discussed in this paper. The order of authorship should not be interpreted as implying a greater contribution on the part of the senior author.

acceptance support and rejection support strategies. In this paper, however, the authors will follow Edwards (1965) and speak of "identifying one's theory with the null hypothesis" and basing support on the acceptance of the null versus "identifying with the alternative hypothesis" and basing support on the rejection of the null and of course, the subsequent acceptance of the alternative.

Binder (1963) has ably pointed out that either strategy may be used effectively under some circumstances. Wilson and Miller (1964a, 1964b) have joined with Grant, however, in arguing that it is generally better to identify with the alternative and, hence, base support for a theory on the rejection of some null hypothesis. These writers point out that while the probability of *rejecting* the null hypothesis wrongly is held constant, for example, at the .05 level, the probability of *accepting* the null hypothesis wrongly varies with the precision of the experiment. To the extent that error is large, the study in question is biased for the null hypothesis and for any theories identified with it. According to their view, the conservative, cautious approach is to identify one's theory with the alternative hypothesis.

This essentially orthodox (Fisherian) view of classical statistical procedures has seemed so reasonable to the present authors that they were surprised to find Edwards, who identifies himself with Bayesian statistics, taking a dramatically opposed view. Edwards' article is only one of several considering the relative virtues of classical versus Bayesian statistics (see Binder, 1964, for an excellent review). Edwards' article seems especially important, however, due to the fact that it strongly urges changes in the tactics of the orthodox classical statistician—changes which might prove to be ill-advised in some cases and impossible in others. Edwards' (1965) paper seems to make or imply the following points: (a) Classical procedures, in fact, "are always violently biased against the null hypothesis [p. 400]." (b) The cautious, conservative approach, therefore, is to identify one's theory with the null hypothesis and, hence, base support for one's theory on the acceptance of the null hypothesis. (c) The ideal solution, however, is to compare the goodness of fit of several models to the same data, thus avoiding the whole problem of null hypothesis testing.

Edwards apparently believes that a Bayesian analysis is always feasible or that if it is not, the experiment in question is not worth doing. The present writers do not agree with this point. They do, however, find much that is admirable in Edwards' position and certainly agree that Bayesian procedures are to be preferred—when they can be used. The purpose of the present paper, therefore, is not to disagree with Edwards so much as to suggest clarification and qualification.

In connection with Point a, it is conceded that Edwards is commenting on differences between classical and Bayesian statistics that really exist. It is suggested, however, that the term "bias" is perhaps not the best way to sum up these differences. A Bayesian analysis typically assumes that the datum comes from a null distribution or from some other distribution. Classical statistics assumes that the datum comes from a null distribution or from some one of all other possible distributions. The assumption that the datum may come from *any* distribution does, indeed, always increase the apparent probability that it comes from some distribution other than the null. The difference lies in the nature of the alternatives which are to be taken into account. Classical procedures happen to assume that all possible alternatives must be taken into account. Granted this assumption, any bias for or against the null is then expressed by the probability level, which, even in the case of the .05 level, clearly favors the null.

In relation to Point *b*, it is suggested that the bias for theories identified with the null hypothesis in imprecise experiments, which Grant talked about, exists independently of and is logically distinct from the bias against the null hypothesis which Edwards talked about. Even when Edwards' bias *against* the null hypothesis exists, it does not imply the absence of Grant's bias *for* the null hypothesis. These considerations make the choice of tactics more complex than Edwards' article indicated. Edwards' (1965) tactical advice was that, "If classical procedures are to be used, a theory identified with a null hypothesis will have several strikes against it. . . . And the more thorough the experiment, the larger that bias becomes [p. 402]." An attempt will be made to show that this advice is valid only in experiments of extreme precision, a type presumably rare in psychology. As experiments become imprecise, just the opposite tactical advice becomes appropriate.

In relation to Point *c*, it is suggested that no matter how many models we may have, many people will still find null hypotheses which will seem to them to need testing.

IS CLASSICAL STATISTICS ALWAYS BIASED AGAINST THE NULL HYPOTHESIS?

Edwards' first point was that classical statistics is always violently biased against the acceptance of the null hypothesis. He argued persuasively for this point in several different ways. He presented, for one thing, the following example, which supposedly illustrates the bias. This example and other points made by Edwards will be considered here in hopes that the discussion will help clarify the circumstances under which classical procedures can and cannot meaningfully be said to be biased, and also give some indication of how often inappropriate rejections of the null hypothesis may, in fact, occur.

The following example gives the flavor of the argument, though it is extremely crude and makes no use of such tools as likelihood ratios. The boiling point of statistic acid is known to be exactly 50°C. [Presumably this number was intended to be 150°C.] You, an organic chemist, have attempted to synthesize statistic acid; in front of you is a beaker full of foul-smelling glop, and you would like to know whether or not it is indeed statistic acid. If it is not, it may be any of a large number of related compounds with boiling points diffusely (for the example, that means uniformly) distributed over the region from 130°C to 170°C. By one of those happy accidents so common in statistical examples, your thermometer is known to be unbiased and to produce normally distributed errors with a standard deviation of 1°. So you measure the boiling point of the glop, once.

The example, of course, justifies the use of the classical critical ratio test with a standard deviation of 1°. Suppose that the glop really is statistic acid. What is the probability that the reading will be 151.96° or higher? Since 1.96 is the .05 level on a two-tailed critical ratio test, but we are here considering only the upper tail, that probability is .025. Similarly, the probability that the reading will be 152.58° or greater is .005. So the probability that the reading will fall between 151.96° and 152.58°, if the glop is really statistical acid, is .025 − .005 = .02.

What is the probability that the reading will fall in that interval if the glop is not statistic acid? The size of the interval is .62°. If the glop is not statistic acid, the boiling points of the other compounds that it might be instead are uniformly distributed over a 40° region. So the probability of any interval within that region is simply the width of the interval divided by the width of the region, .62/40 = .0155. So if the compound is statistic acid, the probability of a reading between 151.96° and 152.58° is .02, while if it is not statistic acid that probability is only .0155. Clearly the occurrence of a reading in that region, especially a reading near its lower end, would favor the null hypothesis, since a reading in that region is more likely if the null hypothesis is true than if it is false. And yet, any such reading would lead to a rejection of the null hypothesis at the .05 level by the critical ratio test [Edwards, 1965, p. 401].

If we follow the mode of analysis Edwards used, that of comparing an area of the null distribution to an area of the alternative distribution, we can note first that the probability of actually rejecting the null when the data in fact favor is quite low, even in this contrived example. It would seem that we would reject the null when the data actually favor it when the observation falls in the interval 151.96–152.40. When the null is true, the data will fall in this interval or in the corresponding lower interval only about 3% of the time.

In addition, the type of analysis Edwards used has two aspects that seem intuitively unappealing: (a) *All* outcomes have a low probability under the alternative hypotheses, and (b) this probability is equally low no matter where an observation occurs. For example, if a datum occurs at 150° the implied inference is that the probability of a hit in this segment of .62 is .62/40 or .0155; likewise, if a datum occurs at 169° the implied inference is that the probability of a hit in this segment of .62 is still .62/40 or .0155.

Another way of looking at this example avoids both of the aspects noted above. Suppose, for simplification, that 1.96 is rounded off to 2.00. The null hypothesis then implies that the observation should fall in a segment of 4° between 148° and 152° 95% of the time. Suppose the actual reading is 152°. The probability of a hit as far as 2° away from 150°, if the null is true, is only .05; therefore, the null is rejected. It would seem that in order to test the alternative hypothesis, one could ask, "What is the probability of a hit as close as 2° to 150° if the acid is not statistic?" The answer would seem to be 4/40 or .10. It would seem, then, that a reading of 151.96° or 152° is more probable if the null hypothesis is false, and rejection of the null would seem appropriate after all. Perhaps, then, the bias is not as prevalent as Edwards would lead us to believe. Indeed, if a person working at the .05 level of confidence is to run any danger of rejecting the null when the data actually support it, it would seem that the width of the distribution under the alternative hypothesis must be more than 80 standard deviations. The present authors will leave it up to Edwards to show that situations of this sort occur frequently enough to justify concern.

The reader may wish to note, however, that the apparent bias in classical procedures can be manipulated at will. If either a broad alternative distribution or a small error term is assumed, the bias will be increased.

Lindley (1957), another Bayesian, presumably had just such considerations in mind when he stated that data can always be invented that highly favor the null hypothesis yet lead to its rejection by a properly applied classical test. Although such data can be invented, it is still meaningful to ask whether such data are inevitable or even likely in reality. By assuming a wide enough alternative, a bias can be created; however, in reality the alternative distribution is supposedly based on some theoretical or empirical consideration, and its width cannot be set arbitrarily. Error terms, on the other hand, can be reduced, even in reality, by the expedient of collecting more cases. However, in order for a reduced error term to lead inevitably to a bias, it is necessary to assume that as the error term is reduced the absolute deviation from chance becomes less, so that the probability of the deviation's occurrence remains constant. This convenient constancy of probability in the face of a shrinking error term cannot be expected in reality. Consider again the statistic acid example, assuming a deviation of 2° and an error term of 1°. As indicated above, the probability of a hit at 152° is .05 under the null and .10 under the alternative. In real life, any observation is more likely than not to be near the true value, so if data are collected until the error term becomes .1°, the mean value of all the observations may still be near 152°. Now the probability under the alternative would still be .10, but the probability under the null would have become far, far less, since the critical ratio would now be 20. Thus, although it is necessary to admit that data can always be invented which will make a classical test look biased, it is also possible to point out that such data are not necessarily obtainable in reality and that certainly it is also easy to invent data that make classical tests look unbiased.

The considerations up to this point seem to support the assertion that classical statistics if blindly applied will sometimes be biased against the null hypothesis. Support for the assertion that classical statistics is always biased seem lacking. Instead, it appears that the extent of such bias depends in great part on the width of the alternative distribution relative to the width of the null distribution.

In further exploring Edwards' position, it may be noted that Edwards conceded that the bias becomes less as the relative width of the alternative distribution decreases. He firmly insisted, however, that no matter how narrow the alternative distribution, the bias will persist. Edwards seems to have come to this conclusion through an inappropriate comparison of probability levels with likelihood ratios.

The likelihood ratio is the ratio of one probability density (at the point of the data) to another. In a case such as the acid example, the Bayesian would apparently rather base his conclusions on the likelihood ratio than on the probability level, and the present authors would have no objection so far as such cases are concerned. In commenting on the relation of likelihood ratios to classical significance tests, Edwards (1965) said:

> Even a single-point alternative hypothesis located exactly where the data fall [the form of alternative distribution that most violently biases the likelihood ratio against the null hypothesis] cannot bias the likelihood ratio against the null hypothesis as severely as classical significance tests are biased [p. 401].

Whether we formally use likelihood ratios (ratios of ordinates) or simply other ratios of conditional probabilities (areas), what Edwards was saying seems to have a certain surface validity. If we consider only the probability (or probability density at a point) under one distribution, the null hypothesis, and do not take into consideration how likely this event might be under some specified alternative distribution, we are biasing ourselves against our null hypothesis by comparing it essentially to all possible alternatives instead of comparing it to just one. It seems to us that this is exactly what the classical statistician *intends* to do. If we have a clearly defined alternative, and we can say that "reality" must be one or the other, then we can justify a likelihood ratio or some similar procedure. If we do not have such alternative models, we cannot invent them to avoid a theoretical bias.

Edwards, Lindman, and Savage (1963) implied that in any inference based on statistics, the decision involved must be a joint function of a prior probability estimate (what you thought before about the likelihood of your hypothesis), the likelihood ratio, and the payoff matrix (the relative rewards and costs of being right and wrong about your hypothesis). Edwards seemed to imply that "classical statisticians" use no such considerations. It is doubtful that this is so. For one thing, a choice of significance level can, under some circumstances, be construed as a prior probability estimate—not the subjective one of the individual scientist but an admittedly arbitrary attempt to standardize a bias against alternative hypotheses (not altogether different from a bias for the null). It appears to be a deliberate attempt to offer a standardized, public method for objectifying an individual scientist's willingness to make an inference. An undisputed goal of science is objectivity—public reproducibility. To introduce, into inferential statistics, a methodology dependent on partially subjective estimates of probability would seem to undermine this goal.

It is apparently impossible to say whether the choice of low probability levels implies a low prior probability or a payoff matrix biased for retention of the null hypothesis, but one or both seem implicit. The natural tendency is for investigators to believe that their hypotheses are correct and that the world can ill afford to ignore them. If such subjective inclinations were allowed full sway, experimentation could become superfluous, and science might well degenerate into controversy. Classical statistics wisely resolves prior probability and payoff considerations in a conservative and standard rather than in a subjective and variable manner.

In answer to the obvious criticism concerning the subjectivity of invented alternative hypotheses, Edwards[3] stated essentially that a scientist always has some information regarding alternatives. Bayesian statistics considers this information; classical statistics does not. It seems that the bias that Edwards was discussing is a function of the failure to use (assume?) this additional information. In point of fact, the choice of one particular alternative distribution with which to compare the null excludes other possible alternatives. It is precisely this exclusion (which may be justifiable under some circumstances) which increases the probability that you will accept the null.

In this section we have attempted to discuss such questions as "Are

---

[3] W. Edwards, personal communication, October 1966.

classical procedures biased against the null hypothesis?" and "How often might such a bias result in rejections of the null when the data actually favor it or discredit it only weakly?" We see no reason to view classical procedures as biased against the null in any absolute sense—the opposite seems to be the case. We see no reason to view classical procedures as biased against the null hypothesis relative to Bayesian statistics since this comparison is at best unsatisfactory due to the difference in the procedures which are used and in the information which is assumed to be available. We can concede only that under special conditions, including presumably a specifiable alternative distribution, blind use of a classical analysis might result in a rejection of the null when a defensible Bayesian analysis, considering only the specifiable alternative, might show that the data actually support the null. We know of not one real-life instance in which the above has been demonstrated. It would seem to us that circumstances making such errors likely are not frequent, and it is suggested that the burden of proof is on those who think that such errors occur frequently—in the literature, as opposed to in the examples used in articles written by Bayesians.

SHOULD THE SCIENTIFIC CONSERVATIVE ALWAYS IDENTIFY HIS THEORY WITH THE NULL HYPOTHESIS?

Edwards' second main point was that if one does use classical statistics, the more conservative strategy is to identify one's theory *not* with the alternative hypothesis—as Wilson, Miller, and Grant advocated —but with the null hypothesis. This point, too, is apparently incorrect. What is more, even if the first point, about classical statistics being always biased, were correct, the second point would not necessarily follow. In order to see the actual independence of the two points, it is helpful to realize that if something is biased, it is biased relative to something else, Edwards presumably meant that classical statistics is always biased relative to Bayesian statistics.

On the other hand, when Grant, Wilson, and Miller said that an insensitive experiment is biased for the acceptance of the null hypothesis and any theory identified with it, they meant that an insensitive experiment is biased relative to a sensitive one. A classical analysis can be biased against the null in comparison to a Bayesian analysis, and an insensitive experiment can still be biased for the null in comparison to a sensitive experiment. The two biases exist independently. What is more, the bias for the null hypothesis in insensitive experiments is just as true with a Bayesian analysis as with a classical analysis.

The following example (see Table 1) is offered as an illustration of the bias in insensitive versus sensitive experiments—a bias which, prior to the Edwards article, the present writers would have considered not in need of illustration. The example views experiments as perfectly sensitive or as completely insensitive; null hypotheses as true or false; and theories as identified with the alternative and supported if the null is rejected, or as identified with the null and supported if the null is not rejected.

Edwards said that if you are a conservative, that is, if you wish to minimize undeserved successes, you should always identify your theory with the null. Wilson, Miller, and Grant said that in the case of insensitive experiments, such as are common in psychology, just the opposite tactic is to be recommended to the conservative. Table 1 indicates the number of deserved and undeserved successes achieved by theorists depending on whether they do identify with the alternative or with the null: Overall, identification with the null clearly promote success for one's theory, a total of 285 (out of 400) successes versus 115 for those who identify with the alternative. The same is true in the case of total undeserved successes. Those who identify with the null get a total of 95 undeserved successes (out of 200) versus 10 for those who identify with the alternative. The most dramatic difference occurs, of course, in the case of insensitive experiments. The score is 95 illegitimate successes for those who test their theories as null hypotheses and only 5 for those who test their theories as alternative hypotheses. Hopefully, no one will wish to attempt to reconcile these outcomes with Edwards' (1965) assertion: "The scientific conservative, eager to make sure that error is scotched at any cost, will therefore [always] prefer to test his theories as null hypotheses—to their detriment [p. 402]."

The authors would like to concede, however, that in recommending continued use of classical statistics combined with identification of theories with alternative hypotheses, they are operating on the basis of several beliefs which are not mathematically demonstrable.

Perhaps the most critical belief in this context is that most experiments in psychology are insensitive. It may be noted that a conservative would not favor identification with the alternative on the basis of Table 1 unless he held this belief. In a perfectly sensitive experiment, only identification with the alternative leads to accepting one's theory when it is false. This would occur because no matter how small the error, t's of 1.96 and greater will occur 5% of the time. These writers would point out, however, that in very sensitive experiments, specious deviations from chance, though technically significant, would be so small that they could hardly mislead anyone. It might also be argued that belief in ESP, for example, survives partly on just such deviations. Such a consideration might justify more concern about such errors, greater use of the .01 or .001 significance level, and greater interest in likelihood ratios when meaningful ones can be computed.

TABLE 1

Limits of Deserved and Undeserved Successes Assuming Identification with the Null versus the Alternative in Perfectly Sensitive and Insensitive Experiments

	Identification of theory with the null		Identification of theory with the alternative	
	Percentage deserved successes	Percentage undeserved successes	Percentage deserved successes	Percentage undeserved successes
Sensitive experiments				
True null	95	0	0	5
False null	0	0	100	0
Insensitive experiments				
True null	95	0	0	5
False null	0	95	5	0

Although the good showing of "acceptance support" in sensitive experiments is in its favor, it should still be noted that, seemingly, grounds are seldom available for deciding if experiments are going to be sensitive. In the face of this inevitable equivocality, investigators are encouraged to identify their theories with the alternative and so put an upper limit on error of 5% rather than 95%. One is not justified, after all, in assuming that favorable circumstances—true theories and sensitive experiments—will occur generally. A sensible strategy must assume the possibility of unfavorable circumstances—false theories and insensitive experiments—and must provide protection against these unfavorable circumstances.

Intuitively, it might seem impossible to base any inference, or express any bias, on the basis of a completely insensitive experiment. This consideration, however, is just the point of the objection to identification with the null. Identification with the null allows one to base positive claims for theoretical confirmation on the acceptance of the null hypothesis, which is, of course, virtually assured ($p = .95$) in the completely insensitive experiment.

Yet another consideration relates to the fact that it makes relatively little difference which approach you use in precise experiments. With great precision, you cannot go too far wrong. As experiments become imprecise, however, the tactical choice makes an increasingly large difference. All the more reason, therefore, for the conservative to choose the tactic that protects him even when experiments are imprecise.

The second belief is that statistics can sometimes reveal facts worth knowing even if they are not apparent to the naked eye. The relationships between smoking and lung cancer and between obesity and heart disease are examples. It is true that statistics may lead one to be overly optimistic about the importance of effects that are more significant than important, and certainly this tendency is to be deplored. On the other hand, although the human observer, unbeguiled by statistics, may indeed discount many trivial effects, he may also infer strong effects when only a trivial effect or no effect at all is present. Many people are, in the opinion of these authors, overly optimistic about the existence of ESP and the efficacy of psychotherapy. This overoptimism exists, however, in spite of statistics, not because of them.

Those who favor the so-called interocular test should realize that situations in which effects are large relative to error will yield their secrets quickly. This consideration suggests that investigators will inevitably spend most of their time on ambiguous situations in which the effects of interest are small relative to the precision of measurement so far achieved. Statistics will be needed in such situations.

The third belief is that false positives are more damaging than false negatives. In the present context, this statement means that it is worse to view a false theory as already proved than to view a true theory as not yet proved. This belief is widespread (see, e.g., Campbell, 1959) and will not be belabored here. Granted the belief that false positives cause more trouble, identification of one's theory with the alternative is only natural. The traditional .05 significance level then limits the investigator to 5% error no matter how insensitive his experiments and no matter how false his hypotheses. When one identifies his theory with the null, however, the natural conservatism inherent in the traditional use of small probability levels works for, rather than against, hasty claims for support. Indeed, if an investigator identifies his theories with the null and if his theories are false, his number of false positives approaches not 5%, but 95%, as his experiments become increasingly imprecise.

The fourth belief is that many null hypotheses are worth testing. Edwards questioned this belief by suggesting that most null hypotheses are obviously false and that the testing of them is, therefore, meaningless. One of several possible replies is that the real question is frequently not whether the data deviate from the null, but whether the deviation is positive or negative. In such a case the testing of the null is obviously meaningful.

If the conservative investigator believes that false positives are the greater threat and that his experiment will be insensitive, he will surely choose to identify his theory with the null hypothesis. On the other hand, none of the beliefs discussed so far necessarily justifies a preference for probability levels over likelihood ratios, and likelihood ratios are, in fact, strongly recommended whenever it is possible to compute them. The last belief, however, is that in most psychological experiments it is not possible to use a Bayesian approach based on likelihood ratios. The calculation of a ratio requires at least one specifiable alternative distribution. In most psychology experiments, no grounds are available for arriving at such a distribution. In such cases, classical statistics appears to be the only alternative, and, under this condition, the possibility of classical statistics being biased relative to Bayesian statistics is a meaningless issue.

Naturally, classical statisticians as well as Bayesian statisticians should consider alternative distributions. The point is that such considerations may or may not yield enough information to clearly justify a Bayesian analysis. So long as such information is often, if not usually, lacking, Edwards' rejection of classical approaches seems premature.

Edwards implied, to be sure, that investigators using classical tests have often rejected true null hypotheses without real evidence even when Bayesian statistics were potentially available. Bayesians could perhaps make a definite contribution by analyzing a number of published experiments and pointing out instances in which classical statistics has led to null hypothesis rejection when a potentially available Bayesian analysis would not have. It seems entirely possible, however, that such examples might be hard to find. At least they seem to be conspicuously absent from Bayesian critiques. It is also suggested that it would be most instructive if Bayesians would supply a precise alternative distribution to accompany any of the following null hypotheses, which are presented as typical psychological problems: (a) partially reinforced subjects extinguish at the same rate as continuously reinforced subjects; (b) patients show no change on tests of adjustment as a function of counseling; (c) punishment does not influence response rate; and (d) students working with teaching machines learn no faster and no better than students reading ordinary textbooks. Furthermore, the likelihood ratio, even if available, is not likely to disagree with the classical test unless the alternative distribution is very broad. If the meaningful alternative is not uniform, as it was in the statistic acid example, but instead has a mode or modes somewhere in the neighborhood of the null value, the likelihood ratio and significance test are even more likely to point in the same direction.

Another tactical point that merits attention is the question of whether a null hypothesis rejection resulting from a sensitive experiment should always be viewed as a legitimate success. As Table 1 indicates, if a null hypothesis is false, a sufficiently sensitive experiment will always reject it, and theories identified with the alternative will always be supported. Table 1 views such successes as legitimate. The possibility always exists, however, that the difference, though real, may be so slight that recording it in the literature is a complete waste of effort. The present writers think that the indiscriminant cataloguing of trivial effects is, in fact, a major problem in psychology today, and they would certainly regret it if their position was in any way interpreted as encouraging this unfortunate practice. On the other hand, as Wilson and Miller have pointed out, one must weigh one problem against another. If investigators identify their theories with the alternative hypothesis, they may be tempted to run *many* subjects and report theories to be true, even though the theories have no predictive utility. On the other hand, if investigators identify their theories with the null hypothesis, they may be tempted to run *few* subjects and report theories to be true, even though they are completely false. The present writers find no difficulty in deciding that they would prefer to confront investigators with the first temptation rather than the second.

A consideration of the problem of accepting trivial effects does not, therefore, greatly modify this paper's inclination towards identification with the alternative hypothesis in combination with the use of classical tests. This strategy may not be the best imaginable, but it is still often the best available, and it is recommended to the scientific conservative.

In summing up this section, we wish to remind the reader that Edwards apparently recommended that investigators switch to Bayesian techniques altogether. He went on to say, however, that if one *does* use classical statistics, the more conservative tactic, that is, the tactic that minimizes erroneous claims for theoretical support, is identification with the null. Our perspective on this advice can be summed up very briefly: If a conservative is a person who wishes to abandon a tactic that puts a ceiling on error of 1 in 20 and adopt instead a tactic that puts a ceiling on error of 19 in 20, then this is excellent advice.

DOES THE DEVELOPMENT OF MULTIPLE MODELS AVOID THE NEED FOR NULL HYPOTHESIS TESTING?

Edwards' last point was that the better tactic is to compare your data to several plausible models, and hence avoid null hypothesis testing altogether. Multiple models seem thoroughly desirable, and it seems worthwhile to note that in a traditional two-tailed test, classical statistics always implies three families of models: one predicting no difference, one predicting a positive difference, and one predicting a negative difference. On the other hand, it is hard to see how multiple models avoid the need for null hypothesis testing. Again an example from Edwards' (1965) paper may be helpful.

> A man from Mars, asked whether or not your suit fits you, would have trouble answering. He could notice the discrepancies between its measurements and yours, and might answer no; he could notice that you did not trip over it, and might answer yes. But give him two suits and ask him which fits you better, and his task starts to make sense [p. 402].

But ask him if he's sure of his decision or if he might reverse it if he saw you model the suits again, and you have a null hypothesis to test. In other words, although on one occasion one model (suit) was judged to fit the data (you) better, one must still ask if the difference in fit is significant relative to the potential sources of error. However stated and however tested, this question still seems to constitute a null hypothesis. Bayesian statistics may offer a meaningful alternative to null hypothesis testing, but it will take more than this example to convince the present authors.

ARE UNDEFINED ALTERNATIVE HYPOTHESES THE FAULT OF CLASSICAL STATISTICS?

From the current vantage point, the classical bias against the null hypothesis does not appear as obvious as Edwards' bias against classical statistics. It is instructive to note one aspect of the rationale behind Edwards' (1965) bias: "The trouble is that in classical statistics the alternative hypothesis is essentially undefined, and so provides no standard by means of which to judge the congruence between datum and null hypothesis [p. 402]." Edwards seemed to see the absence of a specifiable alternative as a short-coming of classical procedure. It seems appropriate to point out that, in fact, the absence of a well-defined alternative is not a vice of classical statistics. The absence of a well-defined alternative is a problem, a problem which is unavoidable in many cases, a problem which Bayesian statistics presumably cannot handle, and also a problem which classical statistics is especially designed to avoid.

SUMMARY

Some of the main points of the position of the present writers may be summed up as follows: (a) Identification with the alternative limits erroneous claim for theoretical support to 5%; identification with the null limits such errors to 95%. The conservative is, therefore, presumably better advised to identify with the alternative. (b) Classical procedures assume that only the null distribution can be specified, and ask if the data are from this distribution or some other. Bayesian procedures assume that two distributions can be specified, and ask if the data are from the null distribution or the alternative. Granted the difference in procedure and in the information assumed, discussion of a bias in one procedure relative to the other seems of dubious meaningfulness. (c) Granted the information assumed by classical procedures, they show an absolute bias in favor of the null hypothesis.

References

Binder, A. Further considerations on testing the null hypothesis and the strategy and tactics of investigating theoretical models. *Psychological Review,* 1963, **70**, 107–115.

Binder, A. Statistical theory. *Annual Review of Psychology,* 1964, **15**, 277–310.

Campbell, D. T. Methodological suggestions from a comparative psychology of knowledge processes. *Inquiry,* 1959, **2**, 152–182.

Edwards, W. Tactical note on the relation between scientific and statistical hypotheses. *Psychological Bulletin,* 1965, **63**, 400–402.

Edwards, W., Lindman, H., & Savage, L. J. Bayesian statistical inference for psychological research. *Psychological Review,* 1963, **70**, 193–242.

Grant, D. A. Testing the null hypothesis and the strategy and tactics of investigating theoretical models. *Psychological Review,* 1962, **69**, 54–61.

Lindley, D. V. A statistical paradox. *Biometrika,* 1957, **44**, 187–192.

Wilson, W., & Miller, H. The negative outlook. *Psychological Reports,* 1964, **15**, 977–978. (a)

Wilson, W. R., & Miller, H. A note on the inconclusiveness of accepting the null hypothesis. *Psychological Review,* 1964, **71**, 238–242. (b)

4.13

The test of significance in psychological research

David Bakan

The vast majority of investigations which pass for research in the field of psychology today entail the use of statistical tests of significance. Most characteristically, when a psychologist finds a problem he wishes to investigate he converts his intuitions and hypotheses into procedures which will yield a test of significance, and will characteristically allow the result of the test of significance to bear the essential responsibility for the conclusions he will draw.

I will attempt to show that the test of significance does not provide the information concerning psychological phenomena characteristically attributed to it; and that, furthermore, a great deal of mischief has been associated with its use. If the test of significance does not yield the expected information concerning the psychological phenomena under investigation, we may well speak of a crisis; for then a good deal of the research of the last several decades must be questioned. What will be said in this paper is hardly original. It is, in a certain sense, what "everybody knows." To say it "out loud" is, as it were, to assume the role of the child who pointed out that the emperor really had no clothes on. Little of what is contained in this paper is not already available in the literature, and the literature will be cited.

Lest what is being said here be misunderstood, some clarification needs to be made at the outset. It is not a blanket criticism of statistics, of mathematics, or, for that matter, even of the test of significance when it can be appropriately used, as in certain decision situations. The argument is, rather, that the test of significance has been carrying too much of the burden of scientific inference. It may well be the case that wise and ingenious investigators can find their way to reasonable conclusions from data because and in spite of their procedures. Too often, however,

David Bakan. "The Test of Significance in Psychological Research." (Chapter One), *On Method: Toward a Reconstruction of Psychological Investigation.* San Francisco: Jossey-Bass, Inc., Publishers, 1967, pp. 1–29.

even wise and ingenious investigators, for varieties of reasons not the least of which are the editorial policies of our major psychological journals, which we will discuss below, tend to credit the test of significance with properties it does not have.

The test of significance has as its aim obtaining information concerning a characteristic of a *population* which is itself not directly observable, whether for practical or more intrinsic reasons. What is observable is the *sample*. The work assigned to the test of significance is that of aiding in making inferences from the observed sample to the unobserved population.

The critical assumption involved in testing significance is that if the experiment is conducted properly, the characteristics of the population have a designably determinative influence on samples drawn from it; that, for example, the mean of a population has a determinative influence on the mean of a sample drawn from it. Thus if P, the population characteristic, has a determinative influence on S, the sample characteristic, then there is some license for making inferences from S to P.

If the determinative influence of P on S could be put in the form of simple logical implication, that P implies S, the problem would be quite simple. For, then we would have the simple situation: if P implies S, and if S is false, P is false. There are some limited instances in which this logic applies directly in sampling. For example, if the range of values in the population is between 3 and 9 (P), then the range of values in any sample must be between 3 and 9 (S). Should we find a value in a sample of, say, 10, it would mean that S is false; and we could assert that P is false.

It is clear from this, however, that, strictly speaking, one can only go from the denial of S to the denial of P; and not from the assertion of S to the assertion of P. It is within this context of simple logical implication that the Fisher school of statisticians have made important contributions—and it is extremely important to recognize this as the context.

In contrast, approaches based on the theorem of Bayes (Edwards, Lindman, & Savage, 1963; Keynes, 1948; Savage, 1954; Schlaifer, 1959) would allow inferences to P from S even when S is not denied, as S adding something to the credibility of P when S is found to be the case. One of the most viable alternatives to the use of the test of significance involves the theorem of Bayes; and the paper by Edwards et al. (1963) is particularly directed to the attention of psychologists for use in psychological research.

The notion of the null hypothesis[1] promoted by Fisher (1947) con-

---

[1] There is some confusion in the literature concerning the meaning of the term "null hypothesis." Fisher used the term to designate any exact hypothesis that we might be interested in disproving, and "null" was used in the sense of that which is to be nullified (see, for example, Berkson, 1942). It has, however, also been used

stituted an advance within this context of simple logical implication. It allowed experimenters to set up a null hypothesis complementary to the hypothesis that the investigator was interested in, and provided him with a way of positively confirming his hypothesis. Thus, for example, the investigator might have the hypothesis that, say, normals differ from schizophrenics. He would then set up the null hypothesis that the means in the population of all normals and all schizophrenics were equal. Thus, the rejection of the null hypothesis constituted a way of asserting that the means of the populations of normals and schizophrenics were different, a seemingly reasonable device whereby to affirm a logical antecedent.

The model of simple logical implication for making inferences from S to P has another difficulty which the Fisher approach sought to overcome. This is that it is rarely meaningful to set up any simple "P implies S" model for parameters that we are interested in. In the case of the mean, for example, it is rather that P has a determinative influence on the frequency of any specific S. But one experiment does not provide many values of S to allow the study of their frequencies. It gives us only one value of S. The sampling distribution is conceived which specifies the relative frequencies of all possible values of S. Then, with the help of an adopted level of significance, we could, in effect, say that S was false; that is, any S which fell in a region whose relative theoretical frequency under the null hypothesis was, say, 5 per cent would be considered false. If such an S actually occurred, we would be in a position to declare P to be false, still within the model of simple logical implication.

It is important to recognize that one of the essential features of the Fisher approach is what may be called the "once-ness" of the experiment; the inference model takes as critical that the experiment has been conducted once. If an S which has a low probability under the null hypothesis actually occurs, it is taken that the null hypothesis is false. As Fisher (1947) put it, why should the theoretically rare event under the null hypothesis actually occur to "us"? If it does occur, we take it that the null hypothesis is false. Basic is the idea that "the theoretically unusual does not happen to me."[2] It should be noted that the referent for all

---

to indicate a parameter of zero (see, for example, Lindquist, 1940): the difference between the population means is zero, or the correlation coefficient in the population is zero, the difference in proportions in the population is zero, etc. Since both meanings are usually intended in psychological research, it causes little difficulty.

[2] I playfully once conducted the following "experiment": Suppose, I said, that every coin has associated with it a "spirit"; and suppose, furthermore, that if the spirit is implored properly, the coin will veer head or tail as one requests of the spirit. I thus invoked the spirit to make the coin fall head. I threw it once; it came up head. I did it again; it came up head again. I did this six times, and got six heads. Under the null hypothesis the probability of occurrence of six heads is $(\frac{1}{2})^6 = .016$, significant at the 2 per cent level of significance. I have never repeated

probability considerations is neither in the population itself nor the subjective confidence of the investigator. It is rather in a hypothetical population of experiments all conducted in the same manner, but only one of which is actually conducted. Thus, of course, the probability of falsely rejecting the null hypothesis if it were true is exactly that value which has been taken as the level of significance. Replication of the experiment vitiates the validity of the inference model, unless the replication itself is taken into account in the model and the probabilities of the model modified accordingly (as is done in various designs which entail replication, where, however, the total experiment, including the replications, is again considered as *one* experiment). According to Fisher (1947), "it is an essential characteristic of experimentation that it is carried out with limited resources." In the Fisher approach, the "limited resources" is not only a making of the best out of a limited situation, but is rather an integral feature of the inference model itself. Lest he be done a complete injustice, it should be pointed out that he did say, "In relation to the test of significance, we may say that a phenomenon is experimentally demonstrable when we know how to conduct an experiment which will rarely fail to give us statistically significant results." However, although Fisher "himself" believes this, it is not built into the inference model.[3]

As already indicated, research workers in the field of psychology place a heavy burden on the test of significance. Let us consider some of the difficulties associated with the null hypothesis.

1. *The a priori reasons for believing that the null hypothesis is generally false anyway.* One of the common experiences of research workers is the very high frequency with which significant results are obtained with large samples. Some years ago, the author had occasion to run a number of tests of significance on a battery of tests collected on about 60,000 subjects from all over the United States. Every test came out significant. Dividing the cards by such arbitrary criteria as east versus west of the Mississippi River, Maine versus the rest of the country, North versus South, etc., all produced significant differences in means. In some instances, the differences in the sample means were quite small,

---

the experiment. But, then, the logic of the inference model does not really demand that I do! It may be objected that the coin, or my tossing, or even my observation was biased. But I submit that such things were in all likelihood not as involved in the result as corresponding things in most psychological research.

[3] Possibly not even this criterion is sound. It may be that a number of statistically significant results which are borderline "speak for the null hypothesis rather than against it" (Edwards et al., 1963). If the null hypothesis were really false, then with an increase in the number of instances in which it can be rejected, there should be some substantial proportion of more dramatic rejections rather than borderline rejections.

but nonetheless, the p values were all very low. Nunnally (1960) has reported a similar experience involving correlation coefficients on 700 subjects. Joseph Berkson (1938) made the observation almost 30 years age in connection with chi-square:

> I believe that an observant statistician who has had any considerable experience with applying the chi-square test repeatedly will agree with my statement that, as a matter of observation, when the numbers in the data are quite large, the P's tend to come out small. Having observed this, and on reflection, I make the following dogmatic statement, referring for illustration to the normal curve: "If the normal curve is fitted to a body of data representing any real observations whatever of quantities in the physical world, then if the number of observations is extremely large—for instance, on an order of 200,000— the chi-square P will be small beyond any usual limit of significance."
>
> This dogmatic statement is made on the basis of an extrapolation of the observation referred to and can also be defended as a prediction from *a priori* considerations. For we may assume that it is practically certain that any series of real observations does not actually follow a normal curve *with absolute exactitude* in all respects, and no matter how small the discrepancy between the normal curve and the true curve of observations, the chi-square P will be small if the sample has a sufficiently large number of observations in it.
>
> If this be so, then we have something here that is apt to trouble the conscience of a reflective statistician using the chi-square test. For I suppose it would be agreed by statisticians that a large sample is always better than a small sample. If, then, we know in advance the P that will result from an application of a chi-square test to a large sample, there would seem to be no use in doing it on a smaller one. But since the result of the former test is known, it is no test at all [pp. 526–527].

As one group of authors has put it, "in typical applications . . . the null hypothesis . . . is known by all concerned to be false from the outset" (Edwards *et al.*, 1963). The fact of the matter is that there is really no good reason to expect the null hypothesis to be true in any population. Why should the mean, say, of all scores east of the Mississippi be identical to all scores west of the Mississippi? Why should any correlation coefficient be exactly .00 in the population? Why should we expect the ratio of males to females to be exactly 50:50 in any population? Or why should different drugs have exactly the same effect on any population parameter (Smith, 1960)? A glance at any set of statistics on total populations will quickly confirm the rarity of the null hypothesis in nature.

The reason the null hypothesis is characteristically rejected with large samples was made patent by the theoretical work of Neyman and Pearson (1933). The probability of rejecting the null hypothesis is a function of five factors: whether the test is one-or two-tailed, the level of significance, the standard deviation, the amount of deviation from the null hypothesis, and the number of observations. The choice of a one- or two-tailed test is the investigator's; the level of significance is also based on the choice of the investigator; the standard deviation is a given of the situation and is characteristically reasonably well estimated; the deviation from the null hypothesis is what is unknown; and the choice of the number of cases in psychological work is characteristically arbitrary or expediential. Should there be any deviation from the null hypothesis in the population, no matter how small—and we have little doubt but that such a deviation usually exists—a sufficiently large number of observations will lead to the rejection of the null hypothesis. As Nunnally (1960) put it,

> If the null hypothesis is not rejected, it is usually because the N is too small. If enough data are gathered, the hypothesis will generally be rejected. If rejection of the null hypothesis were the real intention in psychological experiments, there usually would be no need to gather data [p. 643].

2. Type I error and publication practices. The Type I error is the error of rejecting the null hypothesis when it is indeed true, and its probability is the level of significance. Later in this paper we will discuss the distinction between sharp and loose null hypotheses. The sharp null hypothesis, which we have been discussing, is an exact value for the null hypothesis as, for example, the difference between population means being precisely zero. A loose null hypothesis is one in which it is conceived of as being "around" null. Sharp null hypotheses, as we have indicated, rarely exist in nature. Assuming that loose null hypotheses are not rare, and that their testing may make sense under some circumstances, let us consider the role of the publication practices of our journals in their connection.

It is the practice of editors of our psychological journals, receiving many more papers than they can possibly publish, to use the magnitude of the p values reported as one criterion for acceptance or rejection of a study. For example, consider the following statement made by Arthur W. Melton (1962) on completing twelve years as editor of the *Journal of Experimental Psychology*, certainly one of the most prestigious and scientifically meticulous psychological journals. In listing the criteria by which articles were evaluated, he said:

> The next step in the assessment of an article involved a judgment with respect to the confidence to be placed in the findings—confidence that the results of the experiment would be repeatable under the conditions described. In editing the *Journal* there has been a strong reluctance to accept and publish results related to the principal concern of the

research when those results were significant at the .05 level, whether by one- or two-tailed test. This has not implied a slavish worship of the .01 level, as some critics may have implied. Rather, it reflects a belief that it is the responsibility of the investigator in a science to reveal his effect in such a way that no reasonable man would be in a position to discredit the results by saying that they were the product of the way the ball bounces [pp. 553–554].

His clearly expressed opinion that nonsignificant results should not take up the space of the journals is shared by most editors of psychological journals. It is important to point out that I am not advocating a change in policy in this connection. In the total research enterprise where so much of the load for making inferences concerning the nature of phenomena is carried by the test of significance, the editors can do little else. The point is rather that the situation in regard to publication makes manifest the difficulties in connection with the overemphasis on the test of significance as a principal basis for making inferences.

McNemar (1960) has rightly pointed out that not only do journal editors reject papers in which the results are not significant, but that papers in which significance has not been obtained are not submitted, that investigators select out their significant findings for inclusion in their reports, and that theory-oriented research workers tend to discard data which do not work to confirm their theories. The result of all of this is that "published results are more likely to involve false rejection of null hypotheses than indicated by the stated levels of significance," that is, published results which are significant may well have Type I errors in them far in excess of, say, the 5 per cent which we may allow ourselves.

The suspicion that the Type I error may well be plaguing our literature is given confirmation in an analysis of articles published in the *Journal of Abnormal and Social Psychology* for one complete year (Cohen, 1962). Analyzing seventy studies in which significant results were obtained with respect to the power of the statistical tests used, Cohen found that power, the probability of rejecting the null hypothesis when the null hypothesis was false, was characteristically meager. Theoretically, with such tests, one should not often expect significant results even when the null hypothesis was false. Yet, there they were! Even if deviations from null existed in the relevant populations, the investigations were characteristically not powerful enough to have detected them. This strongly suggests that there is something additional associated with these rejections of the null hypotheses in question. It strongly points to the possibility that the manner in which studies get published is associated with the findings; that the very publication practices themselves are part and parcel of the probabilistic processes on which we base our conclusions concerning the nature of psychological phenomena. Our total research enterprise is, at least in part, a kind of scientific roulette, in which the "lucky," or constant player, "wins," that is, gets his paper or papers published. And certainly, going from 5 per cent to 1 per cent does not eliminate the possibility that it is "the way the ball bounces," to use Melton's phrase. It changes the odds in this roulette, but it does not make it less a game of roulette.

The damage to the scientific enterprise is compounded by the fact that the publication of "significant" results tends to stop further investigation. If the publication of papers containing Type I errors tended to foster further investigation so that the psychological phenomena with which we are concerned would be further probed by others, it would not be too bad. But it does not. Quite the contrary. As Lindquist (1940) has correctly pointed out, the danger to science of the Type I error is much more serious than the Type II error—for when a Type I error is committed, it has the effect of stopping investigation. A highly significant result appears definitive, as Melton's comments indicate. In the twelve years that he edited the *Journal of Experimental Psychology*, he sought to select papers which were worthy of being placed in the "archives," as he put it. Even the strict repetition of an experiment and not getting significance in the same way does not speak against the result already reported in the literature. For failing to get significance, speaking strictly within the inference model, only means that that experiment is inconclusive; whereas the study already reported in the literature, with a low p value, is regarded as conclusive. Thus we tend to place in the archives studies with a relatively high number of Type I errors, or, at any rate, studies which reflect small deviations from null in the respective populations; and we act in such a fashion as to reduce the likelihood of their correction. From time to time the suggestion has arisen that journals should open their pages for "negative results," so called. What is characteristically meant is that the null hypothesis has not been rejected at a conventional level of significance. This is hardly a solution to the problem simply because a failure to reject the null hypothesis is not a "negative result." It is only an instance in which the experiment is inconclusive.

To make this point clearer let us consider the odd case in which the null hypothesis may actually be true; say, the difference between means of a given measure of two identifiable groups in the population is precisely zero. Let us imagine that over the world there are one hundred experimenters who have independently embarked on testing this particular null hypothesis. By the theory under which the whole test of significance is conceived, approximately ninety-five of these experimenters would wind up by not being able to reject the null hypothesis, that is, their results would not be significant. It is not likely that they would write up their experiments and submit them to any journals. However, approximately five of these experimenters would find that their observed difference in means is significant at the 5 per cent level of significance. It is likely that they would write up their experiments and submit them for publication. Indeed, one might imagine interesting quarrels arising among them concerning priority of discovery, if the differences came out in the same direction, and controversy, if the differences came out in different directions. In the former instance, the psychological community might even take it as evidence of "replicability" of the phenomenon, in the latter instance as evidence that the scientific method is "self-corrective." The other ninety-five experimenters would wonder what they did wrong. And this is in the odd instance in which the true difference between means in the population is precisely zero!

The psychological literature is filled with misinterpretations of the nature of the test of significance. One may be tempted to attribute this to such things as lack of proper education, the simple fact that humans may err, and the prevailing tendency to take a cookbook approach in which the mathematical and philosophical framework out of which the tests of significance emerge are ignored; that, in other words, these misinterpretations are somehow the result of simple intellectual inadequacy on the part of psychologists. However, such an explanation is hardly tenable. Graduate schools are adamant with respect to statistical education. Any number of psychologists have taken out substantial amounts of time to equip themselves mathematically and philosophically. Psychologists as a group do a great deal of mutual criticism. Editorial reviews prior to publication are carried out with eminent conscientiousness. There is even a substantial literature devoted to various kinds of "misuse" of statistical procedures, to which not a little attention has been paid.

It is rather that the test of significance is profoundly interwoven with other strands of the psychological research enterprise in such a way that it constitutes a critical part of the total cultural-scientific tapestry. To pull out the strand of the test of significance would seem to make the whole tapestry fall apart. In the face of the intrinsic difficulties that the test of significance provides, we rather attempt to make an "adjustment" by attributing to the test of significance characteristics which it does not have, and overlook characteristics that it does have. The difficulty is that the test of significance can, especially when not considered too carefully, do *some* work; for, after all, the results of the test of significance *are* related to the phenomena in which we are interested. One may well ask whether we do not have here, perhaps, an instance of the phenomenon that learning under partial reinforcement is very highly resistant to extinction. Some of these misinterpretations are as follows:

1. *Taking the p value as a "measure" of significance.* A common misinterpretation of the test of significance is to regard it as a "measure" of significance. It is interpreted as the answer to the question "How significant is it?" A p value of .05 is thought of as less significant than a p value of .01, and so on. The characteristic practice on the part of

psychologists is to compute, say, a t, and then "look up" the significance in the table, taking the p value as a function of t, and thereby a "measure" of significance. Indeed, since the p value is inversely related to the magnitude of, say, the difference between means in the sample, it can function as a kind of "standard score" measure for a variety of different experiments. Mathematically, the t is actually very similar to a "standard score," entailing a deviation in the numerator, and a function of the variation in the denominator; and the p value is a "function" of t. If this use were explicit, it, would perhaps not be too bad. But it must be remembered that this is using the p value as a statistic descriptive of the sample alone, and does not automatically give an inference to the population. There is even the practice of using tests of significance in studies of total populations, in which the observations cannot by any stretch of the imagination be thought of as having been randomly selected from any designable population.[4] Using the p value in this way, in which the statistical inference model is even hinted at, is completely indefensible; for the single function of the statistical inference model is making inferences to populations from samples.

The practice of "looking up" the p value for the t, which has even been advocated in some of our statistical handbooks (e.g., Lacey, 1953; Underwood et al., 1954), rather than looking up the t for a given p value, violates the inference model. The inference model is based on the presumption that one initially adopts a level of significance as the specification of that probability which is too low to occur to "us," as Fisher has put it, in this one instance, and under the null hypothesis. A purist might speak of the "delicate problem . . . of fudging with a posteriori alpha values [levels of significance]" (Kaiser, 1960), as though the levels of significance were initially decided upon, but rarely do psychological research workers or editors take the level of significance as other than a "measure."

But taken as a "measure," it is only a measure of the sample. Psychologists often erroneously believe that the p value is "the probability that the results are due to chance," as Wilson (1961) has pointed out; that a p value of .05 means that the chances are .95 that the scientific hypothesis is correct, as Bolles (1962) has pointed out; that it is a measure of the power to "predict" the behavior of a population (Underwood et al., 1954); and that it is a measure of the "confidence that the results of the experiment would be repeatable under the conditions described," as Melton (1962) put it. Unfortunately, none of these interpretations are within the inference model of the test of significance. Some of our statistical handbooks have "allowed" misinterpretation.

---

[4] It was decided not to cite any specific studies to exemplify points such as this one. The reader will undoubtedly be able to supply them for himself.

For example, in discussing the erroneous rhetoric associated with talking of the "probability" of a population parameter (in the inference model there is no probability associated with something which is either true or false), Lindquist (1940) said, "For most practical purposes, the end result is the same as if the 'level of confidence' type of interpretation is employed." Ferguson (1959) wrote, "The .05 and .01 probability levels are descriptive of our degree of confidence." There is little question but that sizable differences, correlations, etc., in samples, especially samples of reasonable size, speak more strongly of sizable differences, correlations, etc., in the population; and there is little question but that if there is real and strong effect in the population, it will continue to manifest itself in further sampling. However, these are inferences which *we* may make. They are outside the inference model associated with the test of significance. The p value within the inference model is only the value which we take to be as how improbable an event could be under the null hypothesis, which we judge will not take place to "us," in this one experiment. It is not a "measure" of the goodness of the other inferences which we might make. It is an a priori condition that we set up whereby we decide whether or not we will reject the null hypothesis, not a measure of significance.

There is a study in the literature (Rosenthal & Gaito, 1963) which points up sharply the lack of understanding on the part of psychologists of the meaning of the test of significance. The subjects were nine members of the psychology department faculty, all holding doctoral degrees, and ten graduate students, at the University of North Dakota; and there is little reason to believe that this group of psychologists was more or less sophisticated than any other. They were asked to rate their degree of belief or confidence in results of hypothetical studies for a variety of p values, and for N's of 10 and 100. That there should be a relationship between the average rated confidence or belief and p value, as the found, is to be expected. What is shocking is that these psychologists indicated substantially greater confidence or belief in results associated with the larger sample size for the same p values. According to the theory, especially as this has been amplified by Neyman and Pearson (1933), the probability of rejecting the null hypothesis for any given deviation from null and p value increases as a function of the number of observations. The rejection of the null hypothesis when the number of cases is small speaks for a more dramatic effect in the population; and if the p value is the same, the probability of committing a Type I error remains the same. Thus one can be more confident with a small N than a large N. The question is, how could a group of psychologists be so wrong? I believe that this wrongness is based on the commonly held belief that the p value is a "measure" of degree of confidence. Thus, the reasoning behind such a wrong set of answers by these psychologists may well have been something like this: the p value is a measure of confidence; but a larger number of cases also increases confidence; therefore, for any given p value, the degree of confidence should be higher for the larger N. The wrong conclusion arises from the erroneous character of the first premise, and from the failure to recognize that the p value is a function of sample size for any given deviation from null in the population. The author knows of instances in which editors of very reputable psychological journals have rejected papers in which the p values and N's were small on the grounds that there were not enough observations, clearly demonstrating that the same mode of thought is operating in them. Indeed, rejecting the null hypothesis with a small N is indicative of a strong deviation from null in the population, the mathematics of the test of significance having already taken into account the smallness of the sample. Increasing the N increases the probability of rejecting the null hypothesis; and in these studies rejected for small sample size, that task has already been accomplished. These editors are, of course, in some sense the ultimate "teachers" of the profession; and they have been teaching something which is patently wrong.

   2. Automaticity of inference. What may be considered to be a dream, fantasy, or ideal in the culture of psychology is that of achieving complete automaticity of inference. The making of inductive generalizations is always somewhat risky. In Fisher's *The Design of Experiments* (1947), he made the claim that the methods of induction could be made rigorous, exemplified by the procedures which he was setting forth. This is indeed quite correct in the sense indicated earlier. In a later paper (Fisher, 1955), he made explicit what was strongly hinted at in his earlier writing, that the methods which he proposed constituted a relatively complete specification of the process of induction:

> That such a process induction existed and was possible to normal minds, has been understood for centuries; it is only with the recent development of statistical science that an analytic account can now be given, about as satisfying and complete, at least, as that given traditionally of the deductive processes [p. 74].

Psychologists certainly took the procedures associated with the t test, F test, and so on, in this manner. Instead of having to engage in inference themselves, they had but to "run the tests" for the purpose of making inferences, since, as it appeared, the statistical tests were analytic analogues of inductive inference. The "operationist" orientation among psychologists, which recognized the contingency of knowledge on the knowledge-getting operations and advocated their specification could, it would seem, "operationalize" the inferential processes simply by reporting the details of the statistical analysis. It thus removed the burden of responsibility, the chance of being wrong, the necessity for making inductive inferences, from the shoulders of the investigator and placed them on the tests of significance. The contingency of the conclusion upon the experimenter's decision of the level of significance was managed in two ways. The first, by resting on a kind of social agreement that 5 per cent was good, and 1 per cent better. The second in the manner which has already been discussed, by not making a decision of the level of significance, but only reporting the p value as a "result" and a presumably objective "measure" of degree of confidence. But that the probability of getting significance is also contingent upon the number of observations has been handled largely by ignoring it.

A crisis was experienced among psychologists when the matter of the one- versus the two-tailed test came into prominence; for here the contingency of the result of a test of significance on a decision of the investigator was simply too conspicuous to be ignored. An investigator say, was interested in the difference between two groups on some measure. He collected his data, found that Mean A was greater than Mean B in the sample, and ran the ordinary two-tailed t test; and, let us say, it was not significant. Then he bethought himself. The two-tailed test tested against *two* alternatives, that the population Mean A was greater than population Mean B and vice versa. But then, he really wanted to know whether Mean A was greater than Mean B. Thus, he could run a one-tailed test. He did this and found, since the one-tailed test is more powerful, that his difference was now significant.

Now here there was a difficulty. The test of significance is not nearly so automatic an inference process as had been thought. It is manifestly contingent on the decision of the investigator as to whether to run a one- or a two-tailed test. And, somehow, making the decision *after* the data were collected and the means computed seemed like "cheating." How should this be handled? Should there be some central registry in which one registers one's decision to run a one- or two-tailed test before collecting the data? Should one, as one eminent psychologist once suggested to me, send oneself a letter so that the postmark would prove that one had predecided to run a one-tailed test? The literature on ways of handling this difficulty has grown quite a bit in the strain to overcome somehow this particular clear contingency of the results of a test of significance on the decision of the investigator. The author will not attempt here to review this literature, except to cite one very competent paper which points up the intrinsic difficulty associated with this problem, the *reductio ad absurdum* to which one comes. Kaiser (1960), early in his paper, distinguished between the logic associated with the test of significance and other forms of inference, a distinction which, incidentally, Fisher would hardly have allowed: "The arguments developed in this paper are based on logical considerations in statistical inference. (We do not, of course, suggest that statistical inference is the only basis for scientific inference.)" But then, having taken the position that he is going to follow the logic of statistical inference relentlessly, he said (Kaiser's italics): *"we cannot logically make a directional statistical decision or statement when the null hypothesis is rejected on the basis of the direction of the difference in the observed sample means."* One really needs to strike oneself in the head! If sample Mean A is greater than Sample Mean B, and there is reason to reject the null hypothesis, in what other direction can it reasonably be? What kind of logic is it that leads one to believe that it could be otherwise than that Population Mean A is greater than Population Mean B? We do not know whether Kaiser intended his paper as a *reductio ad absurdum,* but it certainly turned out that way.

The issue of the one- versus the two-tailed test genuinely challenges the presumptive "objectivity" characteristically attributed to the test of significance. On the one hand, it makes patent what was the case under any circumstances (at the least in the choice of level of significance, and the choice of the number of cases in the sample), that the conclusion is contingent upon the decision of the investigator. An astute investigator, who foresaw the results, and who therefore predecided to use a one-tailed test, will get one p value. The less astute but honorable investigator, who did not forsee the results, would feel obliged to use a two-tailed test, and would get another p value. On the other hand, if one decides to be relentlessly logical within the logic of statistical inference, one winds up with the kind of absurdity which we have cited above.

3. *The confusion of induction to the aggregate with induction to the general* (see pages 30–36 below). Consider a not atypical investigation of the following sort: A group of, say, twenty normals and a group of, say, twenty schizophrenics are given a test. The tests are scored, and a t test is run, and it is found that the means differ significantly at some level of significance, say 1 per cent. What inference can be drawn? As we have already indicated, the investigator could have insured this result by choosing a sufficiently large number of cases. Suppose we overlook this objection, which we can to some extent, by saying that the difference between the means in the population must have been large enough to have manifested itself with only forty cases. But still, what do we know from this? The only inference which this allows is that the mean of all normals is different from the mean of all schizophrenics in the populations from which the samples have presumably been drawn at random. (Rarely is the criterion of randomness satisfied. But let us overlook this objection too.)

The common rhetoric in which such results are discussed is in the form "Schizophrenics differ from normals in such and such ways." The sense that both the reader and the writer have of this rhetoric is that it has been justified by the finding of significance. Yet clearly it does not mean *all* schizophrenics and *all* normals. All that the test of significance justifies is that measures of central tendency of the aggregates differ in the populations. The test of significance has *not* addressed itself to anything about the schizophrenia or normality which characterizes *each* member of the respective populations. Now it is certainly possible for an investigator to develop a hypothesis about the nature of schizophrenia from which he may infer that there should be differences between the means in the populations; and his finding of a significant difference in the means of his sample would add to the credibility of the former. However, that 1 per cent which he obtained in his study bears only on the means of the populations and is not a "measure" of the confidence that he may have in his hypothesis concerning the nature of schizophrenia. There are two inferences that he must make. One is that of the sample to the population, for which the test of significance is of some use. The other is from his inference concerning the population to his hypothesis concerning the nature of schizophrenia. The p value does not bear on this second inference. The psychological literature is filled with assertions which confound these two inferential processes.

Or consider another hardly atypical style of research. Say an experimenter divides forty subjects at random into two groups of twenty subjects each. One group is assigned to one condition and the other to another condition, perhaps, say, massing and distribution of trials. The subjects are given a learning task, one group under massed conditions, the other under distributed conditions. The experimenter runs a t test on the learning measure and again, say, finds that the difference is significant at the 1 per cent level of significance. He may then say in his report, being more careful than the psychologist who was studying the difference between normals and schizophrenics (being more "scientific" than his clinically interested colleague), that "the mean in the population of learning under massed conditions is lower than the mean in the population of learning under distributed conditions," feeling that he can say this with a good deal of certainty because of his test of significance. But here too (like his clinical colleague) he has made two inferences, and not one, and the 1 per cent bears on the one but not the other. The statistical inference model certainly allows him to make his statement for the population, but only for *that* learning task, and the p value is appropriate only to that. But the generalization to "massed conditions" and "distributed conditions" beyond that particular learning task is a second inference with respect to which the p value is not relevant. The psychological literature is plagued with any number of instances in which the rhetoric indicates that the p value does bear on this second inference.

Part of the blame for this confusion can be ascribed to Fisher who, in *The Design of Experiments* (1947), suggested that the mathematical methods which he proposed were exhaustive of scientific induction, and that the principles he was advancing were "common to all experimentation." What he failed to see and to say was that after an inference was made concerning a population parameter, one still needed to engage in induction to obtain meaningful scientific propositions.

To regard the methods of statistical inference as exhaustive of the inductive inferences called for in experimentation is completely confounding. When the test of significance has been run, the necessity for induction has hardly been completely satisfied. However, the research worker knows this, in some sense, and proceeds, as he should, to make further inductive inferences. He is, however, still ensnarled in his test of significance and the presumption that it is the whole of his inductive activity, and thus mistakenly takes a low p value for the measure of the validity of his other inductions.

The seriousness of this confusion may be seen by again referring back to the Rosenthal and Gaito study and the remark by Berkson which indicate that research workers believe that a large sample is better than a small sample. We need to refine the rhetoric somewhat. Induction consists in making inferences from the particular to the general. It is certainly the case that, as confirming particulars are added, the credibility of the general is increased. However, the addition of observations to a sample is, in the context of statistical inference, not the addition of particulars but the modification of what is one particular in the inference model, the sample aggregate. In the context of statistical inference, it is not necessarily true that "a large sample is better than a small sample." For, as has been already indicated, obtaining a significant result with a small sample suggests a larger deviation from null in the population, and may be considerably more meaningful. Thus more particulars are better than fewer particular in the making of an inductive inference; but not necessarily a larger sample.

In the marriage of psychological research and statistical inference, psychology brought its own reasons for accepting this confusion, reasons which inhere in the history of psychology. Measurement psychology arises out of two radically different traditions, as has been pointed out by Guilford (1936) and Cronbach (1957), and the matter of putting them together raised certain difficulties. The one tradition seeks to find propositions concerning the nature of man in general—propositions of a general nature, with each individual a particular in which the general is manifest. This is the kind of psychology associated with the traditional experimental psychology of Fechner, Ebbinghaus, Wundt, and Titchener. It seeks to find the laws which characterize the "generalized, normal, human, adult mind" (Boring, 1950). The research strategy associated with this kind of psychology is straightforwardly inductive. It seeks inductive generalizations which will apply to every member of a designated class.

A single particular in which a generalization fails forces a rejection of the generalization, calling for either a redefinition of the class to which it applies or a modification of the generalization. The other tradition is the psychology of individual differences, which has its roots more in England and the United States than on the Continent. We may recall that when the young American, James McKeen Cattell, who invented the term "mental test," came to Wundt with his own problem of individual differences, it was regarded by Wundt as *ganz Amerkanisch* (Boring, 1950).

The basic datum for an individual-differences approach is not anything that characterizes each of two subjects, but the difference between them. For this latter tradition, it is the aggregate which is of interest, and not the general. One of the most unfortunate characteristics of many studies in psychology, especially in experimental psychology, is that the data are treated as aggregates while the experimenter is trying to infer general propositions. There is hardly an issue of most of the major psychological journals reporting experimentation in which this confusion does not appear several times, and in which the test of significance, which has some value in connection with the study of aggregates, is not interpreted as a measure of the credibility of the general proposition in which the investigator is interested. Roberts and Wist examined sixty articles from psychological literature from the point of view of the aggregate-general distinction. In twenty-five of the articles it was unambiguous that the authors had drawn general-type conclusions from aggregate-type data.

Thus, what took place historically in psychology is that instead of attempting to synthesize the two traditional approaches to psychological phenomena, which is both possible and desirable, a syncretic combination took place of the methods appropriate to the study of aggregates with the aims of a psychology which sought for general propositions. One of the most overworked terms, which added not a little to the essential confusion, was "error," which was a kind of umbrella term for (at the least) variation among scores from different individuals, variation among measurements for the same individual, and variation among samples.

Let us add another historical note. In 1936, Guilford published his well-known *Psychometric Methods*. In this book, which became a kind of "bible" for many psychologists, he made a noble effort at a "Rapprochement of Psychophysical and Test Methods." He observed, quite properly, that mathematical developments in each of the two fields might be of value in the other, that "Both psychophysics and mental testing have rested upon the same fundamental statistical devices." There is no question of the truth of this. However, what he failed to emphasize sufficiently was that mathematics is so abstract that the same mathematics is applicable to rather different fields of investigation without there being any necessary further identity between them. (One would not, for example, argue that business and genetics are essentially the same because the same arithmetic is applicable to market research and in the investigation of the facts of heredity.) A critical point of contact between the two traditions was in connection with scaling, in which Cattell's principle that "equally often noticed differences are equal unless always or never noticed" (Guilford, 1936) was adopted as a fundamental assumption. The "equally often noticed differences" is, of course, based on aggregates. By means of this assumption, one could collapse the distinction between the two areas of investigation. Indeed, this is not really too bad if one is alert to the fact that it is an assumption, one which even has considerable pragmatic value. As a set of techniques whereby data could be analyzed, that is, as a set of techniques whereby one could describe one's findings, and then make inductions about the nature of the psychological phenomena, what Guilford put together in his book was eminently valuable. However, around this time the work of Fisher and his school was coming to the attention of psychologists. It was attractive for several reasons. It offered advice for handling "small samples." It offered a number of eminently ingenious new ways of organizing and extracting information from data. It offered ways by which several variables could be analyzed simultaneously, away from the old notion that one had to keep everything constant and vary only one variable at a time. It showed how the effect of the "interaction" of variables could be assessed. But it also claimed to have mathematized induction! The Fisher approach was thus "bought," and psychologists got a theory of induction in the bargain, a theory which seemed to exhaust the inductive processes. Whereas the question of the "reliability" of statistics had been a matter of concern for some time before (although frequently very garbled), it had not carried the burden of induction to the degree that it did with the Fisher approach. With the acceptance of the Fisher approach the psychological research worker also accepted, and then overused, the test of significance, employing it as the measure of the significance, in the largest sense of the word, of his research efforts.

Earlier, a distinction was made between sharp and loose null hypotheses. One of the major difficulties associated with the Fisher approach is the problem presented by sharp null hypotheses; for, as we have already seen, there is reason to believe that the existence of sharp null hypotheses is characteristically unlikely. There have been some efforts to correct for this difficulty by proposing the use of loose null hypotheses; in place of a single point, a region being considered null. Hodges and Lehmann (1954) have proposed a distinction between "statistical significance," which entails the sharp hypothesis, and "material significance," in which one tests the hypothesis of a deviation of a stated amount from the null point instead of the null point itself. Edwards (1950) has suggested the notion of "practical significance" in which one takes into account the meaning, in some practical sense, of the magnitude of the deviation from null together with the number of observations which have been involved in getting statistical significance. Binder (1963) has equally argued that a subset of parameters be equated with the null hypothesis. Especially what has been suggested is that the investigator make some kind of a decision concerning "How much, say, of a difference makes a difference?" The difficulty with this solution, which is certainly a sound one technically, is that in psychological research we do not often have very good grounds for answering this question. This is partly due to the inadequacies of psychological measurement, but mostly due to the fact that the answer to the question of "How much of a difference makes a difference?" is not forthcoming outside of some particular practical context. The question calls forth another question, "How much of a difference makes a difference *for what?*"

This brings us to one of the major issues within the field of statistics itself. The problems of the research psychologist do not generally lie within practical contexts. He is rather interested in making assertions concerning psychological functions which have a reasonable amount of credibility associated with them. He is more concerned with "What is the case?" than with "What is wise to do?" (see Rozeboom, 1960).

It is here that the decision-theory approach of Neyman, Pearson, and Wald (Neyman, 1937, 1957; Neyman & Pearson, 1933; Wald, 1939, 1950, 1955) becomes relevant. The decision-theory school, still basing itself on some basic notions of the Fisher approach, deviated from it in several respects:

1. In Fisher's inference model, the two alternatives between which one chose on the basis of an experiment were "reject" and "inconclusive." As he said in *The Design of Experiments* (1947), "the null hypothesis is never proved or established, but is possibly disproved, in the course of experimentation." In the decision-theory approach, the two alternatives are rather "reject" and "accept."

2. Whereas in the Fisher approach the interpretation of the test of significance critically depends on having one sample from a hypothetical population of experiments, the decision-theory approach conceives of, is applicable to, and is sensible with respect to numerous repetitions of the experiment.

3. The decision-theory approach added the notions of the Type II error (which can be made only if the null hypothesis is accepted) and power as significant features of their model.

4. The decision-theory model gave a significant place to the matter of what is concretely lost if an error is made in the practical context, on the presumption that "accept" entailed one concrete action, and "reject" another. It is in these actions and their consequences that there is a basis for deciding on a level of confidence. The Fisher approach has little to say about the consequences.

As it has turned out, the field of application par excellence for the decision-theory approach has been the sampling inspection of mass-produced items. In sampling inspection, the acceptable deviation from null can be specified; both "accept" and "reject" are appropriate categories; the alternative courses of action can be clearly specified; there is a definite measure of loss for each possible action; and the choice can be regarded as one of a series of such choices, so that one can minimize the over-all loss (see Barnard, 1954). Where the aim is only the acquisition of knowledge without regard to a specific practical context, these conditions do not often prevail. Many psychologists who learned about analysis of variance from books such as those by Snedecor (1946) found the examples involving hog weights, etc., somewhat annoying. The decision-theory school makes it clear that such practical contexts are not only "examples" given for pedagogical purposes, but actually are essential features of the methods themselves.

The contributions of the decision-theory school essentially revealed the intrinsic nature of the test of significance beyond that seen by Fisher and his colleagues. They demonstrated that the methods associated with the test of significance constitute not an assertion, or an induction, or a conclusion calculus, but a decision-or risk-evaluation calculus. Fisher (1955) has reacted to the decision-theory approach in polemic style, suggesting that its advocates were like "Russians [who] are made familiar with the ideal that research in pure science can and should be geared to technological performance, in the comprehensive organized effort of a five-year plan for the nation." He also suggested an American "ideological" orientation: "In the U. S. also the great importance of organized technology has I think made it easy to confuse the process appropriate for drawing correct conclusions, with those aimed rather at, let us say, speeding production, or saving money."[5] But perhaps a more reasonable way of looking at this is to regard the decision-theory school to have explicated what was already implicit in the work of the Fisher school.

What then is our alternative, if the test of significance is really of such limited appropriateness? At the very least it would appear that we would be much better off if we were to attempt to estimate the magnitude of the parameters in the populations; and recognize that we then need to make other inferences concerning the psychological phenomena which may be manifesting themselves in these magnitudes. In terms of a statistical approach which is an alternative, the various methods associated with the theorem of Bayes, referred to earlier, may be appropriate; and the paper by Edwards et al. (1963) and the book by Schlaifer (1959) are good starting points. However, what is expressed in the theorem of Bayes alludes to the more general process of inducing propositions concerning

[5] For a reply to Fisher, see Pearson (1955).

the nonmanifest (which is what the population is a special instance of) and ascertaining the way in which what is manifest (of which the sample is a special instance) bears on it. This is what the scientific method has been about for centuries. However, if the reader who might be sympathetic to the considerations set forth in this paper quickly goes out and reads some of the material on the Bayesian approach with the hope that thereby he will find a new basis for automatic inference, this paper will have misfired, and he will be disappointed.

What we have indicated in this paper in connection with the test of significance in psychological research may be taken as an instance of a kind of essential mindlessness in the conduct of research which may be related to the presumption of the nonexistence of mind in the subjects of psychological research. Karl Pearson once indicated that higher statistics was only common sense reduced to numerical appreciation. However, that base in common sense must be maintained with vigilance. When we reach a point where our statistical procedures are substitutes instead of aids to thought, and we are led to absurdities, then we must return to common sense. Tukey (1962) has very properly pointed out that statistical procedures may take our attention away from the data, which constitute the ultimate base for any inferences which we might make. Schlaifer (1959) has dubbed the error of the misapplication of statistical procedures the "error of the third kind," the most serious error which can be made. Berkson has suggested the use of "the interocular traumatic test, you know what the data mean when the conclusion hits you between the eyes" (Edwards et al., 1963). We must overcome the myth that if our treatment of our subject matter is mathematical it is therefore precise and valid. We need to overcome the handicap associated with limited competence in mathematics, a competence that makes it possible for us to run tests of significance while it intimidates us with a vision of greater mathematical competence if only one could reach up to it. Mathematics can serve to obscure as well as reveal.

Most important, we need to get on with the business of generating psychological hypotheses and proceed to do investigations and make inferences which bear on them, instead of, as so much of our literature would attest, testing the statistical null hypothesis in any number of contexts in which we have every reason to suppose that it is false in the first place.

References

Barnard, G. A. Sampling inspection and statistical decisions. *Journal of the Royal Statistical Society* (B), 1954, **16**, 151–165.
Berkson, J. Some difficulties of interpretation encountered in the application of

the chi-square test. *Journal of the American Statistical Association,* 1938, **33**, 526–542.

Berkson, J. Tests of significance considered as evidence. *Journal of the American Statistical Association,* 1942, **37**, 325–335.

Binder, A. Further considerations on testing the null hypothesis and the strategy and tactics of investigating theoretical models. *Psychological Review,* 1963, **70**, 101–109.

Bolles, R. C. The difference between statistical hypotheses and scientific hypotheses. *Psychological Reports,* 1962, **11**, 639–645.

Boring, E. G. *A history of experimental psychology.* (2nd ed.) New York: Appleton-Century-Crofts, 1950.

Cohen, J. The statistical power of abnormal-social psychological research: A review. *Journal of Abnormal and Social Psychology,* 1962, **65**, 145–153.

Cronbach, L. J. The two disciplines of scientific pyschology. *American Psychologist,* 1957, **12**, 671–684.

Edwards, A. L. *Experimental design in psychological research.* New York: Holt, Rinehart and Winston, Inc., 1950.

Edwards, W., Lindman, H., & Savage, L. J. Bayesian statistical inference for psychological research. *Psychological Review,* 1963, **70**, 193–242.

Ferguson, L. *Statistical analysis in psychology and education.* New York: McGraw-Hill, 1959.

Fisher, R. A. *The design of experiments.* (4th ed.) Edinburgh: Oliver & Boyd, 1947.

Fisher, R. A. Statistical methods and scientific induction. *Journal of the Royal Statistical Society* (B), 1955, **17**, 69–78.

Guilford, J. P. *Psychometric methods.* New York: McGraw-Hill, 1936.

Hodges, J. L., & Lehman, E. L. Testing the approximate validity of statistical hypotheses. *Journal of the Royal Statistical Society* (B), 1954, **16**, 261–268.

Kaiser, H. F. Directional statistical decision. *Psychological Review,* 1960, **67**, 160–167.

Keynes, J. M. *A treatise on probability.* London: Macmillan, 1948.

Lacey, O. L. *Statistical methods in experimentation.* New York: Macmillan, 1953.

Lindquist, E. F. *Statistical analysis in educational research.* Boston: Houghton Mifflin, 1940.

McNemar, Q. At random: Sense and nonsense. *American Psychologist,* 1960, **15**, 295–300.

Melton, A. W. Editorial. *Journal of Experimental Psychology,* 1962, **64**, 553–557.

Neyman, J. Outline of a theory of statistical estimation based on the classical theory of probability. *Philosophical Transactions of the Royal Society* (A), 1937, **236**, 333–380.

Neyman, J. "Inductive behavior" as a basic concept of philosophy of science. *Review of the Mathematical Statistics Institute,* 1957, **25**, 7–22.

Neyman, J., & Pearson, E. S. On the problem of the most efficient tests of statistical hypotheses. *Philosophical Transactions of the Royal Society* (A), 1933, **231**, 289–337.

Nunnally, J. The place of statistics in psychology. *Education and Psychological Measurement,* 1960, **20,** 641–650.

Pearson, E. S. Statistical concepts in their relation to reality. *Journal of the Royal Statistical Society* (B), 1955, **17,** 204–207.

Rosenthal, R., & Gaito, J. The interpretation of levels of significance by psychological researchers. *Journal of Psychology,* 1963, **55,** 33–38.

Rozeboom, W. W. The fallacy of the null-hypothesis significance test. *Psychological Bulletin,* 1960, **57,** 416–428.

Savage, L. J. *The foundations of statistics.* New York: Wiley, 1954.

Schlaifer, R. *Probability and statistics for business decisions.* New York: McGraw-Hill, 1959.

Smith, C. A. B. Review of N. T. J. Bailey, *Statistical methods in biology. Applied Statistics,* 1960, **9,** 64–66.

Snedecor, G. W. *Statistical methods.* (4th ed.; orig. publ. 1937) Ames, Iowa: Iowa State College Press, 1946.

Tukey, J. W. The future of data analysis. *Annals of Mathematical Statistics,* 1962, **33,** 1–67.

Underwood, B. J., Duncan, C. P., Taylor, J. A., & Cotton, J. W. *Elementary statistics.* New York: Appleton-Century-Crofts, 1954.

Wald, A. Contributions to the theory of statistical estimation and testing hypotheses. *Annals of Mathematical Statistics,* 1939, **10,** 299–326.

Wald, A. *Statistical decision functions.* New York: Wiley, 1950.

Wald, A. *Selected papers in statistics and probability.* New York: McGraw-Hill, 1955.

Wilson, K. V. Subjectivist statistics for the current crisis. *Contemporary Psychology,* 1961, **6,** 229–231.

4.14

The effects of violations of assumptions underlying the *t* test

C. Alan Boneau[1]

As psychologists who perform in a research capacity are well aware, psychological data too frequently have an exasperating tendency to manifest themselves in a form which violates one or more of the assumptions underlying the usual statistical tests of significance. Faced with the problem of analyzing such data, the researcher usually attempts to transform them in such a way that the assumptions are tenable, or he may look elsewhere for a statistical test. The latter alternative has become popular because of the proliferation of the so-called nonparametric or distribution-free methods. These techniques quite generally, however, couple their freedom from restricting assumptions with a disdain for much of the information contained within the data. For example, by classifying scores into groups above and below the median one ignores the fact that there are intracategory differences between the individual scores. As a result, tests which make no assumptions about the distribution from which one is sampling will tend not to reject the null hypothesis when it is actually false as often as will those tests which do make assumptions. This lack of power of the nonparametric tests is a decided handicap when, as is frequently the case in psychological research, a modicum of reinforcement in the form of an occasional significant result is required to maintain the research response.

C. Alan Boneau, "The Effects of Violations of Assumptions Underlying the *t* Test," *Psychological Bulletin*, **57**, 1960, 49–64. Copyright (1960) by the American Psychological Association, and reproduced by permission.

[1] This project was undertaken while the author was a Public Health Service Research Fellow of the National Institute of Mental Health at Duke University. The computations involved in this study were performed in the Duke University Digital Computing Laboratory which is supported in part by National Science Foundation Grant G-6694. The author wishes to express his appreciation to Thomas M. Gallie, Director of the Laboratory, for his cooperation and assistance.

Confronted with this discouraging prospect and a perhaps equally discouraging one of laboriously transforming data, performing related tests, and then perhaps having difficulty in interpreting results, the researcher is often tempted simply to ignore such considerations and go ahead and run a t test or analysis of variance. In most cases, he is deterred by the feeling that such a procedure will not solve the problem. If a significant result is forthcoming, is it due to differences between means, or is it due to the violation of assumptions? The latter possibility is usually sufficient to preclude the use of the t or F test.

It might be suspected that one could finesse the whole problem of untenable assumptions by better planning of the experiment or by a more judicious choice of variables, but this may not always be the case. Let us examine the assumptions more closely. It will be recalled that both the t test and the closely related F test of analysis of variance are predicated on sampling from a normal distribution. A second assumption required by the derivations is that the variances of the distributions from which the samples have been taken is the same (assumption of homogeneity of variance). Thirdly, it is necessary that scores used in the test exhibit independent errors. The third assumption is usually not restrictive since the researcher can readily conduct most psychological research so that this requirement is satisfied. The first two assumptions depend for their reasonableness in part upon the vagaries inherent in empirical data and the chance shape of the sampling distribution. Certain situations also arise frequently which tend to produce results having intrinsic non-normality or heterogeneity of variance. For example, early in a paired-associate learning task, before much learning has taken place, the modal number of responses for a group will be close to zero and any deviations will be in an upward direction. The distribution of responses will be skewed and will have a small variance. With a medium number of trials, scores will tend to be spread over the whole possible range with a mode at the center, a more nearly normal distribution than before, but with greater variance. When the task has been learned by most of the group, the distribution will be skewed downward and with smaller variance. In this particular case, one would probably more closely approximate normality and homogeneity in the data by using some other measure, perhaps number of trials for mastery. In many situations this option may not be present.

There is, however, evidence that the ordinary t and F tests are nearly immune to violation of assumptions or can easily be made so if precautions are taken (Pearson, 1931; Bartlett, 1935; Welch, 1937; Daniels, 1938; Quensel, 1947; Gayen, 1950a, 1950b; David & Johnson, 1951; Horsnell, 1953; Box, 1954a, 1954b; Box & Anderson, 1955). Journeyman psychologists have been apprised of this possibility by Lindquist (1953) who summarizes the results of a study by Norton (1951). Norton's technique was to obtain samples of *F*s by means of a random sampling procedure from distributions having the same mean but which violated the assumptions of normality and homogeneity of variance in predetermined fashions. As a measure of the effect of the violations, Norton determined the obtained percentage of sample *F*s which exceeded the theoretical 5% and 1% values from the *F* tables for various conditions. If the null hypothesis is true, and if the assumptions are met, the theoretical values are *F* values which would be exceeded by chance exactly 5% or 1% of the time. The discrepancy between these expected percentages and the obtained percentages is one useful measure of the effects of the violations.

Norton's results may be summarized briefly as follows: (*a*) When the samples all came from the same population, the shape of the distribution had very little effect on the percentage of *F* ratios exceeding the theoretical limits. For example, for the 5% level, the percentages exceeding the theoretical limits were 7.83% for a leptokurtic population as one extreme discrepancy and 4.76% for an extremely skewed distribution as another. (*b*) For sampling from populations having the same shape but different variances, or having different shapes but the same variance, there was little effect on the empirical percentage exceeding theoretical limits, the average being between 6.5% and 7.0%. (*c*) For sampling from populations with different shapes and heterogeneous variances, a serious discrepancy between theoretical and obtained percentages occurred in some instances. On the basis of these results, Lindquist (1953, p. 86) concluded that "unless the heterogeneity of either form or variance is so extreme as to be readily apparent upon inspection of the data, the effect upon the *F* distribution will probably be negligible."

This conclusion has apparently had surprisingly little effect upon the statistical habits of research workers (or perhaps editors) as is evident from the increasing reliance upon the less powerful nonparametric techniques in published reports. The purpose of this paper is to expound further the invulnerability of the *t* test and its next of kin the *F* test to ordinary onslaughts stemming from violation of the assumptions of normality and homogeneity. In part, this will be done by reporting results of a study conducted by the author dealing with the effect on the *t* test of violation of assumptions. In addition, supporting evidence from a mathematical framework will be used to bolster the argument.

To temper any imputed dogmatism in the foregoing, it should be emphasized that there are certain restrictions which preclude an automatic utilization of the *t* and *F* tests without regard for assumptions even when these tests are otherwise applicable. It is apparent, for example, that the violation of the homogeneity of variance assumption is drastically disturbing to the distribution of *t*'s and *F*'s if the sample sizes are not the same for all groups, a possibility which was not considered in the Norton study. It also seems clear that in cases of extreme violations, one must have a sample size large enough to allow the statistical effects of averaging to come into play. The need for such considerations will be made apparent in the ensuing discussion. There is abundant evidence, however, that both the t and the F tests are much less affected by extreme violations of the assumptions than has been generally realized.

A SAMPLING EXPERIMENT

At this point we will concern ourselves with the statement of the results of a random sampling study. The procedure is one of computing a large number of t values, each based upon samples drawn at random from distributions having specified characteristics, and constructing a frequency distribution of the obtained t's.

To summarize, random samples were drawn from populations which were either normal, rectangular, or exponential with means equal to 0 and variances of 1 or 4. For several combinations of forms and variances, t tests of the significance of the difference between sample means were computed using combinations of the sample sizes 5 and 15. For each of these combinations, frequency distributions of sample t's were obtained on the IBM 650 Electronic Computer.

RESULTS

The results of the sampling study will be presented in part as a series of frequency distributions in the form of bar graphs of the obtained distribution of t's for a particular condition. Upon these have been superimposed the theoretical t distribution curve for the appropriate degrees of freedom. This furnishes a rapid comparison of the extent to which the empirical distribution conforms to the theoretical.

First we shall consider those combinations possible when both of the samples are from normal distributions but variances and sample sizes may vary. Next will be considered the results of sampling from non-normal distributions, but both samples are from the same type of distribution. Finally we deal with the results of sampling from two different kinds of populations, for example, one sample from the normal distribution, and another from the exponential.

Potentially, a very large number of such combinations are possible. Limitations of the time available on the computer necessitated a paring down to a reasonable number. Although the computer is relatively fast when optimally programmed, it nevertheless required almost an hour, on the average, to complete a frequency distribution of 1000 t's. The combinations presented here are those which seemed most important at the time the study was made.

As a measure of the effect of violation of assumptions, the percentage of obtained *t*'s which exceed the theoretical values delineating the middle 95% of the *t* distribution is used. For 8, 18, and 28 *df* which arise in the present study, the corresponding values are respectively ±2.262, ±2.101, and ±2.048. If the assumptions are met, and if the null hypothesis of equality of means is true, 5% of the obtained *t*'s should fall outside these limits. The difference between this nominal value and the actual value obtained by sampling should be a useful measure of the degree to which violation of assumptions changes the distribution of *t* scores. There is, of course, a random quality to the obtained percentage of *t*'s falling outside the theoretical limits. Hence, the obtained value should be looked upon as an approximation to the true value which should lie nearby.

In the figures and in the text, the various combinations of population, variance, and sample size will be represented symbolically in the following form: $E(0, 1)5\text{-}N(0, 4)15$. Here the letters E, N, and R refer to the population from which the sample was drawn, E for exponential, N for normal, and R for rectangular. The first number in the parenthesis is the mean of the population distribution, in all cases zero, while the second number is the variance. The number following the parenthesis is the sample size for that particular sample. In the example above, the first sample is of Size 5 from an exponential distribution having a variance of 1. The second sample is from a normal distribution with variance of 4 and the sample size is 15.

Sampling from Normal Distributions

In order to justify the random sampling approach utilized in this study, and partly to confirm the faith placed in the tabled values of the mathematical statisticians, the initial comparisons are between the theoretical distributions and the obtained distributions with assumptions inviolate. Figures 1 and 2 exhibit the empirical distributions of *t*'s when both samples are taken from the same normal distribution with zero mean and unit variance—designated $N(0, 1)$. In Figure 1 both samples are of Size 5, while both are 15 in Figure 2. The theoretical curves, one for 8 *df*, the other for 28, represent quite well the obtained distributions. Ordinates approximately two units from the mean of the theoretical distributions mark off the respective 5% limits for rejecting the null hypothesis. In Figure 1,[2] 5.3% of the obtained *t*'s fall outside these bounds, while in Figure 2 only 4.0% of the sample *t*'s are in excess. Since in both cases the expected values is exactly 5%, we must attribute the discrepancy to random sampling fluctuations. The size of these discrepancies should be useful measures in evaluating the discrepancies which will be encountered

---

[2] The numbers in the tails of some of the figures report the number of obtained *t*'s falling outside the boundaries.

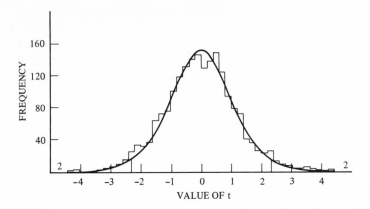

FIGURE 1

Empirical distribution of t's from $N(0, 1)5\text{-}N(0, 1)5$ and theoretical distribution with 28 df.

under other conditions of sampling. For examples of 2000 t's a discrepancy as large as 1% from the nominal 5% value evidently occurs frequently, and for this reason should not be considered as evidence to reject the theoretical distribution as an approximation to the empirical one.

As an initial departure from the simplest cases just presented, Figure 3 compares theoretical and empirical distributions when samples are taken from the same $N(0, 1)$ population, but the first sample size is 5, the second is 15—that is, $N(0, 1)5\text{-}N(0, 1)15$. While this in no sense is a

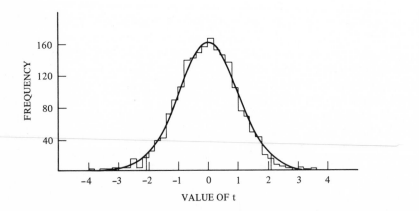

FIGURE 2

Empirical distribution of t's from $N(0, 1)15\text{-}N(0, 1)15$ and theoretical distribution with 28 df.

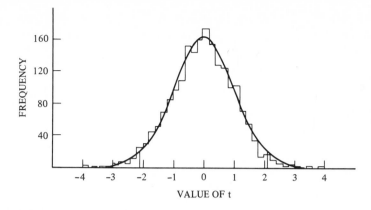

FIGURE 3
Empirical distribution of t's from $N(0, 1)5$-$N(0, 1)15$ and theoretical distribution with 18 df.

violation of the assumptions of the t test, it is interesting to note that again sampling fluctuations have produced an empirical distribution with 4.0% of the t's falling outside the nominal 5% limits.

The violation of the assumption of homogeneity of variance has effects as depicted in Figure 4. Here the obtained distribution is based upon two samples of Size 5, one from $N(0, 1)$ and the other from $N(0, 4)$. The fit is again seen to be close between theoretical and empirical distributions, and 6.4% of the obtained t's exceed the theoretical 5% limits. By increasing the sample size to 15, a distribution results (not shown here) for which only 4.9% of the t's fall outside the nominal limits. It

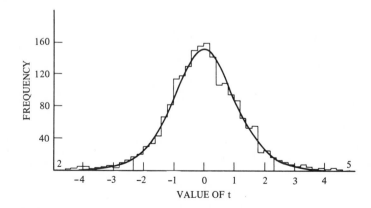

FIGURE 4
Empirical distribution of t's from $N(0, 1)5$-$N(0, 4)5$ and theoretical distribution with 8 df.

would seem that increasing the sample size produces a distribution which conforms rather closely to the t distribution. As will be seen later, this is a quite general result based upon mathematical considerations, the implications of which are important to the argument. For the moment it is evident that differences in variance at least in the ratio of 1 to 4 do not seriously affect the accuracy of probability statements made on the basis of the t test.

This last conclusion is true only so long as the size of both samples is the same. If the variances are different, with the present set of conditions there are two combinations of variance and the sample size possible. In one case the first sample may be of Size 5 and drawn from the population with the smaller variance, while the second sample of Size 15 is drawn from the population having the larger variance—$N(0, 1)5$-$N(0, 4)15$. In the second case the small sample size is coupled with the larger variance, the larger sample size with the smaller variance—$N(0, 4)5$-$N(0, 1)15$. The respective results of such sampling are presented in Figures 5 and 6. The empirical distributions are clearly not approximated by the t distribution. For the distribution of Figure 5, only 1% of the obtained t's exceed the nominal 5% values, while in Figure 6, 16% of the t's fall outside those limits.

There are good mathematical reasons why a difference in sample size should produce such decided discrepancies when the variances are unequal. Recall that $\Sigma(X - M)^2/(N - 1)$ is an estimate of the variance of the population from which the sample is drawn. Hence, $\Sigma(X - M)^2$ will in the long run be equal to $(N - 1)\sigma^2$. The formula used in this study for computing t makes use of this fact and, in addition, under the

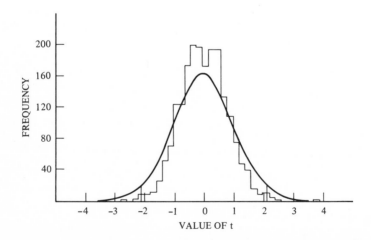

FIGURE 5
Empirical distribution of t's from $N(0, 1)5$-$N(0, 4)15$ and theoretical distribution with 18 df.

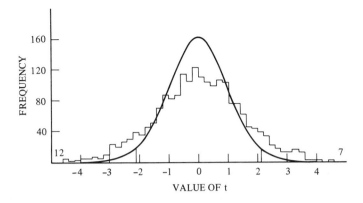

FIGURE 6
Empirical distribution of *t*'s from $N(0, 4)5\text{-}N(0, 1)15$ and theoretical distribution with 18 *df*.

assumption that the variances of the populations from which the two samples are drawn are equal, pools the sum of the squared deviations from the respective sample means to get a better estimate. That is $\Sigma(X_1 - M_1)^2 + \Sigma(X_2 - M_2)^2$ is an estimate of $(N_1 - 1)\sigma_1^2 + (N_2 - 1)\sigma_2^2$. If $\sigma_1^2 = \sigma_2^2 = \sigma^2$ (homogeneity of variance), then the sums estimate $(N_1 + N_2 - 2)\sigma^2$. Hence,

$$\frac{\Sigma(X_1 - M_1)^2 + \Sigma(X_2 - M_2)^2}{N_1 + N_2 - 2} \tag{1}$$

is an estimate of σ^2. If $\sigma_1^2 \neq \sigma_2^2$ the estimating procedure is patently illegitimate, the resulting value depending in a large measure upon the combination of sample size and variance used. For example, the case $N(0, 1)5\text{-}N(0, 4)15$ has $N_1 = 5$, $N_2 = 15$, $\sigma_1^2 = 1$, $\sigma_2^2 = 4$, and $N_1 + N_2 - 2 = 18$. With these values, Formula 1 has an expected value of $[(4 \cdot 1) + (14 \cdot 4)]/18 = 3.33$. Using the appropriate values for the other situation, $N(0, 4)5\text{-}N(0, 1)15$, the result of formula 1 is $[(4 \cdot 4) + (14 \cdot 1)]/18 = 1.67$. This means that on the average, the denominator for the *t* test will be larger for the first case than for the second. If the sample differences between means were of the same magnitude for the two cases, obviously more "significant" *t*'s would emerge when the denominator is smaller. It so happens that when this latter condition exists, the variance of the numerator also tends to be greater than in the other condition, a fact which accentuates the differences between the two empirical distributions.

Welch (1937) has shown mathematically that in the case of sample sizes of 5 and 15, a state which prevails here, the percentage of *t*'s exceeding the nominal 5% value varies as a function of the ratio of the two population variances and can be as low as 0% and as high as 31.3%.

If $N_1 = N_2$ there is never much bias, except perhaps in the case in which the sample sizes are both 2. For $N_1 = N_2 = 10$, the expected value of the percentage of t's exceeding the nominal 5% limits varies between 5% and 6.5% regardless of the difference between the variances. For larger sample sizes, the discrepancy tends to be even less.

Since the pooling procedure for estimating the population variance is used in ordinary analysis of variance techniques, it would seem that the combination of unequal variances and unequal sample sizes might play havoc with F test probability statements. That is, a combination of large variance and large sample size should tend to make the F test more conservative than the nominal value would lead one to expect, and, as with the t test, small variance and large sample size should produce a higher percentage of "significant" F's than expected. These conclusions are based upon a very simple extension to more than two samples of the explanation for the behavior of the t test probabilities with unequal sample sizes.

A more sophisticated mathematical handling of the problem by Box (1954a) reaches much the same conclusions for the simple-randomized analysis of variance. In a table in his article are given exact (i.e., mathematically determined) probabilities of exceeding the 5% point when variances are unequal. In this case, sampling is assumed to be from normal distributions. If the sample sizes are the same, the probability given for equal sample sizes range from 5.55% to 7.42%, for several combinations of variances, and numbers of samples. If, when variances are different, the samples are of different sizes, large discrepancies from the nominal values result. Combining large sample and large variance lessens the probability of obtaining a "significant" result to much less than 5%, just as we have seen for the t test. In a subsequent article, Box (1954b) presents some results from two-way analysis of variance. Since these designs generally have equal cell frequencies the results are not too far from expected. His figures all run within 2% of the 5% value expected if all assumptions were met.

It would seem then that both empirically and mathematically there can be demonstrated only a minor effect on the validity of probability statements caused by heterogeneity of variance, provided the sizes of the samples are the same. This applies to the F as well as the t test. If however, the sample sizes are different, major errors in interpretation may result if normal curve thinking is used.

Sampling from Identical Non-Normal Distributions: (Equal Variances)

Let us now proceed to violate the other main assumption, that of normality of distribution from which sampling takes place. At this time we will consider the t distributions arising when both samples are taken from the same non-normal distribution. The distributions shown here,

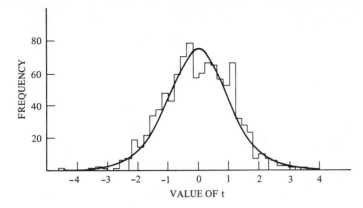

FIGURE 7
Empirical distribution of *t*'s from $E(0, 1)5$-$E(0, 1)5$ and theoretical distribution
with 8 *df*.

and all subsequent ones, are based upon only 1000 *t*'s, and hence will exhibit somewhat more column to column fluctuation than the preceding distributions.

Figure 7 compares the theoretical *t* distribution and the empirical distribution obtained from two samples of Size 5 from the exponential distribution—$E(0, 1)5$-$E(0, 1)5$. The fit is fairly close, but the proportion of cases in the tails seems less for the empirical distribution than for the theoretical. By count, 3.1% of the obtained *t*'s exceed the nominal 5% values—that is, the test in this case seems slightly conservative. If both sample sizes are raised to 15 (distribution not shown here), the corresponding percentage of obtained *t*'s is 4.0%. While this is probably not an appreciably better fit than for samples of Size 5, we shall see later that there are theoretical reasons to suspect that increasing the sample size should better the approximation of the empirical curve by the theoretical no matter what the parent population may be.

If both samples are of Size 5 from the same rectangular distribution —$R(0, 1)5$-$R(0, 1)5$—the result is as depicted in Figure 8. The fit of theoretical curve to empirical data here is as good as any thus far observed. The percentage of obtained *t*'s exceeding the 5% values is 5.1% in this particular case. For the case in which the sample sizes are both 15 (not shown here), the fit is equally good, with 5.0% of the cases falling outside of the nominal 5% bounds.

Sampling from Non-Normal Distributions: (Unequal Variances)

We may assume that if the variances are unequal, and at the same time the sample sizes are different, the resulting distributions from non-normal populations will be effected in the same way as the distributions

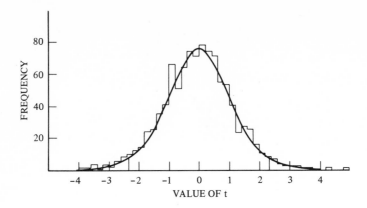

FIGURE 8
Empirical distribution of t's from $R(0, 1)5$-$R(0, 1)5$ and theoretical distribution with 8 df.

derived from normal populations, and for the same reasons. These cases will not be considered.

If sampling is in sizes of 5 from two exponential distributions, one with a variance of 1, and the other of 4, a skewed distribution of obtained t's emerges (not shown here). We shall discover that a skewed distribution of t's generally arises when the sampling is from distributions which are different in degree of skewness or asymmetry. (For an explanation, see discussion of $E(0, 1)5$-$N(0, 1)5$ below.) Apparently, the effect of increasing the variance of the exponential distribution as in the present case—$E(0, 1)5$-$E(0, 4)5$—is to make the negative sample means arising from the distribution with larger variance even more negative than those from the distribution with smaller variance. In terms of per-centage exceeding the nominal 5% limits for this case, the value is 8.3%, of which 7.6% comes from the skewed tail. This combination of variances and distribution was not tested with larger samples, but we shall see when comparing exponential and normal distributions that an increase in the sample size decreases the skew of the obtained t distribution there. Theoretically, this decrease should occur in almost all cases, including the present one.

The result is much less complicated if, while variances are different, the sampling is from symmetrical rectangular distributions—$R(0, 1)$-$R(0, 4)5$. For this small sample situation, (not illustrated), there occurs a distribution of obtained t's having 7.1% of the values exceeding the nominal 5% points. This is roughly the same magnitude as the cor-responding discrepancy from normal distributions. For the normal, it will be recalled that an increase of the sample sizes to 15 decreased the obtained percentage to 4.9%. There is no reason to believe that increasing the size of the rectangular samples would not have the same effect. However, time did not permit the determination of this distribution.

SAMPLING FROM TWO DIFFERENT DISTRIBUTIONS

By drawing the first sample from a distribution having one shape, and by drawing the second from a distribution having another shape (other than shape differences arising from heterogeneity of variance), yet another way has been found to do violence to the integrity of the assumptions underlying the *t* test. Perhaps the least violent of these happenings is that in which at least one of the populations is normal.

When one sample is from the exponential distribution and the other from the normal, the interesting result shown in Figure 9 occurs. This is the small sample case—$E(0, 1)5$-$N(0, 1)5$. It will be recalled that for skewed distributions the mean and median are at different points. In the exponential distributions, for example, the mean is at the 63rd centile. If samples from the exponential distribution are small, there will be a tendency for the sample mean to be less than the population mean, obviously since nearly two thirds of the scores are below that mean. Since the population mean of the present distributions is 0, the result will be a preponderance of negative sample means for small samples. If the other sample is taken from a symmetrical distribution, which would tend to produce as many positive as negative sample means, the resulting distribution of obtained *t*'s would not balance about its zero point, an imbalance exacerbated by small samples. In Figure 9, 7.1% of the obtained cases fall outside the 5% limits, with most 5.6%, lying in the skewed tail. The effect of increasing the sample size to 15 is to

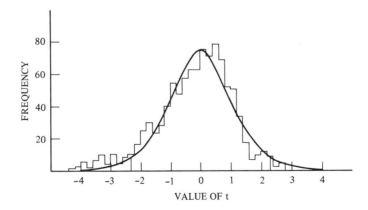

FIGURE 9

Empirical distribution of *t*'s from $E(0, 1)5$-$N(0, 1)5$ and theoretical distribution with 8 *df*.

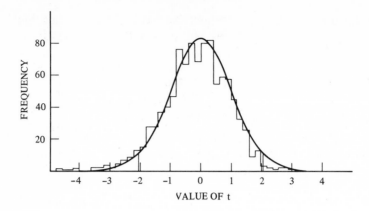

FIGURE 10
Empirical distribution of t's from $E(0, 1)5$-$N(0, 1)15$ and theoretical distribution with **28** df.

normalize the distribution considerably; the resulting curve, Figure 10, is fairly well approximated by the t distribution. One of the tails, however, does contain a disproportionate share of the cases, 4.2% to 0.9% for the other tail, or a total of 5.1% falling outside the nominal 5% limits. Nevertheless, the degree to which the theoretical and empirical distributions coincide under these conditions is striking. It seems likely that if both samples were each of Size **25**, the resulting sample distribution of t's would be virtually indistinguishable from the t distribution for **48** df, or the next best thing, the normal curve itself. To test this hypothesis, an additional empirical t distribution based on sample sizes of 25 from these same exponential and normal populations was obtained (not shown here). The results nicely confirm the presumption. Comparison with the usual 5% values reveals 4.6% of the empirical t's surpassing them. Whereas with the smaller samples the ratio of t's in the skewed tail to those in the other tail is roughly 80:20, the corresponding ratio for the larger case is 59:41. Clearly, the increase in sample sizes has tended to normalize the distribution of t's.

For these conditions, involving rather drastic violation of the mathematical assumptions of the test, the t test has been observed to fare well with an adequate sample size. Such a state of affairs is to be expected theoretically. By invoking a few theorems of mathematical statistics it can be shown that if one samples from any two populations for which the Central Limit Theorem holds, (almost any population that a psychologist might be confronted with), no matter what the variances may be, the use of equal sample sizes insures that the resulting distribution of t's will approach normality as a limit. It would appear from the present results that the approach to normality is rather rapid, since samples of sizes of 15 are generally sufficient to undo most of the damage inflicted by violation of assumptions. Only in extreme cases, such as the last which involves distributions differing in skew, would it seem that slightly larger sizes are prescribed. Thus it would appear that the *t* test is functionally a distribution-free test, providing the sample sizes are sufficiently large (say, 30, for extreme violations) and equal.

The distributions arising when sampling is from the normal and the rectangular distributions—$N(0, 1)5$-$R(0, 1)5$ and $N(0, 1)15$-$R(0, 1)15$ —would further tend to substantiate this claim. The respective percentages exceeding the 5% nominal values are 5.6% and 4.6% from the empirical distributions for these cases, the distributions of *t*'s being symmetrical and close to the theoretical (not shown).

The only other combination examined in the sampling study is the uninteresting case of exponential and rectangular distributions. This distribution (not shown) is again skewed with the effect of increase of sample size from 5 to 15 to cut down the skew and to decrease the percentage of cases falling outside the theoretical 5% values from 6.4% to 5.6%. For those cases falling outside the nominal 5% values, the ratio is 79:21 for the smaller samples. This is changed to 69:31 for the sample size of 15. Here again it would seem that larger sample sizes would be required to insure the validity of probability statements utilizing the *t* distribution as a model.

The results of the total study are summarized in Table 1 which gives for each combination of population, variance, and sample size (*a*) the percentage of obtained *t*'s falling outside the nominal 5% probability limits of the ordinary *t* distribution, and (*b*) the percentage of obtained *t*'s falling outside the 1% limits. The combinations are represented symbolically as before. The table is divided into two parts, the first part presenting information on the empirical distributions which are intrinsically symmetrical. The second part is based upon the intrinsically non-symmetrical distributions, additional information in this section of the table being the percentage of obtained *t*'s falling in the larger of the tails. The percentage for the smaller tail may be obtained by subtraction of the percentage in the larger tail from the total.

Certain implications of the table should be discussed. In the Norton study, more severe distortions sometimes occurred with significance levels of 1% and .1% than appeared with the 5% level. The inclusion in Table 1 of the percentages of obtained *t*'s falling outside the nominal 1% values makes possible the comparison of the 1% and 5% results. The 1% values seem to be approximately what would be expected considering that sampling fluctuations are occurring. It was not felt feasible to determine the results for the .1% level since with only 1000 or 2000

TABLE 1

Obtained Percentages of Cases Falling Outside the Appropriate Tabled t Values for the 5% and 1% Level of Significance

Symmetric Distributions	Obtained Percentage at	
	5% level	1% level
$N(0, 1)5\text{-}N(0, 1)5$	5.3	0.9
$N(0, 1)15\text{-}N(0, 1)15$	4.0	0.8
$N(0, 1)5\text{-}N(0, 1)15$	4.0	0.6
$N(0, 1)5\text{-}N(0, 4)5$	6.4	1.8
$N(0, 1)15\text{-}N(0, 4)15$	4.9	1.1
$N(0, 1)5\text{-}N(0, 4)15$	1.0	0.1
$N(0, 4)5\text{-}N(0, 1)15$	16.0	6.0
$E(0, 1)5\text{-}E(0, 1)5$	3.1	0.3
$E(0, 1)15\text{-}E(0, 1)15$	4.0	0.4
$R(0, 1)5\text{-}R(0, 1)5$	5.1	1.0
$R(0, 1)15\text{-}R(0, 1)15$	5.0	1.5
$R(0, 1)5\text{-}R(0, 4)5$	7.1	1.9
$N(0, 1)5\text{-}R(0, 1)5$	5.6	1.0
$N(0, 1)15\text{-}R(0, 1)15$	5.6	1.1

Asymmetric Distributions	Obtained Percentage at			
	5% level		1% level	
	Total	Larger Tail	Total	Larger Tail
$E(0, 1)5\text{-}N(0, 1)5$	7.1	5.6	1.9	1.9
$E(0, 1)15\text{-}N(0, 1)15$	5.1	4.2	1.4	1.2
$E(0, 1)25\text{-}N(0, 1)25$	4.6	2.7	1.3	1.1
$E(0, 1)5\text{-}R(0, 1)5$	6.4	5.0	3.3	2.5
$E(0, 1)15\text{-}R(0, 1)15$	5.6	3.9	1.6	1.2
$E(0, 1)5\text{-}E(0, 4)5$	8.3	7.6	1.7	1.7

cases the number of obtained t's falling outside prescribed limits was negligible in most cases. It is possible, however, that the distortions in the apparent level of significance are more drastic for the smaller α values.

All the results and discussion have been limited thus far to the two-tailed t test. With notable exceptions, the conclusions we have reached can be applied directly to the one-tailed t test as well. The exceptions involve those distributions which are intrinsically *asymmetric* (see Table 1). In these distributions a preponderance of the obtained t's fall in one tail. Depending upon the particular tail involved in the one-tailed test the use of t should produce too many or too few significant results when sampling is from a combination of populations, from which an asymmetric t distribution is expected. It seems impossible to make any simple statements about the behavior of the tails in the general case of asymmetric *t* distribution except to say that such distributions are expected whenever the skew of the two parent populations is different. The experimenter must determine for each particular instance the direction of skew of the expected distribution and act accordingly. Table 1 gives for the intrinsically asymmetric distributions the total percentage of obtained *t*'s falling outside the theoretical 5% and 1% limits and the percentage in the larger tail. From these values can be assessed the approximate magnitude of the bias incurred when a one-tailed test is used in specific situations.

DISCUSSION AND CONCLUSIONS

Having violated a number of assumptions underlying the *t* test, and finding that, by and large, such violations produce a minimal effect on the distribution of *t*'s, we must conclude that the *t* test is a remarkably *robust* test in the technical sense of the word. This term was introduced by Box (1953) to characterize statistical tests which are only inconsequentially affected by a violation of the underlying assumptions. Every statistical test is in part a test of the assumptions upon which it is based. For example, the null hypothesis of a particular test may be concerned with sample means. If, however, the assumptions underlying the test are not met, the result may be "significant" even though the population means are the same. If the statistical test is relatively insensitive to violations of the assumptions other than the null hypothesis, and, hence, if probability statements refer primarily to the null hypotheses, it is said to be robust. The *t* and *F* tests apparently possess this quality to a high degree.

We may conclude that for a large number of different situations confronting the researcher, the use of the ordinary *t* test and its associated table will result in probability statements which are accurate to a high degree, even though the assumptions of homogeneity of variance and normality of the underlying distributions are untenable. This large number of situations has the following general characteristics: (*a*) the two sample sizes are equal or nearly so, (*b*) the assumed underlying population distributions are of the same shape or nearly so. (If the distributions are skewed they should have nearly the same variance.) If these conditions are met, then no matter what the variance differences may be, samples of as small as five will produce results for which the true probability of rejecting the null hypothesis at the .05 level will more than likely be within .03 of that level. If the sample size is as large as 15, the true probabilities are quite likely within .01 of the nominal value. That is to say, the percentage of times the null hypothesis will be rejected when it is actually true will tend to be between 4% and 6% when the nominal value is 5%.

If the sample sizes are unequal, one is in no difficulty provided the variances are compensatingly equal. A combination of unequal sample sizes and unequal variances, however, automatically produces inaccurate probability statements which can be quite different from the nominal values. One must in this case resort to different testing procedures, such as those by Cochran and Cox (1950), Satterthwaite (1946), and Welch (1947). The Welch procedure is interesting since it has been extended by Welch (1951) to cover the simple randomized analysis of variance which suffers the same defect as the t test when confronted with both unequal variance and unequal sample sizes. The Fisher-Behrens procedure suggested by many psychologically oriented statistical textbooks has had its validity questioned (Bartlett, 1936) and, hence, is ignored by some statisticans (e.g., Anderson & Bancroft, 1952, p. 82).

If the two underlying populations are not the same shape, there seems to be little difficulty if the distributions are both symmetrical. If they differ in skew, however, the distribution of obtained t's has a tendency itself to be skewed, having a greater percentage of obtained t's falling outside of one limit than the other. This may tend to bias probability statements. Increasing the sample size has the effect of removing the skew, and, due to the Central Limit Theorem and others, the normal distribution is approached by this maneuver. By the time the sample sizes reach 25 or 30, the approach should be close enough that one can, in effect, ignore the effects of violations of assumptions except for extremes. Since this is so, the t test is seen to be functionally nonparametric or distribution-free. It also retains its power in some situations (David & Johnson, 1951). There is, unfortunately, no guarantee that the t and F tests are uniformly most powerful tests. It is possible, even probable, that certain of the distribution-free methods are more powerful than the t and F tests when sampling is from some unspecified distributions or combination of distributions. At present, little can be said to clarify the situation. Much more research in this area needs to be done.

Since the t and F tests of analysis of variance are intimately related, it can be shown that many of the statements referring to the t test can be generalized quite readily to the F test. In particular, the necessity for equal sample sizes, if variances are unequal, is important for the same reasons in the F test of analysis of variance as in the t test. A number of the cited articles have demonstrated both mathematically and by means of sampling studies that most of the statements we have made do apply to the F test. It is suggested that psychological researchers feel free to utilize these powerful techniques where applicable in a wider variety of situations, the present emphasis on the nonparametric methods notwithstanding.

References

Anderson, R. L., & Bancroft, T. A. *Statistical theory in research*. New York: McGraw-Hill, 1952.

Bartlett, M. S. The effect of non-normality on the *t*-distribution. *Proceedings of Cambridge Philosophic Society*, 1935, **31**, 223–231.

Bartlett, M. S. The information available in small samples. *Proceedings of Cambridge Philosophic Society*, 1936, **32**, 560–566.

Box, G. E. P. Non-normality and tests on variances. *Biometrika*, 1953, **40**, 318–335.

Box, G. E. P. Some thorems on quadratic forms applied in the study of analysis of variance problems, I. Effect of inequality of variance in the one-way classification. *Annals of Mathematical Statistics*, 1954, **25**, 290–302. (a)

Box, G. E. P. Some theorems on quadratic forms applied in the study of analysis of variance problems, II. Effects of inequality of variance and of correlation between errors in the two-way classification. *Annals of Mathematical Statistics*, 1954, **25**, 484–498. (b)

Box, G. E. P., & Andersen, S. L. Permutation theory in the derivation of robust criteria and the study of departures from assumption. *Journal of the Royal Statistical Society (Series B)*, 1955, **17**, 1–34.

Cochran, W. G., & Cox, G. M. *Experimental designs*. New York: Wiley, 1950.

Daniels, H. E. The effect of departures from ideal conditions other than non-normality on the *t* and *z* tests of significance. *Proceedings of Cambridge Philosophic Society*, 1938, **34**, 321–328.

David, F. N., & Johnson, N. L. The effect of non-normality on the power function of the *F*-test in the analysis of variance. *Biometrika*, 1951, **38**, 43–57.

Gayen, A. K. The distribution of the variance ratio in random samples of any size drawn from non-normal universes. *Biometrika*, 1950, **37**, 236–255. (a)

Gayen, A. K. Significance of difference between the means of two non-normal samples. *Biometrika*, 1950, **37**, 399–408. (b)

Horsnell, G. The effect of unequal group variances on the *F*-test for the homogeneity of group means. *Biometrika*, 1953, **40**, 128–136.

Lindquist, E. F. *Design and analysis of experiments in psychology and education*. Boston: Houghton Mifflin, 1953.

Norton, D. W. An empirical investigation of some effects of non-normality and heterogeneity on the *F*-distribution. Unpublished doctoral dissertation, State University of Iowa, 1952.

Pearson, E. S. The analysis of variance in the case of non-normal. *Biometrika*, 1931, **23**, 114–133.

Quensel, C. E. The validity of the Z-criterion when the variates are taken from different normal populations. *Skandinavisk Aktuarietids*, 1947, **30**, 44–55.

Satterthwaite, F. E. An approximate distribution of estimates of variance components. *Biometrics Bulletin*, 1946, **2**, 110–114.

Welch, B. L. The significance of the difference between two means when the population variances are unequal. *Biometrika,* 1937, **29**, 350–362.

Welch, B. L. The generalization of Student's problem when several different population variances are involved. *Biometrika,* 1947, **34**, 28–35.

Welch, B. L. On the comparison of several mean values: An alternative approach. *Biometrika,* 1951, **38**, 330–336.

4.15

Some consequences when the assumptions for the analysis of variance are not satisfied

W. G. Cochran

1. PURPOSES OF THE ANALYSIS OF VARIANCE

The main purposes are:

(i) To estimate certain treatment differences that are of interest. In this statement both the words "treatment" and "difference" are used in a rather loose sense: e.g., a treatment difference might be the difference between the mean yields of two varieties in a plant-breeding trial, or the relative toxicity of an unknown to a standard poison in a dosage-mortality experiment. We want such estimates to be *efficient.* That is, speaking roughly, we want the difference between the estimate and the true value to have as small a variance as can be attained from the data that are being analyzed.

(ii) To obtain some idea of the accuracy of our estimates, e.g., by attaching to them estimated standard errors, fiducial or confidence limits, etc. Such standard errors, etc., should be reasonably free from bias. The usual property of the analysis of variance, when all assumptions are fulfilled, is that estimated variances are unbiased.

(iii) To perform tests of significance. The most common are the F-test of the null hypothesis that a group of means all have the same true value, and the t-test of the null hypothesis that a treatment dif-

W. G. Cochran, "Some Consequences When the Assumptions for the Analysis of Variance are Not Satisfied," *Biometrics,* **3**, March, 1947, 22–38.

ference is zero or has some known value. We should like such tests to be *valid*, in the sense that if the table shows a significance probability of, say, 0.023, the chance of getting the observed result or a more discordant one on the null hypothesis should really be 0.023 or something near it. Further, such tests should be *sensitive* or *powerful*, meaning that they should detect the presence of real treatment differences as often as possible.

The object of this paper is to describe what happens to these desirable properties of the analysis of variance when the assumptions required for the technique do not hold. Obviously, any practical value of the paper will be increased if advice can also be given on how to detect failure of the assumptions and how to avoid the more serious consequences.

2. ASSUMPTIONS REQUIRED FOR THE ANALYSIS OF VARIANCE

In setting up an analysis of variance, we generally recognize three types of effects:

(a)   treatment effects—the effects of procedures deliberately introduced by the experimenter

(b)   environmental effects (the term is not ideal)—these are certain features of the environment which the analysis enables us to measure. Common examples are the effects of replications in a randomized blocks experiment, or of rows and columns in a Latin square

(c)   experimental errors—this term includes all elements of variation that are not taken account of in (a) or (b).

(1) The treatment effects and the environmental effects must be additive. For instance, in a randomized blocks trial the observation y_{ij} on the ith treatment in the jth replication is specified as

$$y_{ij} = \mu + \tau_i + \rho_j + e_{ij}$$

where μ is the general mean, τ_i is the effect of the ith treatment, p_j is the effect of the jth replication and e_{ij} is the experimental error of that observation. We may assume, without loss of generality, that the e's all have zero means .

(2)   The experimental errors must all be independent. That is, the probability that the error of any observation has a particular value must not depend on the values of the errors for other observations.

(3)   The experimental errors must have a common variance.[1]

(4)   The experimental errors should be normally distributed.

---

[1] This statement, though it applies to the simplest analyses, is an oversimplification. More generally, the analysis of variance should be divisible into parts within each of which the errors have common variance. For instance, in the split-plot design, we specify one error variance for whole-plot comparisons and a different one for sub-plot comparisons.

We propose to consider each assumption and to discuss the consequences when the assumption is not satisfied. The discussion will be in rather general terms, for much more research would be needed in order to make precise statements. Moreover, in practice several assumptions may fail to hold simultaneously. For example, in non-normal distributions there is usually a correlation between the variance of an observation and its mean, so that failure of condition (4) is likely to be accompanied by failure of (3) also.

3. PREVIOUS WORK ON THE EFFECTS OF NON-NORMALITY

Most of the published work on the effects of failures in the assumptions has been concerned with this item. Writing in 1938, Hey (1938) gives a bibliography of 36 papers, most of which deal with non-normality, while several theoretical investigations were outside the scope of his bibliography. Although space does not permit a detailed survey of this literature, some comments on the nature of the work are relevant.

The work is almost entirely confined to a single aspect, namely the effect on what we have called the validity of tests of significance. Further, insofar as the t-test is discussed, this is either the test of a single mean or of the difference between the means of two groups. As will be seen later, it is important to bear this restriction in mind when evaluating the scope of the results.

Some writers, e.g., Bartlett (1935), investigated by mathematical methods the theoretical frequency distribution of F or t, assuming the null hypothesis true, when sampling from an infinite population that was non-normal. As a rule, it is extremely difficult to obtain the distributions in such cases. Others, e.g., E. S. Pearson (1931), drew mechanically 500 or 1000 numerical samples from an infinite non-normal population, calculated the value of F or t for each sample, and thus obtained empirically some idea of their frequency distributions. Where this method was used, the number of samples was seldom large enough to allow more than a chi-square goodness of fit test of the difference between the observed and the standard distributions. A very large number of samples is needed to determine the 5 percent point, and more so the 1 percent point, accurately. A third method, of which Hey's paper contains several examples, is to take actual data from experiments and generate, the F or t distribution by means of randomization similar to that which would be practiced in an experiment. The data are chosen, of course, because they represent some type of departure from normality.

The consensus from these investigations is that no serious error is introduced by non-normality in the significance levels of the F-test or of the two-tailed t-test. While it is difficult to generalize about the range of populations that were investigated, this appears to cover most cases encountered in practice. If a guess may be made about the limits of error, the true probability corresponding to the tabular 5 percent significance level may lie between 4 and 7 percent. For the 1 percent level, the limits might be taken as ½ percent and 2 percent. As a rule, the tabular probability is an underestimate: that is, by using the ordinary F and t tables we tend to err in the direction of announcing too many significant results.

The one-tailed t-test is more vulnerable. With a markedly skew distribution of errors, where one tail is much longer than the other, the usual practice of calculating the significance probability as one-half the value read from the tables may give quite a serious over- or underestimate.

It was pointed out that work on the validity of the t-test covered only the case of a single mean or of the comparison of the means of two groups. The results would be applicable to a randomized blocks experiment if we adopted the practice of calculating a separate error for each pair of treatments to be tested, using only the data from that pair of treatments. In practice, however, it is usual to employ a pooled error for all t-tests in an analysis, since this procedure not only saves labor but provides more degrees of freedom for the estimation of error. It will be shown in section 6 that this use of a pooled error when non-normality is present may lead to large errors in the significance probabilities of individual t-tests. The same remark applies to the Latin square and more complex arrangements, where in general it is impossible to isolate a separate error appropriate to a given pair of treatments, so that pooling of errors is unavoidable.

4. FURTHER EFFECTS OF NON-NORMALITY

In addition to its effects on the validity of tests of significance, non-normality is likely to be accompanied by a loss of efficiency in the estimation of treatment effects and a corresponding loss of power in the F- and t-tests. This loss of efficiency has been calculated by theoretical methods for a number of types of non-normal distribution. While these investigations dealt with the estimation of a single mean, and thus would be strictly applicable only to a paired experiment analyzed by the method of differences, the results are probably indicative of those that would be found for more complex analyses. In an attempt to use these results for our present purpose, the missing link is that we do not know which of the theoretical non-normal distributions that have been studied are typical of the error distributions that turn up in practice. This gap makes speculation hazardous because the efficiency of analysis of variance methods has been found to vary from 100 percent to zero. While I would not wish to express any opinion very forcibly, my impression is that in practice the loss of efficiency is not often great. For instance, in an examination of the Pearson curves, Fisher (1922) has proven that for curves that exhibit only a moderate departure from normality, the efficiency remains reasonably high. Further, an analysis of the logs of the observations instead of the observations themselves has frequently been found successful in converting data to a scale where errors are approximately normally distributed. In this connection, Finney, (1941) has shown that if log x is exactly normally distributed, the arithmetic mean of x has an efficiency greater than 93 percent so long as the coefficient of variation of x is less than 100 percent. In most lines of work a standard error as high as 100 percent per observation is rare, though not impossible.

The effect of non-normality on estimated standard errors is analogous to the effect on the t-test. If a standard error is calculated specifically for each pair of treatments whose means are to be compared, the error variance is unbiased. Bias may arise, however, by the application of a pooled error to a particular pair of treatments.

We now consider how to detect non-normality. It might perhaps be suggested that the standard tests for departure from normality, Fisher (1948), should be applied to the errors in an analysis. This suggestion is not fruitful, however, because for experiments of the size usually conducted, the test would detect, only very violent skewness or kurtosis. Moreover, as is perhaps more important, it is not enough to detect non-normality: in order to develop an improved analysis, one must have some idea of the actual form of the distribution of errors, and for this purpose a single experiment is rarely adequate.

Examination of the distribution of errors may be helpful where an extensive uniformity trial has been carried out, or where a whole series of experiments on similar materials is available. Theoretically, the best procedure would be to try to find the form of the frequency distribution of errors, using, of course, any a priori knowledge of the nature of the data. An improved method of estimation could then be developed by maximum likelihood. This, however, would be likely to lead to involved computations. For that reason, the usual technique in practice is to seek, from a priori knowledge or by trial and error, a transformation that will put the data on a scale where the errors are approximately normal. The hope is that in the transformed scale the usual analysis will be reasonably efficient. Further, we would be prepared to accept some loss in efficiency for the convenience of using a familiar method. Since a detailed account of transformations will be given by Dr. Bartlett in the following paper, this point will not be elaborated.

The above remarks are intended to apply to the handling of a rather extensive body of data. With a single experiment, standing by itself, experience has indicated two features that should be watched for:

(i) evidence of charges in the variance from one part of the experiment to another. This case will be discussed in section 6.

(ii) evidence of gross errors.

5. EFFECTS OF GROSS ERRORS

The effects of gross errors, if undetected, are obvious. The means of the treatments that are affected will be poorly estimated, while if a pooled error is used the standard errors of other treatment means will be over-estimated. An extreme example is illustrated by the data in Table 1, which come from a randomized blocks experiment with four replicates.

TABLE 1
Wheat: Ratio of Dry to Wet Grain

	Nitrogen applied			
Block	*None*	*Early*	*Middle*	*Late*
1	.718	.732	.734	.792
2	.725	.781	.725	.716
3	.704	1.035	.763	.758
4	.726	.765	.738	.781

As is likely to happen when the experimenter does not scrutinize his own data, the gross error was at first unnoticed when the computer carried out the analysis of variance, though the value is clearly impossible from the nature of the measurements. This fact justifies rejection of the value and substitution of another by the method of missing plots, Yates (1933).

Where no explanation can be found for an anomalous observation, the case for rejection is more doubtful. Habitual rejection of outlying values leads to a marked underestimation of errors. An approximate test of significance of the contribution of the suspected observation to the error helps to guard against this bias. First calculate the error sum of squares from the actual observations. Then calculate the error when the suspected value is replaced by the missing-plot estimate: this will have one less degree of freedom and is designated the "Remainder" in the data below. The difference represents the sum of squares due to the suspect. For the data above, the results are

	d.f.	S.S.	M.S.
Actual error	9	.04729	.00525
Suspect	1	.04205	.04205
Remainder	8	.00524	.000655

Alternatively, the contribution due to the suspected observation may be calculated directly and the remainder found by subtraction. If there are t treatments and r replicates, the sum of squares is $(t-1)(r-1)$ d^2/tr, where d is the difference between the suspected observation and the value given by the missing-plot formula. In the present case t and r are 3 and the missing-plot value is 0.7616, so that the contribution is $9(0.2734)^2/16$, or 0.04205.[2]

The F ratio for the test of the suspect against the remainder is 64.2, giving a t value of 8.01, with 8 degrees of freedom. Now, assuming that the suspect had been examined simply because it appeared anomalous, with no explanation for the anomaly, account must be taken of this fact in the test of significance. What is wanted is a test appropriate to the *largest* contribution of any observation. Such a test has not as yet been developed. The following is suggested as a rough approximation. Calculate the significance probability, p, by the ordinary t table. Then use as the correct significance probability np, where n is the number of degrees of freedom in the actual error.[3] In the present case, with $t = 8.01$, p is much less than 1 in a million, and consequently np is less than 1 in 100,000. In general, it would be wise to insist on a rather low significance probability (e.g., 1 in 100) before rejecting the suspect, though a careful answer on this point requires knowledge of the particular types of error to which the experimentation is subject.

6. EFFECTS OF HETEROGENEITY OF ERRORS

If ordinary analysis of variance methods is used when the true error variance differs from one observation to another, there will as a rule be a loss of efficiency in the estimates of treatment effects. Similarly,

[2] This formula applies only to randomized blocks. Corresponding formulas can be found for other types of arrangements. For instance, the formula for a $p \times p$ Latin square is $(p-1)(p-2)d^2/p^2$.

[3] The approximation is intended only to distinguish quickly whether the probability is low or high and must not be regarded as accurate. For a discussion of this type of test in a somewhat simpler case, see E. S. Pearson and C. Chandra Sekar, *Biometrika*, Vol. 28 (1936), pp. 308–320.

there will be a loss of sensitivity in tests of significance. If the changes in the error variance are large, these losses may be substantial. The validity of the F-test for all treatments is probably the least affected. Since, however, some treatment comparisons may have much smaller errors than others, t-tests from a pooled error may give a serious distortion of the significance levels. In the same way the standard errors of particular treatment comparisons, if derived from a pooled error, may be far from the true values.

There is no theoretical difficulty in extending the analysis of variance so as to take account of variations in error variances. The usual analysis is replaced by a weighted analysis in which each observation is weighted in proportion to the inverse of its error variance. The extension postulates, however, a knowledge of the relative variances of any two observations and this knowledge is seldom available in practice. Nevertheless, the more exact theory can sometimes be used with profit in cases where we have good estimates of these relative variances. Suppose, for instance, the situation were such that the observations could be divided into three parts, the error variances being constant within each part. If unbiased estimates of the variances within each part could be obtained and if these were each based on, say, at least 15 degrees of freedom, we could recover most of the loss in efficiency by weighting inversely as the observed variances. This device is therefore worth keeping in mind, though in complex analyses the weighted solution involves heavy computation.

Heterogeneity of errors may arise in several ways. It may be produced by mishaps or damage to some part of the experiment. It may be present in one or two replications through the use of less homogeneous material or of less carefully controlled conditions. The nature of the treatments may be such that some give more variable responses than others. An example of this type is given by the data in Table 2.

The experiment investigated the effects of three levels of chalk

TABLE 2
Mangolds, Plant Numbers per Plot

Block	Control		Chalk			Lime			Total
	0	0	1	2	3	1	2	3	
I	140	49	98	135	117	81	147	130	897
II	142	37	132	151	137	129	131	112	971
III	36	114	130	143	137	135	103	130	928
IV	129	125	153	146	143	104	147	121	1068
Total	447	325	513	575	534	449	528	493	3864
Range	106	88	55	16	26	54	44	18	

TABLE 3

Analysis of Variance for Mangolds Data

	d.f.	S.S.	M.S.
Blocks	3	2,079	
Treatments	6	8,516	
Error	22	18,939	860.9
Total	31	29,534	

dressing and three of lime dressing on plant numbers of mangolds. There were four randomized blocks of eight plots each, the control plots being replicated twice within each block.[4]

Since the soil was acid, high variability might be anticipated for the control plots as a result of partial failures on some plots. The effect is evident on eye inspection of the data. To a smaller extent the same effect is indicated on the plots receiving the single dressing of chalk or lime. If the variance may be regarded as constant within each treatment, there will be no loss of efficiency in the treatment means in this case, contrary to the usual effect of heterogeneity. Any t tests will be affected and standard errors may be biased. In amending the analysis so as to avoid such disturbances, the first step is to attempt to subdivide the error into homogeneous components. The simple analysis of variance is shown [in Table 3].

For subdivision of the error we need the following auxiliary data.

Block	Diff. between Controls	Total–4 (Controls)	(C1–L1)	(C2 + L2 + C3 + L3) − 2(C1 + L1)
1	91	141	17	171
2	105	255	3	9
3	78	328	−5	−17
4	4	52	49	43
Total		776	64	206
Divisor for S.S.	2	24	2	12

The first two columns are used to separate the contribution of the controls to the error. This has 7 d.f. of which 4 represent differences between the two controls in each block. The sum of squares of the first column is divided by 2 as indicated. There remains 3 d.f. which come from a comparison within each block of the total yield of the controls with the total yield of the dressings. Since there are 6 dressed plots to 2 controls per block we take

---

[4] The same data were discussed (in much less detail) in a previous paper, Cochran (1938).

(Dressing total) $-$ 3(Control total) $=$ (Total) $-$ 4(Control total)

Thus $141 = 897 - 4(140 + 49)$.

By the usual rule the divisor for the sum of squares of deviations is 24.

Two more columns are used to separate the contribution of the single dressings. There are 6 d.f. of which 3 compare chalk with line at this level while the remaining 3 compare the single level with the higher levels. The resulting partition of the error sum of squares is shown [in Table 4].

TABLE 4
Partition of Error Sum of Squares

	d.f.	S.S.	M.S.
Total	22	18,939	861
Between controls	4	12,703	3,176
Controls v. Dressings	3	1,860	620
Chalk 1 v. Lime 1	3	850	283
Single v. Higher Dressings	3	1,738	579
Double and Triple Dressings	9	1,788	199

As an illustration of the disturbance to t-tests and to estimated standard errors, we may note that the pooled mean square, 861, is over four times as large as the 9 d.f. error, 199, obtained from the double and triple dressings. Consequently, the significance levels of t and standard errors would be inflated by a factor of two if the pooled error were applied to comparisons within the higher dressings.

In a more realistic approach we might postulate three error variances, σ_c^2 for controls, σ_1^2 for single dressings and σ_h^2 for higher dressings. For these we have unbiased estimates of 3,176, 283, and 199 respectively from Table 4. The mean square for Controls v. Dressings (620) would be an unbiased estimate of $(9\sigma_c^2 + \sigma_1^2 + 2\sigma_h^2)/12$, while that for Single v. Higher Dressings (579) would estimate $(2\sigma_1^2 + \sigma_h^2)/3$.

What one does in handling comparisons that involve different levels depends on the amount of refinement that is desired and the amount of work that seems justifiable. The simplest process is to calculate a separate t-test or standard error for any comparison by obtaining the comparison separately within each block. Such errors, being based on 3 d.f., would be rather poorly determined. A more complex but more efficient approach is to estimate the three variances from the five mean squares given above. Since the error variance of any comparison will be some linear function of these three variances, it can then be estimated.

To summarize, heterogeneity of errors may affect certain treatments or certain parts of the data to an unpredictable extent. Sometimes, as in the previous example, such heterogeneity would be expected in advance from the nature of the experiment. In such cases the data may be inspected carefully to decide whether the actual amount of variation in the error variance seems enough to justify special methods. In fact, such inspection is worthwhile as a routine procedure and is, of course, the only method for detecting heterogeneity when it has not been anticipated. The principal weapons for dealing with this irregular type of heterogeneity are subdivision of the error variance or omission of parts of the experiment. Unfortunately, in complex analyses the computations may be laborious. For the Latin square, Yates (1936) has given methods for omitting a single treatment, row or column, while Yates and Hale (1939) have extended the process to a pair of treatments, rows or columns.

In addition, there is a common type of heterogeneity that is more regular. In this type, which usually arises from non-normality in the distribution of errors, the variance of an observation is some simple function of its mean value, irrespective of the treatment or block concerned. For instance, in counts whose error distribution is related to the Poisson, the variance of an observation may be proportional to its mean value. Such cases, which have been most successfully handled by means of transformations, are discussed in more detail in Dr. Bartlett's paper.

7. EFFECTS OF CORRELATIONS AMONGST THE ERRORS

These effects may be illustrated by a simple theoretical example. Suppose that the errors $e_1, e_2, \ldots, e_r$ of the r observations on a treatment in a simple group comparison have constant variance σ^2 and that every pair has a correlation coefficient ρ. The error of the treatment total, $(e_1 + e_2 \ldots, + e_r)$ will have a variance

$$r\sigma^2 + r(r-1)\rho\sigma^2$$

since there are $r(r-1)/2$ cross-product terms, each of which will contribute $2\rho\sigma^2$. Hence the *true* variance of the treatment mean is

$$\sigma^2\{1 + (r-1)\rho\}/r.$$

Now in practice we would estimate this variance by means of the sum of squares of deviations within the group, divided by $r(r-1)$. But

$$\text{Mean } \Sigma(e_i - \bar{e})^2 = \text{Mean } \Sigma e_i^2 - r\{\text{Mean } \bar{e}^2\} =$$
$$r\sigma^2 - \sigma^2\{1 + (r-1)\rho\} = (r-1)\sigma^2(1-\rho).$$

Hence the *estimated* variance of the treatment mean is $\sigma^2(1-\rho)/r$.

Consequently, if ρ is positive the treatment mean is less accurate than the mean of an independent series, but is estimated to be more accurate. If ρ is negative, these conditions are reversed. Substantial biases in standard errors might result, with similar impairment of t-tests. Moreover, in many types of data, particularly field experimentation, the observations *are* mutually correlated, though in a more intricate pattern.

Whatever the nature of the correlation system, this difficulty is largely taken care of by proper randomization. While mathematical details will not be given, the effect of randomization is, roughly speaking, that we may treat the errors as if they were independent. The reader may refer to a paper by Yates (1933), which presents the nature of this argument, and to papers by Bartlett (1935), Fisher (1949) and Hey (1938), which illustrate how randomization generates a close approximation to the F and t distributions.

Occasionally it may be discovered that the data have been subject to some systematic pattern of environmental variation that the randomization has been unable to cope with. If the environmental pattern obviously masks the treatment effects, resort may be had to what might be called desperate remedies in order to salvage some information.

The data in Table 5 provide an instance. The experiment was a 2^4 factorial, testing the effects of lime (L), fish manure (F) and artificial fertilizers (A). Lime was applied in the first year only; the other dressings were either applied in the first year only (1) or at a half rate every year (2). Two randomized blocks were laid out, the crop being pyrethrum, which forms an ingredient in many common insecticides. The data presented are for the fourth year of the experiment, which was conducted at the Woburn Experimental Farm, England.

TABLE 5
Weights of Dry Heads per Plot (Unit, 10 grams)

Block 1				Block 2			
LA1	LF2	F2	L1	A1	L1	A2	0
84	66	70	81	63	97	56	64
1	1	1	1	1	1	1	1
LF1	A2	A1	FA2	F1	LA2	LA1	LFA1
148	137	146	171	168	158	189	152
0	0	0	0	0	0	0	0
LFA2	F1	LFA1	LA2	LF1	L2	LF2	FA2
179	218	247	228	191	195	189	179
0	0	0	0	0	0	0	0
0	L2	0	FA1	FA1	LFA2	0	F2
124	166	177	153	133	145	141	130
0	0	0	0	0	0	0	0

The weights of dry heads are shown immediately underneath the treatment symbols. It is evident that the first row of plots is of poor fertility—treatments appearing in that row have only about half the yields that they give elsewhere. Further, there are indications that every row differs in fertility, the last row being second worst and the third row best. The fertility gradients are especially troublesome in that the four untreated controls all happen to lie in outside rows. The two replications give practically identical totals and remove none of this variation.

There is clearly little hope of obtaining information about the treatment effects unless weights are adjusted for differences in fertility from row to row. The adjustment may be made by covariance.

For simplicity, adjustments for the first row only will be shown: these remove the most serious environmental disturbance. As x variable we choose a variable that takes the value 1 for all plots in the first row and zero elsewhere. The x values are shown under the weights in Table 5. The rest of the analysis follows the usual covariance technique, Snedecor (1946).

TABLE 6
Sums of Squares and Products (y = weights, z = dummy variates)

	d.f.	y^2	yx	x^2
Blocks	1	657	0.0	0.00
Treatments	13	33,323	-200.2	1.75
Error	17	46,486	-380.0	4.25
Total	31	80,466	-580.2	6.00

Note that there are only 14 distinct treatments, since L1 is the same as L2. The reduction in the error S.S. due to covariance is $(380.0)^2/4.25$, or 33,976. The error mean square is reduced from 2,734 to 782 by means of the covariance, i.e., to less than one-third of its original value. The regression coefficient is $-380.0/4.25$, or -89.4 units.

Treatment means are adjusted in the usual way. For L1, which was unlucky in having two plots in the first row, the unadjusted mean is 89. The mean x value is 1, whereas the mean x value for the whole experiment is 8/32, or ¼. Hence the adjustment increases the L1 mean by (3/4) (89.4), the adjusted value being 156. For L2, which had no plots in the first row, the x mean is 0, and the adjustment reduces the mean from 180 to 158. It may be observed that the unadjusted mean of L2 was double that of L1, while the two adjusted means agree closely, as is reasonable since the two treatments are in fact identical.

If it were desired to adjust separately for every row, a multiple covariance with four x variables could be computed. Each x would take the value 1 for all plots in the corresponding row and 0 elsewhere. It will be realized that the covariance technique, if misused, can lead to an underestimation of errors. It is, however, worth keeping in mind as an occasional weapon for difficult cases. (See Table 6.)

8. EFFECTS OF NON-ADDITIVITY

Suppose that in a randomized blocks experiment, with two treatments and two replicates, the treatment and block effects are multiplicative rather than additive. That is, in either replicate, treatment B exceeds treatment A by a fixed percentage, while for either treatment, replicate 2 exceeds replicate 1 by a fixed percentage. Consider treatment percentages of 20% and 100% and replicate percentages of 10% and 50%. These together provide four combinations. Taking the observation for treatment A in replicate 1 as 1.0, the other observations are shown in Table 7.

TABLE 7
Hypothetical Data for Four Cases Where Effects Are Multiplicative

Rep.	T 20% R 10%		T 20% R 50%		T 100% R 10%		T 100% R 50%	
	A	B	A	B	A	B	A	B
1	1.0	1.2	1.0	1.2	1.0	2.0	1.0	2.0
2	1.1	1.32	1.5	1.8	1.1	2.2	1.5	3.0
d		.02		.10		.10		.50
σ_{na}		.01		.05		.05		.25

Thus, in the first case, 1.32 for B in replicate 2 is 1.2 times 1.1. Since no experimental error has been added, the error variance in a correct analysis should be zero. If the usual analysis of variance is applied to each little table, the calculated error in each case will have 1 d.f. If d is the sum of two corners minus the other two corners, the error S.S. is $d^2/4$, so that the standard error σ_{na} is $d/2$ (taken as positive). The values of d and of σ_{na} are shown below each table.

Consequently, in the first experiment, say, the usual analysis would lead to the statement that the average increase to B is 0.21 units ± 0.01, instead of to the correct statement that the increase to B is 20%. The standard error, although due entirely to the failure of the additive relationship, does perform a useful purpose. It warns us that the actual increase to B over A will vary from replication to replication and measures how much it will vary, so far as the experiment is capable of supplying information on this point. An experimenter who fails to see the correct method of analysis and uses ordinary methods will get less precise information from the experiment for predictive purposes, but if he notes the standard error he will not be misled into thinking that his information is more precise than it really is.

When experimental errors are present, the variance σ_{na}^2 will be added to the usual error variance σ_e^2. The ratio $\sigma_{na}^2/(\sigma_{na}^2 + \sigma_e^2)$ may appropriately be taken as a measure of the loss (fractional) of information due to non-additivity. In the four experiments, from left to right, the values of σ_{na} are respectively 0.9, 3.6, 3.2, and 13.3 percent of the mean yields of the experiments. In the first case, where treatment and replicate effects are small, the loss of information due to non-additivity will be trivial unless σ_e is very small. For example, with $\sigma_e = 5$ percent, the fractional loss is 0.81/25.81 or about 3 percent. In the two middle examples, where either the treatment or the replicate effect is substantial, the losses are beginning to be substantial. With $\sigma_e = 5$ percent in the second case, the loss would be about 30 percent. Finally, when both effects are large the loss is great.

Little study has been made in the literature of the general effects of non-additivity or of the extent to which this problem is present in the data that are usually handled by analysis of variance.[5] I believe, however, that the results from these examples are suggestive of the consequences in other cases. The principal effect is a loss of information. Unless experimental errors are low or there is a very serious departure from additivity, this loss should be negligible when treatment and replication effects do not exceed 20 percent, since within that range the additive relationship is likely to be a good approximation to most types that may arise.

Since the deviations from additivity are, as it were, amalgamated with the true error variance, the pooled error variance as calculated from the analysis of variance will take account of these deviations and should be relatively unbiased. This pooled variance may not, however, be applicable to comparisons between individual pairs of treatments. The examples above are too small to illustrate this point. But, clearly, with three treatments A, B, and C, the comparison (A–B) might be much less affected by non-additivity than the comparison (A–C). Thus non-additivity tends to produce heterogeneity of the error variance.[6]

---

[5] A relevant discussion of this problem for regressions in general with some interesting results, has been given by Jones (1946).

[6] It is an oversimplification to pretend, as in the discussion above, that the deviations from addivity act entirely like an additional component of random error. Discussion of the effects introduced by the systematic nature of the deviations would, however, unduly lengthen this paper.

If treatment or block effects, or both, are large, it will be worth examining whether treatment differences appear to be independent of the block means, or vice versa. There are, of course, limitations to what can be discovered from a single experiment. If relations seem non-additive, the next step is to seek a scale on which effects are additive. Again reference should be made to the paper following on transformations.

9. SUMMARY AND CONCLUDING REMARKS

The analysis of variance depends on the assumptions that the treatment and environmental effects are additive and that the experimental errors are independent in the probability sense, have equal variance, and are normally distributed. Failure of any assumption will impair to some extent the standard properties on which the widespread utility of the technique depends. Since an experimenter could rarely, if ever, convince himself that all the assumptions were exactly satisfied in his data, the technique must be regarded as approximative rather than exact. From general knowledge of the nature of the data and from a careful scrutiny of the data before analysis, it is believed that cases where the standard analysis will give misleading results or produce a serious loss of information can be detected in advance.

In general, the factors that are liable to cause the most severe disturbances are extreme skewness, the presence of gross errors, anomalous behavior of certain treatments or parts of the experiment, marked departures from the additive relationship, and changes in the error variance, either related to the mean or to certain treatments or parts of the experiment. The principal methods for an improved analysis are the omission of certain observations, treatments, or replicates, subdivision of the error variance, and transformation to another scale before analysis. In some cases, as illustrated by the numerical examples, the more exact methods require considerable experience in the manipulation of the analysis of variance. Having diagnosed the trouble, the experimenter may frequently find it advisable to obtain the help of the mathematical statistician.

References

Bartlett, M. S. The effect of non-normality of the t distribution. *Proceedings of the Cambridge Philosophical Society*, 1935, **31**, 223–231.

Cochran, W. G. Some difficulties in the statistical analysis of replicated experiments. *Empire Journal of Experimental Agriculture*, 1938, **6**, 157–175.

Finney, D. J. On the distribution of a variate whose logarithm is normally distributed. *Journal of The Royal Statistical Society, Supplement* 1941, **7**, 155–161.

Fisher, R. A. On the mathematical foundations of theoretical statistics. *Philosophical Transactions of the Royal Society of London, A,* 1922, **222**, 309–368.

Fisher, R. A. *Statistical methods for reasearch workers.* Edinburgh: Oliver and Boyd, 1948.

Fisher, R. A. *The design of experiments.* Edinburgh: Oliver and Boyd, 1949.

Hey, G. B. A new method of experimental sampling illustrated on certain non-normal populations. *Biometrika,* 1938, **30**, 68–80.

Jones, H. L. Linear regression functions with neglected variables. *Journal of the American Statistical Association,* 1946, **41**, 356–369.

Pearson, E. S. The analysis of variance in cases of non-normal variation. *Biometrika,* 1931, **23**, 114.

Snedecor, G. W. *Statistical methods* (4th ed.), Ames, Iowa: Iowa State College Press, 1946, Chaps. 12 and 13.

Yates, F. The analysis of replicated experiments when the field results are incomplete. *Empire Journal of Experimental Agriculture,* 1933, **1**, 129–142.

Yates, F. Incomplete latin squares. *Journal of Agricultural Science,* 1936, **26**, 301–315.

Yates, F. The formation of latin squares for use in field experiments. *Empire Journal of Experimental Agriculture,* 1933, **1**, 235–244.

Yates, F., & Hale, R. W. The analysis of latin squares when two or more rows, columns or treatment are missing. *Journal of the Royal Statistical Society, Supplement,* 1939, **6**, 67–79.

5

Potpourri

INTRODUCTION

One of the most central, and most hard to grasp, of all concepts in statistical theory is that of "degrees of freedom." Nevertheless, most textbooks provide very little information on degrees of freedom. It has been my experience that students know how to arrive at value of the degrees of freedom in a formula or when they must use a statistical table, but they do not know why they are using degrees of freedom in the first place. Fortunately, Helen M. Walker has provided an edifying article about degrees of freedom. Her article fulfills its purpose, ". . . to state as simply as possible what degrees of freedom represent, why the concept is important, and how the appropriate number may be readily determined" (p. 350).

The article by J. W. Tukey, "Conclusions vs. Decisions," was based on a talk given by Tukey to the Section of the Physical and Engineering Sciences of the American Statistical Association. The article argues for the place of statistical analysis in the development of knowledge, and differentiates decisions and decision procedures from conclusions and conclusion procedures. Tukey views a decision as a choice made to maximize gain in a specified situation with or without formal data and with or without logical   consistency with other decisions.

Conclusions are statements derived from data that are logically consistent with existing knowledge and are "accepted" but subject to future revision or rejection, given strong evidence against them. Conclusions are accepted as factual until proven otherwise. Thus, we may *decide* to cross a street against a red light but *conclude* that it is safe to cross the street when the light is green. Conclusions are not judged by one outcome from one situation. Tukey argues that decisions and conclusions both play a role in our accumulation of knowledge, but asks whether the major aim of science is to make decisions, to make conclusions, or both.

The last two articles in this chapter are somewhat different than most of those in this collection. The article by William F. Dukes, "N = 1," is a provocative discussion of the use of only one subject in an experimental study, something we would scarely think of after a course in statistics. Dukes offers historical evidence for the importance of research using only one subject, reasoning that "if uniqueness is involved a sample of one exhausts the population" (p. 382) and "if complete population generality exists," then a sample of one is sufficient. Certainly there are instances when one subject does provide evidence of generality, but before using an N = 1, we must remember to ask whether the one subject is representative.

Sir R. A. Fisher was one of the most eminent statisticians of the twentieth century. Fisher possessed the unusual ability to take complex mathematical statistical procedures and make them comprehensible to the research worker. He was the early link between the statistician and the research scientist. His book, *Statistical Methods for Research Workers* (Oliver and Boyd, Edinburgh, 1925) was first published in 1925 and has gone through twelve editions. Perhaps Fisher's greatest contribution was, as J. Neyman said, his "actual founding of an entirely novel discipline related to scientific research which I like to call the theory of experimentation" (p. 388). Fisher's work was often challenged—one of his most prominent antagonists was J. Neyman—yet the contribution he made to science is attested to by the fact that Neyman wrote the last article in this collection: "R. A. Fisher (1890–1902): An Appreciation."

5.1

Degrees of freedom

Helen M. Walker

A concept of central importance to modern statistical theory which few textbooks have attempted to clarify is that of "degrees of freedom." For the mathematician who reads the original papers in which statistical theory is now making such rapid advances, the concept is a familiar one needing no particular explanation. For the person who is unfamiliar with N-dimensional geometry or who knows the contributions to modern sampling theory only from secondhand sources such as textbooks, this concept often seems almost mystical, with no practical meaning.

Tippett, one of the few textbook writers who attempt to make any general explanation of the concept, begins his account (1931, p. 64) with the sentence, "This conception of *degrees of freedom* is not altogether easy to attain, and we cannot attempt a full justification of it here; but we shall show its reasonableness and shall illustrate it, hoping that as a result of familiarity with its use the reader will appreciate it." Not only do most texts omit all mention of the concept but many actually give incorrect formulas and procedures because of ignoring it.

In the work of modern statisticians, the concept of degrees of freedom is not found before "Student's" paper of 1908, it was first made explicit by the writings of R. A. Fisher, beginning with his paper of 1915 on the distribution of the correlation coefficient, and has only within the decade or so received general recognition. Nevertheless the concept was familiar to Gauss and his astronomical associates. In his classical work on the *Theory of the Combination of Observations* (Theoria Combinationis Observationum Erroribus Minimis Obnoxiae) and also in a work generalizing the theory of least squares with reference to the combination of observations (Ergänzung zur Theorie der den kleinsten Fehlern unter-

Helen M. Walker, "Degrees of Freedom," *Journal of Educational Psychology*, 31, 1940, 253–269. Copyright (1940) by the American Psychological Association, and reproduced by permission.

worfen Combination der Beobachtungen, 1826), he states both in words and by formula that the number of observations is to be decreased by the number of unknowns estimated from the data to serve as divisor in estimating the standard error of a set of observations, or in our terminology $\sigma^2 = \Sigma x^2/N - r$ where r is the number of parameters to be estimated from the data.

The present paper is an attempt to bridge the gap between mathematical theory and common practice, to state as simply as possible what degrees of freedom represent, why the concept is important, and how the appropriate number may be readily determined. The treatment has been made as non-technical as possible, but this is a case where the mathematical notion is simpler than any non-mathematical interpretation of it. The paper will be developed in four sections: (I) The freedom of movement of a point in space when subject to certain limiting conditions, (II) The representation of a statistical sample by a single point in N-dimensional space, (III) The import of the concept of degrees of freedom, and (IV) illustrations of how to determine the number of degrees of freedom appropriate for use in certain common situations.

I. THE FREEDOM OF MOVEMENT OF A POINT IN SPACE WHEN SUBJECT TO CERTAIN LIMITING CONDITIONS

As a preliminary introduction to the idea, it may be helpful to consider the freedom of motion possessed by certain familiar objects, each of which is treated as if it were a mere moving point without size. A drop of oil sliding along a coil spring or a bead on a wire has only one degree of freedom for it can move only on a one-dimensional path, no matter how complicated the shape of that path may be. A drop of mercury on a plane surface has two degrees of freedom, moving freely on a two-dimensional surface. A mosquito moving freely in three-dimensional space, has three degrees of freedom.

Considered as a moving point, a railroad train moves backward and forward on a linear path which is a one-dimensional space lying on a two-dimensional space, the earth's surface, which in turn lies within a three-dimensional universe. A single coördinate, distance from some origin, is sufficient to locate the train at any given moment of time. If we consider a four-dimensional universe in which one dimension is of time and the other three dimensions of space, two coördinates will be needed to locate the train, distance in linear units from a spatial origin and distance in time units from a time origin. The train's path which had only one dimension in a space universe has two dimensions in a space-time universe.

A canoe or an automobile moves over a two-dimensional surface which lies upon a three-dimensional space, is a section of a three-dimensional space. At any given moment, the position of the canoe, or auto, can be given by two coördinates. Referred to a four-dimensional space-time universe, three coördinates would be needed to give its location, and its path would be a space of three dimensions, lying upon one of four.

In the same sense an airplane has three degrees of freedom in the usual universe of space, and can be located only if three coördinates are known. These might be latitude, longitude, and altitude; or might be altitude, horizontal distance from some origin, and an angle; or might be direct distance from some origin, and two direction angles. If we consider a given instant of time as a section through the space-time universe, the airplane moves in a four-dimensional path and can be located by four coördinates, the three previously named and a time coördinate.

The degrees of freedom we have been considering relate to the motion of a point, or freedom of translation. In mechanics freedom of *rotation* would be equally important. A point, which has position only, and no size, can be translated but not rotated. A real canoe can turn over, a real airplane can turn on its axis or make a nose dive, and so these real bodies have degrees of freedom of rotation as well as of translation. The parallelism between the sampling problems we are about to discuss and the movement of bodies in space can be brought out more clearly by discussing freedom of translation, and disregarding freedom of rotation, and that has been done in what follows.

If you are asked to choose a pair of numbers (x, y) at random, you have complete freedom of choice with regard to each of the two numbers, have two degrees of freedom. The number pair may be represented by the coördinates of a point located in tht x, y plane, which is a two-dimensional space. The point is free to move anywhere in the horizontal direction parallel to the xx' axis, and is also free to move anywhere in the vertical direction, parallel to the yy' axis. There are two independent variables and the point has two degrees of freedom.

Now suppose you are asked to choose a pair of numbers whose sum is 7. It is readily apparent that only one number can be chosen freely, the second being fixed as soon as the first is chosen. Although there are two variables in the situation, there is only one independent variable. The number of degrees of freedom is reduced from two to one by the imposition of the condition $x + y = 7$. The point is not now free to move anywhere to the xy plane but is constrained to remain on the line whose graph is $x + y = 7$, and this line is a one-dimensional space lying in the original two-dimensional space.

Suppose you are asked to choose a pair of numbers such that the sum of their squares is 25. Again it is apparent that only one number can be chosen arbitrarily, the second being fixed as soon as the first is chosen. The point represented by a pair of numbers must lie on a circle with center at the origin and radius 5. This circle is a one-dimensional space lying in the original two-dimensional plane. The point can move only forward or backward along this circle, and has one degree of freedom only. There were two numbers to be chosen $(N = 2)$ subject to one limiting relationship $(r = 1)$ and the resultant number of degrees of freedom is $N - r = 2 - 1 = 1$.

Suppose we simultaneously impose the two conditions $x + y = 7$ and $x^2 + y^2 = 25$. If we solve these equations algebraically we get only two possible solutions, $x = 3$, $y = 4$, or $x = 4$, $y = 3$. Neither variable can be chosen at will. The point, once free to move in two directions, is now constrained by the equation $x + y = 7$ to move only along a straight line, and is constrained by the equation $x^2 + y^2 = 25$ to move only along the circumference of a circle, and by the two together is confined to the intersection of that line and circle. There is no freedom of motion for the point. $N = 2$ and $r = 2$. The number of degrees of freedom is $N - r = 2 - 2 = 0$.

Consider now a point (x, y, z) in three-dimensional space $(N = 3)$. If no restrictions are placed on its coördinates, it can move with freedom in each of three directions, has three degrees of freedom. All three variables are independent. If we set up the restriction $x + y + z = c$, where c is any constant, only two of the numbers can be freely chosen, only two are independent observations. For example, let $x - y - z = 10$. If now we choose, say, $x = 7$ and $y = 9$, then z is forced to be -12. The equation $x - y - z = c$ is the equation of a plane, a two-dimensional space cutting across the original three-dimensional space, and a point lying on this space has two degrees of freedom. $(N - r = 3 - 1 = 2.)$ If the coördinates of the (x, y, z) point are made to conform to the condition $x^2 + y^2 + z^2 = k$, the point will be forced to lie on the surface of a sphere whose center is at the origin and whose radius is $\sqrt{k}$. The surface of a sphere is a two-dimensional space. $(N = 3, r = 1, N - r = 3 - 1 = 2.)$

If both conditions are imposed simultaneously, the point can lie only on the intersection of the sphere and the plane, that is, it can move only along the circumference of a circle, which is a one-dimensional figure lying in the original space of three dimensions. $(N - r = 3 - 2 = 1.)$ Considered algebraically, we note that solving the pair of equations in three variables leaves us a single equation in two variables. There can be complete freedom of choice for one of these, no freedom for the other. There is one degree of freedom.

The condition $x = y = z$ is really a pair of independent conditions, $x = y$ and $x = z$, the condition $y = z$ being derived from the other two. Each of these is the equation of a plane, and their intersection gives a straight line through the origin making equal angles with the three axes. If $x = y = z$, it is clear that only one variable can be chosen arbitrarily, there is only one independent variable, the point is constrained to move along a single line, there is one degree of freedom.

These ideas must be generalized for N larger than 3, and this generalization is necessarily abstract. Too ardent an attempt to visualize the outcome leads only to confusion. Any set of N numbers determine a single point in N-dimensional space, each number providing one of the N coördinates of that point. If no relationship is imposed upon these numbers, each is free to vary independently of the others, and the number of degrees of freedom is N. Every necessary relationship imposed upon them reduces the number of degrees of freedom by one. Any equation of the first degree connecting the N variables is the equation of what may be called a hyperplane (Better not try to visualize it!) and is a space of $N - 1$ dimensions. If, for example, we consider only points such that the sum of their coördinates is constant, $\Sigma X = c$, we have limited the point to an $N - 1$ space. If we consider only points such that $\Sigma(X - M)^2 = k$, the locus is the surface of a hypersphere with center at the origin and radius equal to $\sqrt{k}$. This surface is called the locus of the point and is a space of $N - 1$ dimensions lying within the original N space. If r such conditions should be imposed simultaneously, the point would be confined to the intersection of the various loci, which would be a space of $N - r$ dimensions lying within the original N space. The number of degrees of freedom would be $N - r$.

II. THE REPRESENTATION OF A STATISTICAL SAMPLE BY A POINT IN N-DIMENSIONAL SPACE

If any N numbers can be represented by a single point in a space of N dimensions, obviously a statistical sample of N cases can be so represented by a single sample point. This device, first employed by R. A. Fisher in 1915 in a celebrated paper ("Frequency distribution of the values of the correlation coefficient in samples from an indefinitely large population") has been an enormously fruitful one, and must be understood by those who hope to follow recent developments.

Let us consider a sample space of N dimensions, with the origin taken at the true population mean, which we will call μ, so that $X_1 - \mu = x_1$, $X_2 - \mu = x_2$, etc., where X_1, X_2, ... X_N are the raw scores of the N individuals in the sample. Let M be the mean and s the standard deviation of a sample of N cases. Any set of N observations determines a single sample point, such as S. This point has N degrees of freedom if no conditions are imposed upon its coördinates.

All samples with the same mean will be represented by sample points lying on the hyper-plane $(X_1 - \mu) + (X_2 - \mu) + \ldots + (X_N - \mu) = N(M - \mu)$, or $\Sigma X = NM$, a space of $N - 1$ dimensions.

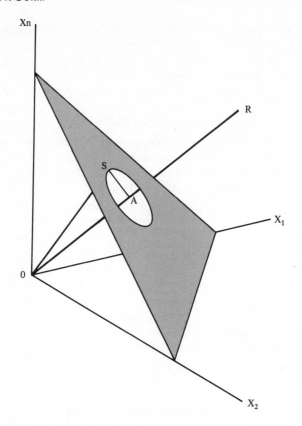

FIGURE 1

If all cases in a sample were exactly uniform, the sample point would lie upon the line $X_1 - \mu = X_2 - \mu = X_3 - \mu = \ldots = X_N - \mu = M - \mu$ which is the line OR in Figure 1, a line making equal angles with all the coördinate axes. This line cuts the plane $\Sigma X = NM$ at right angles at a point we may call A. Therefore, A is a point whose coördinates are each equal to $M - \mu$. By a well-known geometric relationship,

$$\overline{OS^2} = (X_1 - \mu)^2 + (X_2 - \mu)^2 + \cdots + (X - \mu)^2$$

$$\overline{OA^2} = N(M - \mu)^2$$

$$\overline{OS^2} = \overline{OA^2} + \overline{AS^2}$$

$$\overline{AS^2} = \Sigma(X - \mu)^2 - N(M - \mu)^2 = \Sigma X^2 - NM^2 = Ns^2$$

Therefore, $OA = (M - \mu)\sqrt{N}$ and $AS = s\sqrt{N}$.

The ratio OA/AS is thus $M - \mu/s$ and is proportional to the ratio of the amount by which a sample mean deviates from the population mean to its own standard error. The fluctuation of this ratio from sample to sample produces what is known as the t-distribution.

For computing the variability of the scores in a sample around a population mean which is known *a priori*, there are available N degrees of freedom because the point S moves in N-dimensional space about O; but for computing the variability of these same scores about the mean of their own sample, there are available only $N - 1$ degrees of freedom, because one degree has been expended in the computation of that mean, so that the point S moves about A in a space of only $N - 1$ dimensions.

Fisher has used these spatial concepts to derive the sampling distribution of the correlation coefficient. The full derivation is outside the scope of this paper but certain aspects are of interest here. When we have N individuals each measured in two traits, it is customary to represent the N pairs of numbers by a correlation diagram of N points in two-dimensional space. The same data can, however, be represented by two points in N-dimensional space, one point representing the N values of X and the other the N values of Y. In this frame of reference the correlation coefficient can be shown to be equal to the cosine of the angle between the vectors to the two points, and to have $N - 2$ degrees of freedom.

III. THE IMPORT OF THE CONCEPT

If the normal curve adequately described all sampling distributions, as some elementary treatises seem to imply, the concept of degrees of freedom would be relatively unimportant, for this number does not appear in the equation of the normal curve, the shape of the curve being the same no matter what the size of the sample. In certain other important sampling distributions—as for example the Poisson—the same thing is true, that the shape of the distribution is independent of the number of degrees of freedom involved. Modern statistical analysis, however, makes much use of several very important sampling distributions for which the shape of the curve changes with the effective size of the sample. In the equations of such curves, the number of degrees of freedom appears as a parameter (called n in the equations which follow) and probability tables built from these curves must be entered with the correct value of n. If a mistake is made in determining n from the data, the wrong probability value will be obtained from the table, and the significance of the test employed will be wrongly interpreted. The Chi-square distribution, the t-distribution, and the F and z distributions are now commonly used even in elementary work, and the table for each of these must be entered with the appropriate value of n.

Let us now look at a few of these equations to see the rôle played in them by the number of degrees of freedom. In the formulas which follow, C represents a constant whose value is determined in such a way as to make the total area under the curve equal to unity. Although this constant involves the number of degrees of freedom, it does not need to be considered in reading probability tables because, being a constant multiplier, it does not affect the proportion of area under and given segment of the curve, but serves only to change the scale of the entire figure.

Normal Curve

$$y = C_1 e^{-\frac{x^2}{2\sigma^2}}$$

The number of degrees of freedom does not appear in the equation, and so the shape of the curve is independent of it. The only variables to be shown in a probability table are x/σ and y or some function of y such as a probability value.

Chi-square

$$y = C_2(\chi^2)^{\frac{n-2}{2}} e^{-\frac{x^2}{2}}$$

The number of degrees of freedom appears in the exponent. When $n = 1$, the curve is J-shaped. When $n = 2$, the equation reduces to $y = C_2 e^{-\frac{x^2}{2}}$ and has the form of the positive half of a normal curve. The curve is always positively skewed, but as n increases it becomes more and more nearly like the normal, and becomes approximately normal when n is 30 or so. A probability table must take account of three variables, the size of Chi-square, the number of degrees of freedom, and the related probability value.

t-Distribution

$$y = C_3 \left(1 + \frac{t^2}{n}\right)^{-\frac{(n+1)}{2}}$$

The number of degrees of freedom appears both in the exponent and in the fraction t^2/n. The curve is always symmetrical, but is more peaked than the normal when n is small. This curve also aproaches the normal form as n increases. A table of probability values must be entered with the computed value of t and also with the appropriate value of n. A few selected values will show the comparison between estimates of significance read from a table of the normal curve and a t-table.

For a normal curve, the proportion of area in both tails of the curve beyond 3σ is .0027. For a t-distribution the proportion is as follows:

n	1	2	5	10	20
p	.204	.096	.030	.014	.007

Again, for a normal curve, the point such that .01 of the area is in the tails, is 2.56σ from the mean.

For a t-distribution, the position of this point is as follows:

n	1	2	3	5	10	20	30
x/σ	63.6	9.9	5.8	4.0	3.2	2.8	2.75

F-Distribution and z-Distribution

$$y = C_4 \frac{F^{\frac{n_1-2}{2}}}{(n_1 F + n_2)^{\frac{n_1+n_2}{2}}}$$

and

$$y = C_5 \frac{e^{n_1 z}}{(n_1 e^{2z} + n_2)^{\frac{n_1+n_2}{2}}}$$

In each of these equations, which provide the tables used in analysis of variance problems, there occurs not only the computed value of F (or of z), but also the two parameters n_1 and n_2, n_1 being the number of degrees of freedom for the mean square in the numerator of F and n_2 the number of degrees of freedom for that in the denominator. Because a probability table must be entered with all three, such a table often shows the values for selected probability values only. The tables published by Fisher give values for $p = .05$, $p = .01$, and $p = .001$; those by Snedecor give $p = .05$ and $p = .01$.

Sampling Distribution of r

This is a complicated equation involving as parameters the true correlation in the population, p; the observed correlation in the sample, r; and the number of degrees of freedom. If $\rho = 0$ the distribution is symmetrical. If $\rho = 0$ and n is large, the distribution becomes normal. If $\rho \neq 0$ and n is small the curve is definitely skewed. David's *Tables of the Correlation Coefficient* (Issued by the Biometrika Office, University College, London, 1938) must be entered with all three parameters.

IV. DETERMINING THE APPROPRIATE NUMBER OF DEGREES OF FREEDOM IN CERTAIN TYPICAL PROBLEMS

A universal rule holds: The number of degrees of freedom is always equal to the number of observations minus the number of necessary relations obtaining among these observations. In geometric terms, the number of observations is the dimensionality of the original space and each relationship represents a section through that space restricting the sample point to a space of one lower dimension. Imposing a relationship upon the observations is equivalent to estimating a parameter from them. For example, the relationship $\Sigma X = NM$ indicates that the mean of the population has been estimated from observations. The number of degrees of freedom is also equal to the number of independent observations, which is the number of original observations minus the number of parameters estimated from them.

Standard Error of a Mean

This is $\sigma_{mean} = \sigma/\sqrt{N}$ when σ is known for the population. As σ is seldom known *a priori*, we are usually forced to make use of the observed standard deviation in the sample, which we will call s. In this case $\sigma_{mean} \doteq s/\sqrt{N-1}$, one degree of freedom being lost because deviations have been taken around the sample mean, so that we have imposed one limiting relationship, $\Sigma X = NM$, and have thus restricted the sample point to a hyperplane of $N - 1$ dimensions.

Without any reference to geometry, it can be shown by an algebraic solution that $s\sqrt{N} \doteq \sigma\sqrt{N-1}$. (The symbol $\doteq$ is to be read "tends to equal" or "approximates.")

Goodness of Fit of Normal Curve to a Set of Data

The number of observations is the number of intervals in the frequency distribution for which an observed frequency is compared with the frequency to be expected on the assumption of a normal distribution. If this normal curve has an arbitrary mean and standard deviation agreed upon in advance, the number of degrees of freedom with which we enter the Chi-square table to test goodness of fit is one less than the number of intervals. In this case one restriction is imposed; namely $\Sigma f = \Sigma f'$, where f is an observed and f' a theoretical frequency. If, however, as is more common, the theoretical curve is made to conform to the observed data in its mean and standard deviation, two additional restrictions are imposed; namely $\Sigma fX = \Sigma f'X$ and $\Sigma f(X - M)^2 = \Sigma f'(X - M)^2$, so that the number of degrees of freedom is three less than the number of intervals compared. It is clear that when the curves are made to agree in mean and standard deviation, the discrepancy between observed and theoretical frequencies will be reduced, so the number of degrees of freedom in relation to which that discrepancy is interpreted should also be reduced.

Relationship in a Contingency Table

Suppose we wish to test the existence of a relationship between trait A, for which there are three categories, and trait B, for which there are five, as shown in Figure 2. We have fifteen cells in the table, giving us fifteen observations, inasmuch as an "observation" is now the frequency in a single cell. If we want to ask whether there is sufficient evidence to believe that in the population from which this sample is drawn A and B are independent, we need to know the cell frequencies which would be expected under that hypothesis. There are then fifteen comparisons to be made between observed frequencies and expected frequencies. But are all fifteen of these comparisons independent?

If we had *a priori* information as to how the traits would be distributed theoretically, then all but one of the cell comparisons would be independent, the last cell frequency being fixed in order to make up the proper total of one hundred fifty, and the degrees of freedom would be $15 - 1 = 14$. This is the situation Karl Pearson had in mind when he first developed his Chi-square test of goodness of fit, and Table XII in Vol. I of his *Tables for Statisticians and Biometricians* is made up on the assumption that the number of degrees of freedom is one less than the number of observations. To use it when that is not the case we merely readjust the value of n with which we enter the table.

In practice we almost never have *a priori* estimates of theoretical frequencies, but must obtain them from the observations themselves, thus imposing restrictions on the number of independent observations and reducing the degrees of freedom available for estimating reliability. In this case, if we estimate the theoretical frequencies from the data, we would estimate the frequency $f'_{11} = (20)(40)/150$ and others in similar fashion. Getting the expected cell frequencies from the observed marginal frequencies imposes the following relationships:

(a) $f_{11} + f_{21} + f_{31} + f_{41} + f_{51} = 40$
$f_{12} + f_{22} + f_{32} + f_{42} + f_{52} = 60$
$f_{13} + f_{23} + f_{33} + f_{43} + f_{53} = 50$

(b) $f_{11} + f_{12} + f_{13} = 20$
$f_{21} + f_{22} + f_{23} = 20$
$f_{31} + f_{32} + f_{33} = 35$
$f_{41} + f_{42} + f_{43} = 30$
$f_{51} + f_{52} + f_{53} = 50$

(c) $f_{11} + f_{21} + \cdots + f_{51} + f_{12} + \cdots + f_{53} = 150$

At first sight, there seem to be nine relationships, but it is immediately apparent that (c) is not a new one, for it can be obtained either by adding the three (a) equations or the five (b) equations. Also any one of the remaining eight can be obtained by appropriate manipulation of the other seven. There are then only seven independent necessary relationships imposed upon the cell frequencies by requiring them to add up to the observed marginal totals. Thus $n = 15 - 7 = 8$, and if we compute Chi-square, we must enter the Chi-square table with eight

	A_1	A_2	A_3	
B_1	12	3	5	20
B_2	3	6	11	20
B_3	3	30	2	35
B_4	9	14	7	30
B_5	13	7	25	45
	40	60	50	150

FIGURE 2

Observed joint frequency distribution of two traits A and B

	A_1	A_2	A_3	
B_1	f_{11}	f_{12}	f_{13}	20
B_2	f_{21}	f_{22}	f_{23}	20
B_3	f_{31}	f_{32}	f_{33}	35
B_4	f_{41}	f_{42}	f_{43}	30
B_5	f_{51}	f_{52}	f_{53}	45
	40	60	50	150

FIGURE 3

Observed marginal frequencies of two traits A and B

degrees of freedom. The same result can be obtained by noting that two entries in each row and four in each column can be chosen arbitrarily and there is then no freedom of choice for the remaining entries.

In general in a contingency table, if $c =$ number of columns and $r =$ number of rows, the number of degrees of freedom is $n = (c - 1)(r - 1)$ or $n = rc - (r + c - 1)$.

Variance in a Correlation Table

Suppose we have a scatter diagram with c columns, the frequencies in the various columns being $n_1, n_2, \ldots n_c$, the mean values of Y for the columns being $m_1, m_2, \ldots m_c$, and the regression values of Y estimated from X being $\tilde{Y}_1, \tilde{Y}_2, \ldots \dot{Y}_c$. Thus for any given column, the sum of the Y's is

$$\sum_1^{n_i} fY = n_i m_i.$$

For the entire table

$$N = n_1 + n_2 + \cdots + n_c, NM = \sum_1^c \sum_1^{n_i} fY,$$

so that $NM = n_1 m_1 + n_2 m_2 + \ldots n_c m_c.$

Now we may be interested in the variance of all the scores about the total mean, of all the scores about their own column means, of all the scores about the regression line, of regressed values about the total mean, of column means about the total mean, or of column means about the regression line, and we may be interested in comparing two such variances. It is necessary to know how many degrees of freedom are available for such comparisons.

Total variance. For the variance of all scores about the total mean, this is

$$s^2 = \frac{1}{N} \sum_1^N (Y - M)^2,$$

we have N observations and only one restriction; namely, $\Sigma f Y = NM$. Thus there are $N - 1$ degrees of freedom.

Variance of regressed values about total mean. The equation for the regressed values being

$$\tilde{Y} - M_y = r \frac{s_x}{s_y} (X - M_x),$$

it is clear that as soon as x is known, y is also known. The sample point can move only on a straight line. There is only one degree of freedom available for the variance of regressed values.

Variance of scores about regression line. There are N residuals of the form $Y - \tilde{Y}$ and their variance is the square of the standard error of estimate, or $s^2{}_y(1 - r^2{}_{xy})$. There are N observations and two restrictions; namely,

$$\Sigma f(Y - \tilde{Y}) = 0$$

and

$$\Sigma f(Y - \tilde{Y})^2 = N s_y^2 (1 - r_{xy}^2).$$

Thus there are $N - 2$ degrees of freedom available.

Variance of scores about column means. If from each score we subtract not the regression value but the mean of the column in which it stands, the variance of the residuals thus obtained will be $s_y{}^2(1 - E^2)$ where E is the correlation ratio obtained from the sample. There are N such residuals. For each column we have the restriction

$$\sum_1^{n_i} f Y = n_i m_i$$

making c restrictions in all. The number of degrees of freedom for the variance within columns is therefore $N - c$.

Variance of column means about total means. To compute this variance we have c observations, *i.e.*, the means of c columns, restricted by the single relation

$$NM = \sum_1^c n_i m_i,$$

and therefore have $c - 1$ degrees of freedom. The variance itself can be proved to be $s_y{}^2 E^2$, and represents the variance among the means of columns.

Variance of column means about regression line. If for each column we find the difference $m_i - \tilde{Y}_i$ between the column mean and the regression value, and then find

$$\frac{1}{N} \sum_1^c f_i(m_i - \tilde{Y}_i)^2,$$

the result will be $s_y^2(E^2 - r^2)$ which is a variance representing the departure of the means from linearity. There is one such difference for each column, giving us c observations, and these observations are restricted by the two relationships,

$$\sum_1^c f_i(m_i - \tilde{Y}_i) = 0$$

and

$$\sum_1^c f_i(m_i - \tilde{Y}_i)^2 = N s_y^2(E^2 - r^2).$$

Therefore, we have $c - 2$ degrees of freedom.

The following scheme shows these relationships in summary form:

Source of variation	Formula	Degrees of freedom
(d) Scores about column means	$s^2(1 - E^2)$	$N - c$
(e) Means about total mean	$s^2 E^2$	$c - 1$
(a) Total	s^2	$N - 1$
(c) Scores about regression line	$s^2(1 - r^2)$	$N - 2$
(b) Regressed values about total mean	$s^2 r^2$	1
(a) Total	s^2	$N - 1$
(d) Scores about column means	$s^2(1 - E^2)$	$N - c$
(f) Column means about regression line	$s^2(E^2 - r^2)$	$c - 2$
(c) Scores about regression line	$s^2(1 - r^2)$	$N - 2$
(b) Regressed values about total mean	$s^2 r^2$	1
(f) Column means about regression line	$s^2(E^2 - r^2)$	$c - 2$
(e) Column means about total mean	$s^2 E^2$	$c - 1$
(b) Regressed values about total mean	$s^2 r^2$	1
(f) Column means about regression line	$s^2(E^2 - r^2)$	$c - 2$
(d) Scores about column means	$s^2(1 - E^2)$	$N - c$
(a) Total	s^2	$N - 1$

It is apparent that these variances have additive relationships and that their respective degrees of freedom have exactly the same additive relationships.

Tests based on ratio of two variances. From any pair of these additive variances, we may make an important statistical test. Thus, to test whether linear correlation exists in the population or not, we may divide $s^2r^2/1$ by $s^2(1-r)^2/N-2$ obtaining $r^2(N-2)/1-r^2$. To test whether a relationship measureable by the correlation ratio exists in the population, we may divide $s^2E^2/c-1$ by $s^2(1-E^2)/N-c$ obtaining $E^2/1-E^2 \cdot N - c/c - 1$. To test whether correlation is linear, we may divide $s^2(E^2-r^2)/c-2$ by $s^2r^2/1$ obtaining $E^2-r^2/r^2(c-2)$ or may divide $s^2(E^2-r^2)/c-2$ by $s^2(1-E^2)/N-c$ obtaining $E^2-r^2/1-E^2 \cdot N - c/c - 2$. In each case, the resulting value is referred to Snedecor's F-table which must be entered with the appropriate number of degrees of freedom for each variance. Or we may find the logarithm of the ratio to the base e, take half of it, and refer the result to Fisher's z-table, which also must be entered with the appropriate number of degrees of freedom for each variance.

Partial Correlation

For a coefficient of correlation of zero order, there are $N-2$ degrees of freedom. This is obvious, since a straight regression line can be fitted to any two points without residuals, and the first two observations furnish no estimate of the size of r. For each variable that is held constant in a partial correlation, one additional degree of freedom is lost, so that for a correlation coefficient of the pth order, the degrees of freedom are $N-p-2$. This places a limit upon the number of meaningful interrelationships which can be obtained from a small sample. As an extreme illustration, suppose twenty-five variables have been measured for a sample of twenty-five cases only, and all the intercorrelations computed, as well as all possible partial correlations—the partials of the twenty-third order will of necessity be either $+1$ or -1, and thus are meaningless. Each such partial will be associated with $25-23-2$ degrees of freedom. If the partial were not $+1$ or -1 the error variance $\sigma^2(1-r^2)/N-p-2$ would become infinite, a fantastic situation.

References

Fisher, R. A. "Frequency distribution of the values of the correlation coefficient in samples from an indefinitely large population." *Biometrika*, Vol. x, 1915, pp. 507–521. First application of n-dimensional geometry to sampling theory.

Tippett, L. H. C. *The Methods of Statistics.* Williams and Norgate, Ltd., 1931. One of the few attempts to treat the concept of degrees of freedom in general terms, but without geometric background, is made on pages 64–65.

5.2

Conclusions vs. decisions[1]

John W. Tukey

With the exception of appendices 2 and 3, the following is based
on the after dinner talk given by Professor John W. Tukey at the
first meeting of the Section of the Physical and Engineering Sciences of the American Statistical Association held in New York
City on May 26, 1955. This talk was repeated at a later date
before a dinner meeting of the Metropolitan Section of the
American Society for Quality Control. On both occasions considerable discussion ensued. The talk is published here both for
the record, and in the hope that some readers may be stimulated
to prepare written rejoinders.

INTRODUCTION

My subject tonight should be both interesting and professionally
relevant, and yet should not involve formulas or a blackboard. Of the
topics most professionally relevant to statisticians, I must choose between human relations, as between statistician and client, and statistical
philosophy, both subjects where our practices often outshine our formal
philosophy, both subjects were more discussion and better understanding
are needed if our practices are to improve as fast as they should.

It is especially important that our discussion and understanding of
statistical philosophy be firm and well-balanced. For one-sided development, no matter how important the single aspect may be, will ultimately
deflect some, if not all, of our practices into unwise bypaths.

I have been concerned for a number of years with the tendency of
decision theory to attempt the conquest of all statistics. This concern
has been founded, in large part, upon my belief that science does not

John W. Tukey, "Conclusions vs. Decisions," *Technometrics*, **2**, 1960, 424–433.

[1] Prepared in part in connection with research sponsored by the Office of
Naval Research.

live by decisions alone—that its main support is a different sort of inference.

Effective discussion of this problem, and a real start toward the development of a consensus of opinion, has been retarded by the absence of a word for this other sort of inference, a word which could be contrasted with "decisions." For me, there is now a word. (Some dislike it, but no one suggested a better choice.) The word is "conclusions." Conclusion theory is intended, not to replace decision theory, but to stand firm beside it.

Because I believe that conclusions are even more important to science than decisions, it is particularly appropriate that I am able to speak to the first meeting of the ASA's new Section on Physical and Engineering Sciences about the relations, and the differences, between decisions and conclusions. I know of no better way to wish the Section well than to encourage its membership to thought and discussion on a topic which I believe will remain important to the carrying out of the functions of all of its members.

DECISIONS, WHAT ARE THEY?

Some of us have read about decision theory, most of us have heard of it, and all of us make decisions. But do we have a clear idea of what a decision-theorist's decision is? Have the books made the essential situation clear? Or have they discussed only the externals of a single formulation? In fact, there has been so little discussion of essentials that I have had to formulate my own idea of what a "decision," in the sense of modern decision theory, really is.

The decisions of practice are far more nearly of the form "let us decide to act for the present as if" than of the form long conventional in treatments of decision theory—"we accept." The distinction is important and too often neglected. The restrictions "act · · · as if" and "for the present" convey two separate and important ideas, ideas which serve to distinguish conclusions from decisions, ideas which epitomize much of what I wish to say.

When an engineer must choose at once between two ways of building a bridge, or a doctor must choose which of two treatments to apply to a patient who is critically ill, or when a businessman must choose between two policies for the season that is now upon him, each must *weigh* alternative A against alternative B in this *immediate* situation, and strive to select the alternative that will yield the bigger reward, whether this reward be a cheaper safe bridge, a better chance of recovery for the patient, or a more profitable season. The possible actions are defined, their consequences in various "states of nature" are understood, and some evidence about these states of nature is at hand. In each instance the individual must judge whether to act as if the reward from alternative A will indeed prove to be greater than that from alternative B, (which we may abbreviate "$A > B$"), or whether the opposite is true ("$A < B$").

The three alternative decisions:

(1) to act in the present situation as if $A > B$,
(2) to act in the present situation as if $A = B$,
(3) to act in the present situation as if $A < B$

seem to me reasonably stated, while the conventional statements of the alternatives:

(1′) to accept $A > B$,
(2′) to accept $A = B$,
(3′) to accept $A < B$

seem to have been (unconsciously) well calculated to mislead the reader or student.

When we say "act as if $A > B$," we have made *no* judgment as to the "truth" or "certainty beyond a reasonable doubt" of the statement "$A > B$." When we say "for the present," we are referring only to the particular situation under consideration at present. Thus what we have done is weigh both the evidence concerning the relative merits of A and B and also *the probable consequences in the present situation* of various actions (actions, not decisions!). Finally, we have decided that the particular course of action which would be appropriate if A were truly $>B$ is the most reasonable one to adopt *in the specific situation* that faces us.

When we say "act as if $A > B$" and "in the present situation," we assert *no* judgment as to the "truth" or "certainty beyond a reasonable doubt" of the statement "$A > B$," and we make no judgment about the wisdom of choosing among actions in all, or even many, of the situations in which a knowledge that A was truly $> B$ would determine a wise man's choice. The consequences in *other* situations of acting as if $A > B$ have not been considered. It is important that we have not done these things; it is perhaps even more important that we know that we have not done them.

What has been done is simple and specific. The *evidence* concerning the relative rewards from the alternatives has been weighed: The *consequences in the present situation* of various actions (not decisions!) have been assessed. We have decided that, in this single specific situation, the particular action that would be appropriate if A were truly $> B$ is the most reasonable action to take.

Two sorts of special cases may help to tie down these remarks: It is often necessary to make a decision on the basis of no formal data at all. (Consider the hen crossing the road!) It may be reasonable to make two opposite decisions at the same time with regard to different actions. (How many of us both save for *our own* future and carry life insurance, perhaps even in a single policy? One is a decision to act as if we will live, the other a decision to act as if we will die!)

Decisions to "act for the present as if" are attempts to do as well as possible in specific situations, to choose wisely among the available gambles.

CONCLUSIONS, WHAT MAY THEY BE?

Like any other human endeavor science involves many decisions, but it progresses by the building up of a fairly well established body of knowledge. (One whose relevance is supposed to be broad.) This body grows by the reaching of conclusions—by acts whose essential characteristics differ widely from the making of decisions. Conclusions are established with careful regard to evidence, but without regard to consequences of specific actions in specific circumstances. (They are, of course, based on specific experiments or observations.) Conclusions are withheld until adequate evidence has accumulated.

A conclusion is a statement which is to be accepted as applicable to the conditions of an experiment or observation unless and until unusually strong evidence to the contrary arises. This definition has three crucial parts; two explicit, and the third implicit. It emphasizes "acceptance," in the original, strong sense of that word; it speaks of "unusually strong evidence"; and it implies the possibility of later rejection.

First, the conclusion is to be *accepted*. It is taken into the body of knowledge, not just into the guidebook of advice for immediate action, as would be the case with a decision. It is something of lasting value extracted from the data.

Indeed, the conclusion is to remain accepted, unless and until *unusually strong* evidence to the contrary arises. This implies that only a small percentage of all conclusions will, in due course, be upset.

Third, a conclusion is accepted subject to future *rejection*, when and if the evidence against it becomes strong enough. (Only a small proportion of conclusions will be rejected.) It is taken to be of lasting value, but not necessarily of everlasting value.

These characteristics are very different from those of a decision-theorist's decision. The differences are extremely important.

It has been wisely said that "science is the use of alternative working hypotheses." Wise scientists use great care and skill in selecting the bundle of alternative working hypotheses they use. Conclusions typically reduce the spread of the bundle of those working hypotheses which are regarded as still consistent with the observations. Hence conclusions must be reached cautiously, firmly, not too soon and not too late. And they must be judged by their long run effects, by their "truth," not by specific consequences of specific actions.

STATISTICAL VS. EXPERIMENTER'S CONCLUSIONS

As statisticians we must insist upon more than one kind of conclusion, upon the difference between "statistical conclusions" and "experimenter's conclusions." A "statistical conclusion" applies to the *actual* conditions of the experiment. If a consistent blunder were made, if the instruments or measurements yield substantial systematic errors (they will always have some systematic errors, though we may hope that these are small), if the measurements were reduced according to a theory which is incomplete in some important way (it will always be incomplete to a certain extent), if the conditions or measurements were incorrectly recorded, if the importance of important variables were not recognized (so that their values were not recorded or reported), the stated conclusions are likely to be wrong. Errors for such reasons are not to be charged against *statistical* conclusions.

But experimenter's conclusions, be they physical conclusions, chemical conclusions, biological conclusions or engineering conclusions, must take account of all these possibilities. In most areas of experiment or observation it will be either desirable or necessary for the experimenter to make specific allowance, beyond the statistically recognizable uncertainty, for such deviations of the actual situation from the supposed situation. For this reason, his conclusions will be weaker than the statistical ones.

This difference, which arises from what may loosely be called the problem of systematic error, is an important challenge to the statistician. Both the statistician's morale and his integrity are tested when, for example, he has to face the possibility of a really substantial systematic error just after he has used all his skill to reduce, in the same experiment, the effects of fluctuating errors to 95% of their former value. It challenges his relationship to his clients in two opposite ways. When his client is quantitatively sophisticated, as many physical and engineering scientists are, he must face the systematic errors or lose his client's respect. When his client is not quantitatively sophisticated, as is often the case in other fields, he must educate the client at the proper rate, not too rapidly and not too slowly—first, perhaps, about fluctuating errors, but eventually about systematic errors, too!

ASYMMETRY CAN BE ESSENTIAL

We have emphasized the most important differences between decisions and conclusions. There is another difference which is not quite among the most important, but which yet deserves a place of its own. This is the treatment of doing nothing.

In most accounts of decision theory, the decision to do nothing is either ignored (which is probably the worst thing to do in practice) or treated on a par with all the other decisions. In conclusion theory, on the other hand, not coming to a conclusion plays a very special role. Three instances may help us to reflect on this distinction:

(1) All of us who were originally brought up in physical or biological science feel quite clearly, I am sure, that "to be not yet certain" is very different from other attitudes about a question.

(2) We may be surprised to find a related attitude among administrators —Barnard, 1938, p. 194, says (his italics) *"The fine art of executive decision consists in not deciding questions that are now not pertinent, in not deciding prematurely, in not making decisions that cannot be made effective, and in not making decisions that others should make."*

(3) An active worker in decision theory told me recently that the decision to do nothing was "the only decision without a loss function."

Each of these emphasizes, in a different way, the distinctive character of "doing nothing." Each deserves further examination. (Appendix 2 will treat (1).)

Barnard's statement implies that the "decisions" of the executive are much more nearly what we have called *conclusions* than what we have called *decisions*. They are not to be entered upon lightly, and there is a clear implication that, once reached, they are to be referred to for some time as part of a growing body of doctrine.

The decision-theorist's statement in (3) reveals him, it seems to me, as one who is really in search of conclusions. Why else is "doing nothing" so different? It is an action, one that can, in particular, lose money.

Decision theory ought to be *symmetrical* with regard to the action "do nothing." Conclusion theory must be *unsymmetrical* with regard to the action "conclude nothing."

TESTS OF SIGNIFICANCE

The prototype of modern experimental statistics was the test of significant difference. It came first as a tool of analysis and inference, not as a tool of mere description. When we examine its purport in the framework we are describing, we find that it is a *qualitative conclusion* procedure. Its purpose is to answer the question "Dare we conclude that this difference is not zero?"

We may, on the basis of a test of significance, *conclude* that $A \neq B$, or even more specifically that $A < B$ or $A > B$. But failure to attain significance is not, of itself, intended to produce a conclusion, is not intended to be accepted, in that strong sense of the word "accept" which is relevant to conclusions.

Where do we stand when the difference between A and B has not reached "significance"? Some would like to wield Occam's razor and say that "We have shown that $A = B$." Surely we have not *concluded* that $A = B$. For no quantitative evidence can establish that A is not just a very little different from B. Perhaps we have *decided* that $A = B$, but if so, for what specific situation, on what evidence, and with what assessment of consequences?

To interpret appropriately a failure to attain significance, it is necessary to know something about the precision of the comparison, to know how close there is reason to believe A is to B. Only by advancing into the use of confidence techniques (about which more anon) can a negative statement about significance be converted into a positive conclusion of established smallness of difference.

TESTS OF HYPOTHESIS

Symmetry, and mathematical simplicity, seemed to lead along a straight path from tests of significance to tests of hypothesis. As the procession traversed this path, few if any stopped to see where they had gone—to notice that they had left a qualitative conclusion procedure and had come to what was suspiciously like a qualitative decision procedure.

The choice between two simple hypotheses can be viewed in two quite different ways:

(1) as an attempt to choose the best risk, without regard to certainty—which is surely a decision procedure, or

(2) as an attempt to control, often by a sequential procedure, both kinds of error (both the error of accepting the hypothesis when it is false, and the error of rejecting it when it is true) at suitably low levels—which is, on the face of it, a conclusion procedure.

The aim of (2) can be expressed as follows: "We will take enough observations to allow us to dare to conclude either that the first hypothesis is false, or that the second hypothesis is false, but we shall not try to conclude that both are false, even if the observations prove adequate to do this." The form of this statement is clearly that of a conclusion procedure, though it is natural to wonder at the presence of its last proviso.

If, on the other hand, the aim is really (1), to choose the best risk, then there is no real place in the procedure for the artificial limitations of 5%, of 1%, or of any of the conventional significance or confidence levels. If nothing is to be concluded, only something decided, there is no need to control the probability of error. (Only the mathematical expectation of gain needs to be positive to make a small gamble profitable. There is no need for high confidence in winning individual bets. A coin which comes heads 60% of the time will win more money safely than one that comes heads 95% or 99% of the time.)

Until we go through the accounts of the testing of hypotheses, separating decision elements from conclusion elements, the intimate mixture of disparate elements will be a continual source of confusion. The writer looks forward to the day when the history and status of tests of hypotheses will have been disentangled. (See also Appendix 3.)

ESTIMATION

Older by far than any other statistical techniques are point estimates, each a simple indication of the value the data seem to point out or suggest. They are quantitative, where the other classical procedures we have so far discussed are qualitative. They are attempts to do the best that we can, not to do only what we can be certain about. Hence they are decision procedures, more specifically, quantitative decision procedures. So far as our classification goes, they offer no problems.

Probably the greatest ultimate importance, among all types of statistical procedures we now know, belongs to *confidence procedures* which, by making interval estimates, attempt to reach as strong conclusions as are reasonable by pointing out, not single likely values, but rather whole classes (intervals, regions, etc.) of *possible* values, so chosen that there can be high confidence that the "true" value is *somewhere among them*. Such procedures are clearly quantitative conclusion procedures. They make clear the essential "smudginess" of experimental knowledge.

THE TWIN DICHOTOMIES

Keeping the varied sorts of statistical inference procedures separate, and yet properly related to one another, is important to every statistician. Hopefully, the distinction between decisions and conclusions, as well as the distinction between qualitative and quantitative, are now clear.

The writer has found, and continues to find, these twin dichotomies (qualitative-quantitative and conclusion-decision) most helpful in organizing the procedure of statistics into a pattern which is useful both for application and reflection.

Surely the quantitative is preferable to the qualitative whenever both are equally available and equally relevant. Thus most qualitative statistical procedures are interim measures, introduced to serve until equally relevant quantitative procedures become available.

If we use the phrases "to do one's best" and "to state only that which is certain" as typifying decisions on the one hand, and conclusions on the other, we can see that there is a real place for both. And in particular situations we can usually tell what these places are.

To sum things up: the case of qualitative vs. quantitative should have a mixed verdict, granting "qualitative" squatters rights, but only until "quantitative" is ready to move in; while the case of conclusion vs. decisions should be settled out of court, with an understanding that cooperation is vital to both parties. There is a place for both "doing one's best" and "saying only what is certain," but it is important to know, in each instance, both which one is being done, and which one ought to be done.

Appendix 1

SOME CONFUSING RELATIONS

So long as we lacked clearly contrasted words, confusion between decisions and conclusions was very easy, especially since they are so thoroughly combined, both so frequently, and in almost every possible way.

Both decisions and conclusions are required in almost every field of human endeavor, yet the proportions, mutual relations and relative dominance which are appropriative vary greatly from one field to another. The aim and purpose of pure science lies in the *conclusions* which build up knowledge. Yet these conclusions are reached because individual scientists *decide* to attack certain problems in certain ways. (They rarely, if ever, know enough to *conclude* which problems they should attack, or how.) In most fields of engineering much must depend on the wisdom of experience, on engineering judgment, on engineering decisions. Yet these *decisions* are built upon the *conclusions* of pure and applied science. Engineering uses decisions fortified with conclusions, just as science uses decisions to reach conclusions.

In statistics, too, conclusions and decisions are interrelated and intertwined. It is not infrequent that we come to *conclusions* about *decision procedures*. What may prove to be one of the greatest monuments to Abraham Wald's memory is the notion of admissibility. And one of its more important elements is the fact that we may *conclude* (in this instance purely from theory and presuppositions) that one *decision procedure* is always worse than another.

We have seen that point estimates may reasonably be regarded as decisions. If we have a situation in which alternative point estimates are investigated by experimental sampling, and if the sampling is continued until the effects of sampling fluctuations fall below a prechosen standard of smallness, we are really experimenting until we can reach a *conclusion* about competing *decision procedures*. A third instance, closer in feeling to the first, is provided by [Fisher (1920)] one of many objective comparisons of estimators.

On the other hand, all of us make *decisions* about *conclusion procedures*. Some of us do it every day. "How is it best to analyze this data?" is a question which cannot be left to the experimenter alone, which the statistician is bound by his profession to try to answer. If the answer should clearly be a procedure to provide a conclusion, then he must do something about a *conclusion procedure*. Does he *decide* about it, or *conclude* about it? As an inherent conservative (professionally, anyway!), he would like to conclude. But will he have enough firm evidence? Often he will not!

When a transformation is chosen, whether for an analysis of variance, for quantal response assay, or for some other statistical procedure, how often does the chooser know what is *the* best transformation? In a theoretical sense, the answer is "never," for he will have only a finite amount of information—since his estimate of "best" will have a finite standard deviation—and transformations can be varied in arbitrarily small steps. In practice, it must be recognized that *exactly the best* transformation is not required, so that such an argument is not compelling. Yet, even in a practical sense, the answer is "not nearly often enough," for adequate information is often, or even usually, lacking. Who knows of an instance, to take a concrete example, where the choice between probits and logits for a quantal response assay was a conclusion and not a decision?

In handling complex data by analysis of variance, how shall we set up the analysis? How detailed shall be our computations? On what orthogonal functions shall we calculate regressions? Can any of you recall situations where the answer to any of these was a conclusion?

Appendix 2

CONCLUSION THEORY AS AN ACTION SYSTEM

Insofar as man's organized activities can be regarded as striving toward at least dimly recognized goals, it is easy to argue that the individual actions which make up these activities should be guided by some appropriate form of decision theory. Actions are to be taken in specific instances, and the gains or losses resulting from specific combinations of actions and states of nature can, in principle, be at least roughly assessed. Why, then, should there be a place for conclusion theory which seems from such a broad viewpoint to be a poor substitute for what is really needed? At least four classes of important reasons loom up over the horizon; problems of communication, problems of assessment of gain and loss, problems of assessment of the *a priori*, problems of adequate mathematical treatment.

Most human affairs are not conducted by a single individual, nor even by a single executive hierarchy. Science, in the broadest sense, is both one of the most successful of human affairs, and one of the most decentralized. In principle, each of us puts his evidence (his observations, experimental or not, and their discussion) before all the others, and in due course an adequate consensus of opinion develops. In the early decades of the Royal Society of London, this was indeed very nearly how things were. But the number of working scientists has doubled, and redoubled very many times since then. As a consequence, problems of communication have probably come to dominate the problems of scientific method. And the practices of science have developed to meet the challenge. Outstanding among these practices is the use of conclusions. A scientist is helped little to know that another, given different evidence and facing a different specific situation, decided (even decided wisely) to act as if so-and-so were the true state of nature. The communication (for information, not as directives) of decisions is often inappropriate, and usually inefficient. A scientist is helped much to know that another reached a certain conclusion, that he felt that the correctness of so-and-so was established with high confidence. In order to replace conclusions as the basic means of communication, it would be necessary to rearrange and replan the entire fabric of science. No statistician should dare to attempt such a task on the basis of his limited area of specialized knowledge.

But suppose a new fabric of science were to be developed. How could the old be compared with the new? Let us admit for simplicity that rapidity of progress is what is desired. (To do this for argument's sake alone does not mean that the intellectual and artistic aspects of science are being neglected in comparison with its pragmatic aspects.) Can one judge now how far science will progress (using the old fabric) in twenty years? And if not, how could we judge whether twenty years' use of a new fabric had done better or worse? If twenty years of trial would not be adequate for evaluation, how can an advance assessment give any useful idea of the gains and losses to be expected from a change to a new fabric?

If there were to be a change to a new fabric of science, one based more explicitly on decision-theoretic principles, how would the choice among many such fabrics be made? There would be a need to choose something like an *a priori* state of the whole world, more precisely to choose an *a priori* distribution of probability over all the possible states of the whole world since just as the admissable decision procedures are the Bayesian solutions, (those solutions which are optimum for suitable assumptions about the *a priori* probabilities of all "states of nature" considered) so too the admissible decision fabrics are to be expected to be Bayesian fabrics. And it is a little too much to ask of those who have learned to study certain limited aspects of the world, and who are striving to learn a little more of these aspects, that they envisage all possible worlds and distribute probability among them.

Finally, there are problems of adequate mathematical treatment. Statistics can solve a few vastly over-simplified problems in great generality or great detail, but it has barely begun to chew out a few little entrances into many problems of moderate difficulty. Problems of the order of difficulty of finding a Bayesian fabric, given the gains and losses, are wholly outside its present grasp. Today it has not provided even a beginning of an answer for such vastly simpler problems as: given samples of moderate size from each of two populations, given that the populations are so nearly normal (i.e., Gaussian) that samples of 1000 have no more than an even chance of detecting (at 5% significance) that the populations are not normal, and (even) given that the populations are symmetrical, what is the safest way to compare the centers of the two populations on the basis of the samples, where safety combines (1) reasonable reliability of significance or confidence percentages and (2) avoidance of procedures which are relatively very wasteful for particular population shapes. (Notes: (a) Even these many words, of course, have not completely specified a problem. (b) Adding a probability distribution over shapes to the hypotheses seems unlikely to make the problem easier.)

There are four types of difficulty, then, ranging from communication through assessment to mathematical treatment, each of which by itself will be sufficient for a long time, to prevent the replacement, in science, of the system of conclusions by a system based more closely on today's decision theory. Once these four have been examined, the natural question becomes: "How did the conclusion system escape the parallel sets of difficulties?" The answer is simple and clear. It grew. This means that it evolved; that many minor alternatives were tried, often unconsciously, that most were found wanting, and were discarded; that this process of trial and selection went through cycle after cycle. The strength of the *process of science* today comes from *e*xperience rather than *in*sight, and this state of affairs may be expected to continue for a long time. Indeed, it will not be easy to gain the limited insight required to understand how the present processes of science do as well as they do.

Appendix 3

WHAT OF TESTS OF HYPOTHESIS?

In view of Neyman's continued insistence on "inductive behavior," words which relate more naturally to decisions than to conclusions, it is reasonable to suppose that Neyman-Pearson theory of testing hypotheses was, at the very least, a long step in the direction of decision theory, and that the appearance of 5%, 1% and the like in its development and discussion was a carryover from the then dominant qualitative conclusion theory, the theory of tests of significance. If this view is correct, Wald's decision theory now does much more nearly what tests of hypothesis were intended to do. Indeed, there are three ways in which it does better. First, it has given up a fixed probability for errors of the first kind, and has focussed on gains, losses, or regrets (be they average or minimax). Secondly, it has made it somewhat easier to consider a much wider variety of specifications, to make much less stringent assumptions. And finally it has shown that one should expect mathematics to provide, not a single best procedure, but rather an assortment of good procedures (e.g. a complete class of admissible procedures) from which judgment and insight into a particular instance (perhaps expressed in the form of an *a priori* distribution) must be used to select the "best" procedure.

If one aspect of the theory of testing hypotheses has been embodies in modern decision theory, what of its other aspects? The notion of the power function of a test, which is of course strictly analogous to the notion of the operating characteristic of a sampling plan, is just as applicable to tests of significance (conclusions) as to tests of hypotheses (decisions). And, indeed, its natural generalization to confidence procedures (conclusions) seems more natural and reasonable than such conventional criteria as the average length of confidence intervals.

Conclusion theory can take over these nondecisional aspects of the theory of testing hypotheses. Its main concern in so doing must be caution about over-narrow specifications. To know that a certain confidence procedure is optimum, so long as the underlying observations follow a normal (i.e., Gaussian) distribution precisely, is not enough, if the procedure is poor for distributions whose shapes are very difficult to distinguish (in practice) from normality. (And such situations exist, at least in large samples; Tukey, 1960.)

In the long run, then, the theory of testing hypotheses can be absorbed into the contrasted bodies of decision theory and conclusion theory. (And the Neyman-Pearson lemma can serve, in its proper place, in both.)

ACKNOWLEDGMENT

Like most of the writer's papers, the present one owes much to conversations with, and careful readings of drafts by many friends and colleagues. Outstanding in this instance are Dr. Edgar Anderson and Dr. Jayarajan Chanmugan. My thanks go to all who have helped.

References

Barnard, C. I. *The function of the executive.* Cambridge, Mass.: Harvard University Press, 1938.

Fisher, R. A. A mathematical examination of methods of determining the accuracy of an observation by the mean error, and by the mean square error. *Monthly Notices of the Royal Astronomical Society,* 1920, **80** (8), 758–770. Reprinted in R. A. Fisher, *Contributions to mathematical statistics.* New York: Wiley, 1950.

Tukey, J. W. A survey of sampling from contaminated distributions. In Oklin et al. (Eds.), *Contributions to probability and statistics: Essays in honor of Harold Hotelling.* Stanford, Calif.: Stanford University Press, 1960.

5.3

N = 1

William F. Dukes

In the search for principles which govern behavior, psychologists generally confine their empirical observations to a relatively small sample of a defined population, using probability theory to help assess the generality of the findings obtained. Because this inductive process commonly entails some knowledge of individual differences in the behavior involved, studies employing only one subject (N = 1) seem somewhat anomalous. With no information about intersubject variability in performance, the general applicability of finding is indeterminate.

Although generalizations about behavior rest equally upon adequate sampling of both subjects and situations, questions about sampling most often refer to subjects. Accordingly, the term "N = 1" is used throughout the present discussion to designate the *reductio ad absurdum* in the sampling of subjects. It might, however, equally well (perhaps better, in terms of frequency of occurrence) refer to the limiting case in the sampling of situations—for example, the use of one maze in an investigation of learning, or a simple tapping task in a study of motivation. With the respect to the two samplings, Brunswik (1956), foremost champion of the representative design of experiments, speculated:

> In fact, proper sampling of situations and problems may in the end be more important than proper sampling of subjects, considering the fact that individuals are probably on the whole more alike than are situations among one another [p. 39].

As a corollary, the term N = 1 might also be appropriately applied to the sampling of experiments. Long recognized as a potential source of variance in interview data (e.g., Cantril, 1944; Katz, 1942), the investigator has recently been viewed as a variable which may also influence laboratory results (e.g., McGuigan, 1963; Rosenthal, 1963).

Except to note these other possible usages of the term N = 1, the present paper is not concerned with one-experimenter or one-situation treatments, but is devoted, as indicated previously, to single-subject studies.

Despite the limitation stated in the first paragraph, N = 1 studies cannot be dismissed as inconsequential. A brief scanning of general and historical accounts of psychology will dispel any doubts about their importance, revealing, as it does, many instances of pivotal research in which the observations were confined to the behavior of only one person or animal.

SELECTIVE HISTORICAL REVIEW

Foremost among N = 1 studies is Ebbinghaus' (1885) investigation of memory. Called by some authorities "a landmark in the history of psychology . . . a model which will repay careful study [McGeoch & Irion, 1952, p. 1]," considered by others "a remedy . . . at least as bad as the disease [Bartlett, 1932, p. 3]," Ebbinghaus' work established the pattern for much of the research on verbal learning during the past 80 years. His principal findings, gleaned from many self-administered learning situations consisting of some 2,000 lists of nonsense syllables and 42 stanzas of poetry, are still valid source material for the student of memory. In another well-known pioneering study of learning, Bryan and Harter's (1899) report on plateaus, certain crucial data were obtained from only one subject. Their letter-word-phrase analysis of learning to receive code was based on the record of only one student. Their notion of habit hierarchies derived in part from this analysis is, nevertheless, still useful in explaining why plateaus may occur.

Familiar even to beginning students of perception is Stratton's (1897) account of the confusion from and the adjustment to wearing inverting lenses. In this experiment according to Boring (1942), Stratton, with only himself as subject,

> settled both Kepler's problem of erect vision with an inverted image, and Lotze's problem of the role of experience in space perception, by showing that the "absolute" localization of retinal positions—up-down and right-left—are learned and consist of bodily orientation as context to the place of visual excitation [p. 237].

The role of experience was also under scrutiny in the Kelloggs' (1933) project of raising one young chimpanzee, Gus, in their home. (Although observations of their son's behavior were also included in their report, the study is essentially of the N = 1 type, since the "experimental group" consisted of one.) This attempt to determine whether early experience may modify behavior traditionally regarded as instinctive was for years a standard reference in discussions of the learning-maturation question.

Focal in the area of motivation is the balloon-swallowing experiment of physiologists Cannon and Washburn (1912) in which kymographic recordings of Washburn's stomach contractions were shown to coincide with his introspective reports of hunger pangs. Their findings were widely incorporated into psychology textbooks as providing an explanation of hunger. Even though in recent years greater importance has been attached to central factors in hunger, Cannon and Washburn's work continues to occupy a prominent place in textbook accounts of food-seeking behavior.

In the literature on emotion, Watson and Rayner's study (1920) of Albert's being conditioned to fear a white rat has been hailed as "one of the most influential papers in the history of American psychology" (Miller, 1960, p. 690). Their experiment, Murphy (1949) observes,

> immediately had a profound effect on American psychology; for it appeared to support the whole conception that not only simple motor habits, but important, enduring traits of personality, such as emotional tendencies, may in fact be 'built into' the child by conditioning. [p 261].

Actually the Albert experiment was unfinished because he moved away from the laboratory area before the question of fear removal could be explored. But Jonas (1924) provided the natural sequel in Peter, a child who, through a process of active reconditioning, overcame a non-laboratory-produced fear of white furry objects.

In abnormal psychology few cases have attracted as much attention as Prince's (1905) Miss Beauchamp, for years the model case in accounts of multiple personality. An excerpt from the Beauchamp case was recently included, along with selection from Wundt, Pavlov, Watson, and others, in a volume of 36 classics in psychology (Shipley, 1961). Perhaps less familiar to the general student but more significant in the history of psychology is Breuer's case (Breuer & Freud, 1895) of Anna O., the analysis of which is credited with containing "the kernel of a new system of treatment, and indeed a new system of psychology [Murphy, 1949, p. 307]." In the process of examining Anna's hysterical symptoms, the occasions for their appearance, and their origin, Breuer claimed that with the aid of hypnosis these symptoms were "talked away." Breuer's young colleague was Sigmund Freud (1910), who later publicly declared the importance of this case in the genesis of psychoanalysis.

There are other instances, maybe not so spectacular as the preceding, of influential $N = 1$ studies—for examples, Yerkes' (1927) exploration of the gorilla Congo's mental activities; Jacobson's (1931) study of neuromuscular activity and thinking in an amputee; Culler and Mettler's (1934) demonstration of simple conditioning in a decorticate dog; and

TABLE 1

Total Distribution of N = 1 Studies (1939–1963)

Category	f	Examples
Maturation, development	29	Sequential development of prehension in a macaque (Jensen, 1961); smiling in a human infant (Salzen, 1963)
Motivation	7	Differential reinforcement effects of true, esophagal, and sham feeding in a dog (Hull, Livingston, Rouse, & Barker, 1951)
Emotion	12	Anxiety levels associated with bombing (Glavis, 1946)
Perception, sensory processes	25	Congenital insensitivity to pain in a 19-year-old girl (Cohen, Kipnis, Kunkle, & Kubzansky, 1955); figural aftereffects with a stabilized retinal image (Krauskopf, 1960)
Learning	27	Delayed recall after 50 years (Smith, 1963); imitation in a chimpanzee (Hayes & Hayes, 1952)
Thinking, language	15	"Idealess" behavior in a chimpanzee (Razran, 1961); opposite speech in a schizophrenic patient (Laffal & Ameen, 1959)
Intelligence	14	Well-adjusted congenital hydrocephalic with IQ of 113 (Teska, 1947); intelligence after lobectomy in an epileptic (Hebb, 1939)
Personality	51	Keats' personality from his poetry (McCurdy, 1944); comparison in an adult of P and R techniques (Cattell & Cross, 1952)
Mental health psychotherapy	66	Multiple personality (Thigpen & Cleckley, 1954); massed practice as therapy for patient with tics (Yates, 1958)
Total	246	

Burtt's (1932) striking illustration of his son's residual memory of early childhood.

Further documentation of the significant role of N = 1 research in psychological history seems unnecessary. A few studies, each in impact like the single pebble which starts an avalanche, have been the impetus for major developments in research and theory. Others, more like missing pieces form nearly finished jigsaw puzzles, have provided timely data on various controversies.

This historical recounting of "successful" cases is, of course, not an exhortation for restricted subject samplings, nor does it imply that their greatness is independent of subsequent related work.

FREQUENCY AND RANGE OF TOPICS

In spite of the dated character of the citations—the latest being 1934—N = 1 studies cannot be declared the product of an era unsophisticated in sampling statistics, too infrequent in recent psychology to merit attention. During the past 25 years (1939–1963) a total of 246 N = 1 studies, 35 of them in the last 5-year period, have appeared in the following psychological periodicals: the *American Journal of Psychology, Journal of Genetic Psychology, Journal of Abnormal and Social Psychology, Journal of Educational Psychology, Journal of Comparative and Physiological Psychology, Journal of Experimental Psychology, Journal of Applied Psychology, Journal of General Psycholgy, Journal of Social Psychology, Journal of Personality,* and *Journal of Psychology.* These are the journals, used by Bruner and Allport (1940) in their survey of 50 years of change in American psychology, selected as significant for and devoted to the advancement of psychology as science. (Also used in their survey were the *Psychological Review, Psychological Bulletin,* and *Psychometrika,* excluded here because they do not ordinarily publish original empirical work.) Although these 246 studies constitute only a small percent of the 1939–1963 journal articles, the absolute number is noteworthy and is sizable enough to discount any notion that N = 1 studies are a phenomenon of the past.

When furthermore, these are distributed, as in Table 1, according to subject matter, they are seen to coextend fairly well with the range of topics in general psychology. As might be expected, a large proportion of them fall into the clinical and personality areas. One cannot, however, explain away N = 1 studies as case histories contributed by clinicians and personologists occupied less with establishing generalizations than with exploring the uniqueness of an individual and understanding his total personality. Only about 30% (74) are primarily oriented toward the individual, a figure which includes not only works in the "understanding" tradition, but also those treating the individual as a universe of responses and applying traditionally nomothetic techniques to describe and predict individual behavior (e.g., Cattell & Cross, 1952; Yates, 1958).

In actual practice, of course, the two orientations—toward uniqueness or generality—are more a matter of degree than of mutual exclusion, with the result that in the literature surveyed purely idiographic research is extremely rare. Representative of that approach are Evans' (1950) novel-like account of Miller who "spontaneously" recovered his sight after more than 2 years of blindness, Rosen's (1949) "George X: A self-analysis by an avowed fascist," and McCurdy's (1944) profile of Keats.

RATIONALE FOR N = 1

The appropriateness of restricting an idiographic study to one individual is obvious from the meaning of the term. If uniqueness is involved, a sample of one exhausts the population. At the other extreme, an N of 1 is also appropriate if complete population generality exists (or can reasonably be assumed to exist). That is, when between-individual variability for the function under scrutiny is known to be negligible or the data from the single subject have a point-for-point congruence with those obtained from dependable collateral sources, results from a second subject may be considered redundant. Some $N = 1$ studies may be regarded as approximations of this ideal case, as for example, Heinemann's (1961) photographic measurement of retinal images and Bartley and Seibel's (1954) study of entoptic stray light, using the flicker method.

A variant on this typicality theme occurs when the researcher, in order to preserve some kind of functional unity and perhaps to dramatize a point, reports in depth one case which exemplifies many. Thus Eisen's (1962) description of the effects of early sensory deprivation is an account of one quondam hard-of-hearing child, and Bettelheim's (1949) paper on rehabilitation a chronicle of one seriously delinquent child.

In other studies an N of 1 is adequate because of the dissonant character of the findings. In contrast to its limited usefulness in *established* generalizations from "positive" evidence, an N of 1 when the evidence is "negative," is as useful as an N of 1,000 in *rejecting* an asserted or assumed universal relationship. Thus Krauskopf's (1960) demonstration with one stopped-image subject eliminates motion of the retinal image as necessary for figural aftereffects; and Lenneberg's (1962) case of an 8-year-old boy who lacked the motor skills necessary for speaking but who could understand language makes it "clear that hearing oneself babble is not a necessary factor in the acquisition of understanding . . . [p. 422]." Similarly Teska's (1947) case of a congenital hydrocephalic, 6½ years old, with an IQ of 113, is sufficient evidence to discount the notion that prolonged congenital hydrocephaly results in some degree of feeblemindedness.

While scientists are in the long run more likely to be interested in knowing *what is* than *what is not* and more concerned with how many exist or in what proportion they exist than with the fact that at least one exists, one negative case can make it necessary to revise a traditionally accepted hypothesis.

Still other $N = 1$ investigations simply reflect a limited opportunity to observe. When the search for lawfulness is extended to infrequent "nonlaboratory" behavior, individuals in the population under study may be so sparsely distributed spatially or temporally that the psychologist can observe only one case, a report of which may be useful as a part of a cumulative record. Examples of this include cases of multiple personality (Thigpen & Cleckley, 1954), unilateral color blindness (Graham, Sperling, Hisa, & Coulson, 1961) congenital insensitivity to pain (Cohen et al., 1955), and mental deterioration following carbon monoxide poisoning (Jensen, 1950). Situational complexity as well as subject sparsity may limit the opportunity to observe. When the situation is greatly extended in time, requires expensive or specialized training for the subject, or entails intricate and difficult to administer controls, the investigator may, aware of their exporatory character, restrict his observations to one subject. Projects involving home-raising a chimpanzee (Hayes & Hayes, 1952) or testing after 16 years for retention of material presented during infancy (Burtt, 1941) would seem to illustrate this use of an N of 1.

Not all N = 1 studies can be conveniently fitted into this rubric; nor is this necessary. Instead of being oriented either toward the person (uniquenesss) or toward a global theory (universality), researchers may sometimes simply focus on a problem. Problem-centered research on only one subject may, by clarifying questions, defining variables, and indicating approaches, make substantial contributions to the study of behavior. Besides answering a specific question, it may (Ebbinghaus' work, 1885, being a classic example) provide important groundwork for the theorists.

Regardless of rationale and despite obvious limitations, the usefulness of N = 1 studies in psychological research seems, from the preceding historical and methodological considerations, to be fairly well established. (See Shapiro, 1961, for an affirmation of the value of single-case investigations in fundamental clinical psychological research.) Finally, their status in research is further secured by the statistician's assertion (McNemar, 1940) that:

> The statistician who fails to see that important generalizations from research on a single case can ever be acceptable is on a par with the experimentalist who fails to appreciate the fact that some problems can never be solved without resort to numbers [p. 361].

References

Bartlett, F. C. *Remembering*. Cambridge, England: University Press, 1932.

Bartley, S. H., & Seibel, Jean L. A further study of entoptic stray light. *Journal of Psychology*, 1954, **38**, 313–319.

Bettelheim, B. H. A study in rehabilitation. *Journal of Abnormal and Social Psychology*, 1949, **44**, 231–265.

Boring, E. G. *Sensation and perception in the history of experimental psychology*. New York: Appleton-Century, 1942.

Breier, J., & Freud, S. Case histories. (Orig. publ. 1895; trans. by J. Strachey). In J. Strachey (Ed.), *The standard edition of the complete psychological works of Sigmund Freud*. Vol. 2. London: Hogarth Press, 1955. Pp. 19–181.

Bruner, J. S., & Allport, G. W. Fifty years of change in American psychology. *Psychological Bulletin*, 1940, **37**, 757–776.

Brunswik, E. *Perception and the representative design of psychological experiments*. Berkeley: University California Press, 1956.

Bryan, W. L., & Harter, N. Studies on the telegraphic language. The acquisition of a hierarchy of habits. *Psychological Review*, 1899, **6**, 345–375.

Burtt, H. E. An experimental study of early childhood memory. *Journal of Genetic Psychology*, 1932, **40**, 287–295.

Burtt, H. E. An experimental study of early childhood memory: Final report. *Journal of Genetic Psychology*, 1941, **58**, 435–439.

Cannon, W. B., & Washburn, A. L. An explanation of hunger. *American Journal of Physiology*, 1912, **29**, 441–454.

Cantril, H. *Gauging public opinion*. Princeton: Princeton University Press, 1944.

Cattell, R. B., & Cross, K. P. Comparison of the ergic and self-sentiment structures found in dynamic traits by R- and P- techniques. *Journal of Personality*, 1952, **21**, 250–271.

Cohen, L. D., Kipnis, D., Kunkle, E. C., & Kubzansky, P. E. Observations of a person with congenital insensitivity to pain. *Journal of Abnormal and Social Psychology*, 1935, **51**, 333–338.

Culler, E., & Mettler, F. A. Conditioned behavior in a decorticate dog. *Journal of Comparative Psychology*, 1934, **18**, 291–303.

Ebbinghaus, H. *Über das Gedächtnis*. Leipzig: Duncker & Humblot, 1885.

Eisen, N. H. Some effects of early sensory deprivation on later behavior: The quondem hard-of-hearing child. *Journal of Abnormal and Social Psychology*, 1962, **65**, 338–342.

Evans, Jean. Miller. *Journal of Abnormal and Social Psychology*, 1950, **45**, 359–379.

Freud, S. The origin and development of psychoanalysis. *American Journal of Psychology*, 1910, **21**, 181–218.

Glavis, L. R., Jr. Bombing mission number fifteen. *Journal of Abnormal and Social Psychology*, 1946, **41**, 189–198.

Graham, C. H., Sperling, H. G., Hsia, Y., & Coulson, A. H. The determination of some visual functions of a unilaterally color-blind subject. *Journal of Psychology*, 1961, **51**, 3–32.

Hayes, K. J., & Hayes, Catherine. Imitation in a home-raised chimpanzee. *Journal of Comparative and Physiological Psychology*, 1952, **45**, 450–459.

Hebb, D. O. Intelligence in man after large removals of cerebral tissue: Defects following right temporal lobectomy. *Journal of General Psychology*, 1939, **21**, 437–446.

Heinemann, E. G. Photographic measurement of the retinal image. *American Journal of Psychology*, 1961, **74**, 440–445.

Hull, C. L., Livingston, J. R., Rouse, R. O., & Barker, A. N. True, sham, and esophageal feeding as reinforcements. *Journal of Comparative and Physiological Psychology*, 1951, **44**, 236–245.

Jacobson, E. Electrical measurements of neuromuscular states during mental activities: VI. A note on mental activities concerning an amputated limb. *American Journal of Physiology*, 1931, **96**, 122–125.

Jensen, G. D. The development of prehension in a macaque. *Journal of Comparative and Physiological Psychology*, 1961, **54**, 11-12.

Jensen, M. B. Mental deterioration following carbon monoxide poisoning. *Journal of Abnormal and Social Psychology*, 1950, **45**, 146–153.

Jones, Mary C. A laboratory study of fear: The case of Peter. *Journal of Genetic Psychology*, 1924, **31**, 308–315.

Kayz, D. Do interviewers bias poll results? *Public Opinion Quarterly*, 1942, **6**, 248–268.

Kellogg, W. N., & Kellogg, Luella. *The ape and the child.* New York: McGraw-Hill, 1933.

Krauskopf, J. Figural after-effects with a stabilized retinal image. *American Journal of Psychology*, 1960, **73**, 294–297.

Laffal, J., & Ameen, L. Hypotheses of opposite speech. *Journal of Abnormal and Social Psychology*, 1959, **58**, 267–269.

Lenneberg, E. H. Understanding language without ability to speak: A case report. *Journal of Abnormal and Social Psychology*, 1962, **65**, 419–425.

McCurdy, H. G. *La belle dame sans merci. Character and Personality*, 1944, **13**, 166–177.

McGeoch, J. A., & Irion, A. L. *The psychology of human learning.* New York: Longmans, Green, 1952.

McGuigan, F. J. The experimenter: A neglected stimulus object. *Psychological Bulletin*, 1963, **60**, 421–428.

McNemar, Q. Sampling in psychological research. *Psychological Bulletin*, 1940, **37**, 331–365.

Miller, D. R. Motivation and affect. In Paul H. Mussen (Ed.), *Handbook of research methods in child development.* New York: Wiley, 1960. Pp. 688–769.

Murphy, G. *Historical introduction to modern psychology.* New York: Harcourt, Brace, 1949.

Prince, M. *The dissociation of a personality.* New York: Longmans, Green, 1905.

Razran, C. Raphael's "idealess" behavior. *Journal of Comparative and Physiological Psychology*, 1961, **54**, 366–367.

Rosen, E. George X: The self-analysis of an avowed fascist. *Journal of Abnormal and Social Psychology*, 1949, **44**, 528–540.

Rosenthal, R. Experimenter attributes as determinants of subjects' responses. *Journal of Projection Techniques*, 1963, **27**, 324–331.

Salzen, E. A. Visual stimuli eliciting the smiling response in the human infant. *Journal of Genetic Psychology.* 1963, **102**, 51–54.

Shapiro, M. B. The single case in fundamental clinical psychological research. *British Journal of Medical Psychology*, 1961, **34**, 255–262.

Shipley, T. (Ed.) *Classics in psychology.* New York: Philosophical Library, 1961.

Smith, M. E. Delayed recall of previously memorized material after fifty years. *Journal of Genetic Psychology*, 1963, **102**, 3–4.

Stratton, G. M. Vision without inversion of the retinal image. *Psychological Review*, 1897, **4**, 341–360, 463–481.

Teska, P. T. The mentality of hydrocephalics and a description of an interacting case. *Journal of Psychology*, 1947, **23**, 197–203.

Thigpen, C. H., & Cleckley, H. A case of multiple personality. *Journal of Abnormal and Social Psychology*, 1954, **49**, 135–151.

Watson, J. B., & Rayner, Rosalie. Conditioned emotional reactions. *Journal of Experimental Psychology*, 1920, **3**, 1–14.

Yates, A. J. The application of learning theory to the treatment of tics. *Journal of Abnormal and Social Psychology*, 1958, **56**, 175–182.

Yerkes, R. M. The mind of a gorilla. *Genetic Psychology Monographs*, 1927, **2**, 1–193.

5.4

R. A. Fisher (1890-1962): An appreciation

Jerzy Neyman

R. A. Fisher was a great scholar, typifying Cambridge, the wonderful melting pot of ideas, where astronomers rub shoulders with historians, neurophysiologists with layers, and mathematicians with geneticists, statisticians, and others. The appraisal of Fisher's scholarly activity should, then, be made from at least three points of view, the point of view of statistics, that of mathematics, and that of empirical science.

As a statistician, R. A. Fisher appears to me to be a direct descendant of Karl Pearson, with side influences of G. Udny Yule and of F. Y. Edgeworth. From the point of view of mathematics, the situation is more complicated. Fisher's early days at Cambridge were, roughly, on the dividing line between two epochs, the earlier epoch of manipulative skills and the subsequent period of conceptual developments. Fisher seems to have belonged to the epoch of manipulative skills, in which he was supreme. Also, I remember his declaring that the change symbolized by the names of Hardy and Littlewood was a disaster in English

Jerzy Neyman, "R. A. Fisher (1890–1962): An Appreciation," *Science*, **156**, 1967, 1456–1460. Copyright (1967) by the American Association for the Advancement of Science.

The author is professor of statistics and director of the Statistical Laboratory, University of California, Berkeley. This is the text of an address delivered 29 December 1966 at the Washington meeting of the AAAS.

mathematics. Nevertheless, some of the most important writings of Fisher appear to be influenced by what was then the incipient era of conceptual mathematics.

The part of Fisher's work that I admire most is one that must have resulted from the general Cambridge atmosphere, from his contacts with representatives of empirical sciences, astronomers (in particular with A. S. Eddington), geologists, and biologists. Here I have in mind not only Fisher's direct contributions to science, especially to population genetics, but also Fisher's actual founding of an entirely novel discipline related to scientific research which I like to call the theory of experimentation. Even though Fisher's close contact with experimentation began after he left Cambridge and assumed a position at the Rothamsted Experimental Station, I rather think that the roots of experimental designs started growing while he was still at his alma mater.

My contacts with Fisher go back to 1932, the second of the two occasions known to me on which he and Karl Pearson agreed. The first occasion occurred in or slightly before 1915, when Pearson agreed to publish Fisher's paper giving the distribution of the coefficient of correlation (Fisher, 1915). The occasion of 1932, memorable to me, was when Karl Pearson communicated to the Royal Society a joint paper on the theory of testing hypotheses by E. S. Pearson and myself and on which Fisher wrote a favorable review. Generally, the reviews are confidential. However, on this occasion Fisher deviated from the customary anonymity in order to call our attention to an oversight: Moments do not always determine uniquely the corresponding distributions. Fisher's kindly suggestion is duly acknowledged on page 315 of our paper (Neyman & Pearson, 1933).

My subsequent, already personal, contacts with Fisher began in April 1934 when I came to London, first on a temporary and, later on, a tenure appointment at University College. Karl Pearson, the founder of the famous institute, had retired a year or so before and his institute was divided into two departments: the department of statistics with Egon S. Pearson as its head, and the department of eugenics under Fisher. My position was under E. S. P.

There was a sharp feud raging between Karl Pearson and R. A. Fisher with, seemingly, the population of the earth divided into two categories: K. P.'s school and R. A. F.'s school. Both E. S. Pearson and I did not like the situation and did our best to avoid becoming involved. To begin with, my own relations with Fisher were excellent and not infrequently I was a guest at Fisher's home in Harpenden. However, after a year or so, a break occurred between Fisher and me, an individual break, not a part of the K. P.–R. A. F. feud. While personalities were involved, the break had a scholarly background and was the start of a dispute lasting a quarter of a century (Fisher, 1934, 1935, 1960; Neyman, 1961). Except for the ready admission that Fisher's writings, including his polemical articles, influenced my own work, the dispute itself and the issues involved are not the subject of the present paper.

FOUR PRINCIPAL DIRECTIONS OF FISHER'S WORK

Any classification of empirical material, be it plants, inhabitants of a country, stars, or works of a particular scholar must be somewhat fuzzy, involving borderline cases. Also, any such classification is unavoidably subjective, reflecting the background and the attitudes of the person attempting the classification. The classification of Fisher's works given below has both these characteristics: It is somewhat fuzzy and it is subjective.

As I see it, the many research papers published by Fisher fall under four principal headings: (i) conceptual papers dealing with foundations of theory; (ii) papers giving exact distributions of various statistics; (iii) papers and books dealing with the theory of experimentation; and (iv) papers on stochastic models of natural phenomena.

As mentioned earlier, in the history of mathematics Fisher belongs to the era of generations preceding his own. Contrary to this, in the history of statistics Fisher's role was that of a pioneer. In order that his role may be more easily understood, a brief historical background might be helpful.

TRANSITION FROM "KOLLEKTIVMASSLEHRE" TO ANALYTICAL STATISTICS

The era in the history of statistics that preceded Fisher's may be labeled by the excellent German term *Kollektivmasslehre* invented, I think, by Bruns. Its development followed the realization, at the end of the 19th century, that the treatment of the then novel subjects of scientific investigations, namely studies of what we now call "populations," be it populations of stars or of molecules, of plants or of humans, require a new "collective" mathematical discipline. A population is characterized by the distribution of one or of several "individual' 'characteristics of this population's members. Thus, the mathematical problem was to devise flexible formulas which could be considered as idealization of empirical distributions, such as the distribution of magnitudes of stars or of anthropological measurements of various kinds. As is well known, the efforts in this direction resulted in several systems of frequency curves connected with the names of Burns, Charlier, Edgeworth, and Karl Pearson.

The next era in the history of mathematical statistics, the era that might be labeled analytical, began with a considerable temporal overlap with the *Kollektivmasslehre* period. The basic question typical of the analytical period is: What is the chance mechanism operating within a given population that generated the particular distribution we observe? For example: given that the number of industrial accidents per worker per year follows, approximately, the negative binomial distribution, what might be the chance mechanism underlying this phenomenon? (Actually, this particular question was asked, and answered, in the late 1920's and in the early 1930's. It is used here for its excellent illustrative quality.)

A very clear and compact definition of the problem typical of the analytical period has been given by Émile Borel in his book *Éléments de la Théorie des Probabilités* first published in 1909. As described by Borel, this problem is *the* problem of mathematical statistics (Borel, 1924).

> Le problème général de la statistique mathématique est le suivant: déterminer un système de tirages effectués, dans des urnes de composition fixe, de telle manière que le résultats d'une serie de tirages, interprétés à l'aide de coefficients fixes convenablement choisis, puissent avec une très grande vraisemblance conduire à un tableau indentique au tableau des observations.

In modern terminology, and with an appropriate extension, this statement may be reformulated in the following ways:

1) The general problem of mathematical statistics is that of stochastic models: given a distribution—what is the chance mechanism (Borel's system of draws of balls) that generated it?

2) Admitting tentatively that an empirical distribution has been generated by a specified chance mechanism involving several unknown parameters, what are the values that one should ascribe to these parameters? (These are the Borel's *"coefficients, convenablement choisis."*)

These two questions are inevitably followed by a third:

3) Given a set S of observations and a tentatively proposed chance mechanism M, how to decide whether M is consistent with S?

To my knowledge, apart from the definition quoted and apart from a somewhat inconclusive discussion in *Le Hasard* (Borel, 1914) of the possibility of solving problem 3 (above) though the use of *"fonctions en quelque sorte remarquables,"* Borel did not contribute much to the development of mathematical statistics as he understood the term. On the other hand, Fisher's contribution was tremendous. Even though the analytical period of the history of statistics began before Fisher, the relevant papers were few and far between and, apart from a few exceptions [for example, Karl Pearson's paper (1900) on χ^2 published in 1900 and "Student's" paper (1908) published a few years later], the analytical element in them was rather tentative and indistinct. From the point of view of generality of interest, the analytical era of statistics began with Fisher.

FISHER'S ATTACK ON DISTRIBUTION PROBLEMS

Curiously, the first important series of Fisher's papers, clearly belonging to the analytical period, was concerned with certain technical details and conceptualizations came later. The first sequence of Fisher's papers, the sequence that shook the statistical community, was concerned with the distribution problems involved in the general problem 3 (above), which we now label the problem of testing statistical hypotheses. This series began with the paper on the distribution of the coefficient of correlation, published in *Biometrika* in 1915 when Fisher was 25 years of age (Fisher, 1915). Compared with distribution problems that were treated earlier, the problem of the correlation coefficient was emphatically very much more difficult and Fisher deserved general applause. Yet, he seems to have had difficulties. In fact, his subsequent papers, giving the distributions of partial correlation, of multiple correlation, of the correlation ratio, and of the F statistic, were published not in the same journal, where they obviously belonged, but in many different journals where, occasionally, they appeared out of place.

This section of Fisher's activity exercised, and continues to exercise, a considerable influence on statistical literature, with such contributors as Harold Hotelling, S. S. Wilks, and J. W. Wishart. Later, Harold Cramér's book was published, popularizing matrix theory, and serving as predecessor of the works by T. W. Anderson, R. C. Bose, S. N. Roy, Henry Scheffé, and others.

Quite apart from the unusual manipulative skill exhibited by Fisher in his early distribution papers, they contain a very important conceptual contribution. As of now, this contribution may seem trivial. However, the mere fact that at the time it appeared to have been a novelty indicates the heavy weight of routine of thought that Fisher managed to shake off. The particular contribution consists in introducing a clear distinction between the value of a parameter characterizing a population (for example, of the population correlation coefficient ρ) on the one hand, and the value of the same parameter (which we now call "statistic," following Fisher), computed from the sample (for example, the sample correlation coefficient r). Distinct as ρ and r are, some of the earlier studies indicate a degree of most embarrassing confusion which Fisher helped to dispel.

CONCEPTUAL PAPERS BY FISHER

The strictly conceptual papers of Fisher began to appear after a substantial interval since his first groundbreaking paper of 1915. Here I have in mind mainly the paper, "On the Mathematical Foundations of

Theoretical Statistics," published in 1922, and the paper, "Theory of Statistical Estimation," that appeared in 1925, both reprinted in the collection of Fisher's papers published by Wiley (Fisher, 1950).

It is here that Fisher's ideological descendance from Karl Pearson and, partly, from Edgeworth, is apparent. It is also here that I sense the influence on Fisher of the contemporary devolopment of conceptual mathematics in Cambridge. At the time the first of the two papers was written, the Pearson system of frequency curves was fully developed, as well as the method of fitting them by moments, complete with a set of the necessary tables. Also in 1908–09, Edgeworth came up with two novel ideas. One was that the method of moments may not be the most advantageous method of fitting all the Pearson curves, and the other, that the most advantageous method of fitting is the one which we now call the method of maximum likelihood. Edgeworth formulated a conjecture (1908, 1909) that the a priori asymptotic variance of the maximum likelihood estimate (which he termed the a posteriori most probable value) cannot be greater than that of any other estimate. Edgeworth admitted his inability to prove this proposition in its full generality, but provided proofs for several particular cases.

In his two papers just quoted, Fisher divested himself of all considerations of a posteriori probabilities and attacked the two problems with considerable vigor and ingenuity. In fact, a strong attempt was made to build up a general theory of statistics, at least the theory of point estimation, as a balanced architectural whole. Compared to what has been done for probability by Kolmogorov (1933), this attempt cannot be considered successful, which must have been apparent to the editors of the two journals. In fact, the introductory sections to the two papers have the appearance of having been written as addenda to the rest of the texts, possibly in response to objections by the referees. In particular, the "Prefactory Note" to the second paper begins with the sentence (Fisher, 1950, p 11.700): "It has been pointed out to me that some of the statistical ideas employed in the following investigation have never received a strictly logical definition and analysis." Clearly the word "never" covers the earlier paper by Fisher. "On Mathematical Foundations. . . ."

Nevertheless, in spite of the lack of mathematical rigor, not only were the two papers published, as they should have been, but they also exercised a very considerable influence on the generations of statisticians that came on the scene after their publication.

Of the concepts formulated by Fisher, the following appear to be the most fruitful: consistency of an estimate, its efficiency, and the concept of sufficiency. The number of papers dealing with these concepts is tremendous and their enumeration is an impossibility. However, the authors who were inspired by Fisher's two papers are a good illustration of the importance of Fisher's ideas. These authors include Darmois,

Dugué, and Fréchet in France, Harold Cramér in Sweden, Hotelling, Doob, Wald, and Wolfowitz in the United States, C. R. Rao and Mahalanobis in India, Pitman in Australia, Dynkin, Linnik, and Kolmogorov in the Soviet Union, not to mention many British authors.

A fair description of the situation seems to be as follows: Even though Fisher failed to construct a theory of statistics, or even a theory of estimation, as an internally consistent system of concepts, he did through persistent work on a great number of particular problems manage to bring out several recurring phenomena of prime importance which inspired generations of other research workers. In addition, Fisher was a fighter and, after reaching a result which satisfied him, he would struggle for the general acceptance of this result. One example of this is Fisher's series of papers on a subject brought out by Yule, concerned with the loss of degrees of freedom in χ^2 due to the use of estimates of unknown parameters. At times there were "no holds barred" in the disputes that developed, and in the process it was inevitable for Fisher to step heavily on the toes of some generally recognized contemporary authorities. This led to feuds and to several spectacular developments. One was a long series of lengthy articles in Italy, in which an offended authority and his students repeatedly claimed to have "annihilated" the Anglo-Saxon theory of statistics. Incidentally, after having delat thus with Fisher, the same individuals dealt similarly with me. Another and perhaps even more spectacular occurrence was the December 1934 meeting of the Royal Statistical Society. This was the first and, so far as I know, the only meeting of the Society at which Fisher was invited to present a paper (1935). The subsequent motion of the "vote of thanks" (Bowley, et al., 1935) (the quotes are intentional and fitting the situation) and the following discussion, all duly recorded in the Society's journal, have few parallels in the scholarly literature known to me. However, the violent attacks Fisher sustained on this occasion were harmless. At the time Fisher had easy access to the printing press, both through the *Annals of Eugenics*, of which he was editor, and by being a fellow of the Royal Society of London. With these advantages, Fisher could ignore the displeasure of the leaders of the Royal Statistical Society. Besides, the angry outbursts of several "oldsters," countered by manifestations of Fisher's high polemical talent, impressed the audience as evidence of his originality. His following grew and the problems he was concerned with attracted more attention.

THEORY OF EXPERIMENTATION

As the research in science inexorably drifted to subjects exhibiting more and more variability from one experimental unit to the next, the problem of designs of experiments, taking this variability into account explicitly, became more and more urgent. However, the urgency of a problem in a given domain is not always recognized by the rank and file of workers in this domain. This is particularly true with problems that are interdisciplinary in character. Also, if an active research worker W in a domain A is requested to do something with reference to problems that are pressing in another domain B, the very frequent response of the worker W is unwillingness to cooperate. When the authorities of the Rothamsted Experimental Station sought the cooperation of Fisher, he not only agreed to cooperate, but put his heart and soul into this cooperation. Both sides deserved compliments—the authorities of Rothamsted for their foresightedness and Fisher for his ability to become genuinely and actively interested in the problems at Rothamsted.

Fisher's contributions to the theory of experimentation are many and his now famous books, *Statistical Methods for Research Workers* (1950) and *The Design of Experiments* (1949) are really composed of items of his own finding. Also, the third book (1963), of which Frank Yates is a coauthor, *Statistical Tables for Biological, Agricultural and Medical Research Workers* represents a compendium of a number of findings either by the two authors themselves or by other members of what may be called the Rothamsted School of Statistics.

These books were followed by a long series of other books by other authors. In this country alone the literature on experimentation is very extensive, from Snedecor, to Cochran and Cox, to Brownless, to the manuals of Youden, each starting where Fisher left off and then extending the principles of scientific experimentation to further and further domains of scientific research. Since about 1950, I also joined this general trend with an effort to introduce Fisher's experimental principles into research on weather control.

Important as the books on experimentation are, my favorite publications are two little papers, one under the title "The Arrangement of Field Experiments," originally published in 1925, and the other, a joint paper by Fisher and S. Barbacki, "A Test of the Supposed Precision of Systematic Arrangements," published in 1936. Both are reprinted in the volume (1950) of Fisher's *Contributions to Mathematical Statistics*. My particular preference for these papers is due to the emphasis they place on randomization as a prerequisite to soundness of an experiment. Fisher's own argument in favor of randomization is that it is necessary for a valid estimation of the error variance. While this is undoubtedly true, I prefer to formulate essentially the same statement somewhat differently: Without randomization there is no guarantee that the experimental data will be free from a bias that no test of significance can detect.

Fisher must be credited not only with a clear statement of principles of experimentation, but also with the great success he achieved in propagandizing these principles so that now they are generally accepted and adhered to in many domains of science. Also, in other domains where experiments are going on without randomization, the particular experimenters feel compelled to present excuses for not randomizing. Thus it is likely that, in a generation or so, sound experimentation will spread to these domains also.

The development of sound principles of experimentation is a great achievement per se. However, Fisher did more. Here I have in mind his analysis of variance and the development of a system of relevant tests, both the parametric tests based on assumptions of normality and independence, and of nonparametric randomization tests. When speaking of a system of tests I have in mind prescriptions like that of first deciding (using the F test) whether there is any significant effect in an experiment at all. Further analysis, leading to decisions as to particular treatments tested is only justified if the F test indicates significance. As a further development of the same idea, Fisher introduced the familiar χ^2 procedure of combining the results of several independent trials; each trial taken separately may fail to indicate significance, but jointly these results may provide evidence that the treatment studied did, in fact, have some effects. Alternatively, it may be that the true effects of treatments are really zero and the occasional significance observed in a few independent experiments out of a substantial number of them is the result of the unavoidable random variation. Fisher's summary test, for which Egon S. Pearson proved a property of optimality (1938, 1966), is a means of controlling these two sources of error.

The influence of Fisher's theory of experimentation on further developments in statistics seems to have no limits. Fisher's summary treatment of several already completed experiments appears to me as a predecessor of the achievements of Tukey and Scheffé, now nicely summarized by Rupert Miller, concerned with multiple decision problems. Fisher's experimental designs, particularly those involving incomplete blocks and confounding, brought to the fore delicate combinatorial problems and appear to have inspired R. C. Bose who, in his turn, generated a special branch of literature. Finally, Fisher's study of experiments influenced Abraham Wald and, thereby, a long series of outstanding scholars who follow Wald.

There is an interesting classification of problems of experimentation born out of a conversation I once had with M. G. Kendall. As we saw it, the original problems of Fisher, exemplified by single experiments to be conducted on given pieces of land, might be called problems of experimental tactics. Contrary to this, the problems of Wald visualized sequences of experiments and the possibility of a variety of decisions after each member of the sequence. One decision could be to discontinue the sequence with some sort of "terminal" substantive decision. Another possibility is to continue experimentation, perhaps with some novel design. Problems of this kind, obviously different from Fisher's, might be called problems of experimental strategy. Wald's ideas as introduced in his *Theory of Statistical Decision Functions* (1950), are obvious predecessors of the more modern works of Blackwell, Girshick, and their followers.

It is now appropriate for me to mention a problem which obviously belongs to the category of experimental tactics, but is missing in Fisher's writings. This is the problem of evaluating the probability that an experiment with a tentatively fixed design will detect the effects that it is designed to detect, if such effects are real and have preassigned magnitudes. The problem is that of the power of the tests contemplated for the treatment of the given experiment. The consideration of power is occasionally implicit in Fisher's writings, but I would have liked to see it treated explicitly.

I do not believe that Fisher ever thought of his work on experimental designs with reference to Borel's definition of the typical problem of mathematical statistics. However, the correspondence is unambiguous. Consider, for example, the randomized blocks experiment designed to compare some s varieties. Let n be the number of blocks. Let u_{ijk}, for $i = 1,2$, . . . , n and $j,k = 1,2, . . . , s$, denote the potential yield of the kth variety if grown on the jth plot of the ith block. The Borel type scheme of draws of balls visualizes n groups of urns of s urns each. The ith group corresponds to the ith block of plots. The kth urn of each group corresponds to the kth variety tested. It contains s numbered balls, the jth of them corresponding to the jth plot on the ith block and carrying the number u_{ijk} written on it. The s urns of the ith group have the magic property that, if the jth ball is removed from any one of them, the jth balls disappear from all the other urns of the same group. Under this system, the randomized n-blocks experiment with s varieties is equivalent to extracting just one ball out of each of the ns urns. The number written on the ball extracted from the kth urn of the ith group is a random variable, say x_{ik}, representing the yield of the kth variety on the randomly selected plot within the ith block.

The above schematization may be considered as the structural part of the stochastic model of the randomized blocks experiment; no numerical assumptions are involved. This structure may then be supplemented by other particularizing assumptions, such as the assumption of additivity of block and varietal effects and assumptions regarding the values of the u_{ijk}, and others. The final urn model to test against the observations is that described by all such assumptions and, in addition, by the assumption that the block means, say $u_{i.k}$ do not depend upon k.

The peculiarity of this situation consists in the fact that the structural part of the phenomenon, namely the randomization of the n blocks of s plots each, is the result of a deliberate choice by the experimenter and is known for certain. The only freedom that Nature is allowed is the values of the numbers u_{ijk}. This is in contrast with the frequent situation where the statistician is confronted with a truly natural phenomenon, such as the phenomenon of inheritance, where Nature plays a game of chance constructed by herself and the problem is to guess the underlying mechanism, including its structural part.

Here again we are confronted with remarkable achievements of Fisher. While these concern several domains of science, including earth magnetism, Fisher's preference seems to have been genetics and evolution, both out of my usual bailiwick. The general impression I formed from occasional reading is that the modern discipline of population genetics, including such authors as S. Karlin, O. Kempthorne, M. Kiumra, R. G. Lewontin, G. Malécot, K. Mather, P. A. P. Moran, and a number of others, is a development that grew out of the works of essentially only three scholars of the earlier generation: R. A. Fisher, J. B. S. Haldane, and Sewall Wright. The provenance of the ideas that underlay the population genetical studies of these three research workers is likely to be quite complex and, probably, very different. However, the relevant works of Karl Pearson (1909) and, perhaps unexpectedly, a little note by G. H. Hardy (1908), one of the purest of pure mathematicians, seem to have been a common inspiration.

In his very interesting book (1962), Moran refers to 28 contributions to population genetics by R. A. Fisher, either alone or with some coauthors, extending from 1918 (Fisher, 1918) to 1943, and probably this list is not complete. There is no doubt in my mind that in this domain also Fisher's role was that of the founder, at least that of one of the founders, of a fruitful novel domain of human thought and inquiry.

CONCLUDING REMARKS

As stated at the outset, the present appreciation of Fisher's scholarly work is subjective. Also it is one-sided. Both the strict subjectivity and one-sidedness are intentional.

The subjectivity of my account of Fisher's work depends on my personal scientific past and on my personal perspective. No doubt, other scholars will view the same developments differently. Also, I rather expect that Fisher himself would have disagreed with my views on a number of points. One example is the connection between Fisher's own work on experimental tactis, on the one hand, and Wald's work on experimental strategy, on the other. In fact, soon after the apperance of Wald's book, Fisher published an article emphasizing his view that Wald's theory of decision functions has no relation with Fisher's designs of experiments. In a sense, I agree. Wald's work was original work on his own, not on Fisher's problems. My point is that, if Fisher's theory of experimentation did not exist, then, probably, Wald's theory of statistical decision functions would not have been developed as it was developed. As stated by Wald himself, his thinking was stimulated by Fisher's.

Another point on which Fisher is likely to have disagreed with me is my calling him a "descendant" of Karl Pearson. Here a few comments might be useful. A "descendant" does not necessarily mean either a follower or even a student. What I mean here is that, in the early phase of his scholarly activities, Fisher was preoccupied with problems immediately suggested by Karl Pearson's writings. In fact, Fisher seems to have picked up where Karl Pearson left off, and for the history of human thought, it is this link that is significant, not the feelings that the two great scholars had for each other.

The one-sided character of the present article results from my opinion as to how an individual's scholarly activity should be judged. In several earlier writings I have pointed out that certain of Fisher's conceptual developments, not mentioned here, are erroneous. Lest there be a misunderstanding on this point, I emphasize that I continue to maintain this view. However, to err is a part of human nature and I feel that a scholar's activity should be judged by his positive achievements and, particularly, by the influence he exercised on subsequent generations. The purpose of the above outline of Fisher's work is to emphasize my personal views on his record, which is second to none.

References

Borel, É. *Le Hasard*. Paris: Alcan, 1914.

Borel, É. *Eléments de la théorie des probabilitiés* (3rd ed.). Paris: Hermann, 1924.

Bowley, A. L., Irwin, J. O., Isserlis, L., Pearson, B. S., Jeffreys, H., Bartlett, M. S., Neyman, J., and Fisher, R. A. Discussion on professor Fisher's paper, *Journal of the Royal Statistical Society*, 1935, **98**, 55–82.

Edgeworth, F. Y. *Journal of the Royal Statistical Society*, 1908, **71**, 381, 409.

Edgeworth, F. Y. On the probable errors of frequency-constants. *Journal of the Royal Statistical Society*, 1909, **72**, 81–90.

Fisher, R. A. The frequency distribution of the values of the correlation correlation coefficient. *Biometrika*, 1915, **10**, 507–521.

Fisher, R. A. Discussion on Dr. Neyman's paper [On the two different aspects of the representative method]. *Journal of the Royal Statistical Society*, 1934, **97**, 614–619.

Fisher, R. A. The logic of inductive inference. *Journal of the Royal Statistical Society*, 1935, **98**, 39–54.

Fisher, R. A. Discussion of Dr. Neyman's paper. *Journal of the Royal Statistical Society*, Supplement, 1935, **2**, 154–157.

Fisher, R. A. *The design of experiments*. (5th ed.). Edinburgh: Oliver and Boyd, 1949.

Fisher, R. A. *Contributions to mathematical statistics*. New York: Wiley, 1950.

Fisher, R. A. Statistical methods for research workers. (11th ed.). Edinburgh: Oliver and Boyd, 1950.

Fisher, R. A. Scientific thought and the refinement of human reasoning. *Journal of Operations Research Society, Japan.* 1960, **3**, 1–10.

Fisher, R. A. *Statistical tables for biological, agricultural and medical research workers*. (6th ed.). New York: Hafner, 1963.

Kalmogorov, A. N. *Grundbegriffe der Wahrscheinlickeitsrechnung*. Berlin: Springer, 1933.

Moran, P. A. P. *The statistical processes of evolutionary theory*. Oxford: Clarendon Press, 1962.

Neyman, J. Silver jubilee of my dispute with Fisher. *Journal of the operations research society. Japan*, 1961, **3**, 145–154.

Neyman, J. and Pearson, E. S. On the problem of the most efficient tests of statistical hypotheses. *Philosophic Transactions of the Royal Society of London, Series A*, 1933, **231**, 289–337.

Pearson, E. S. The probability integral transformation for testing goodness of fit and combining independent tests of significance. *Biometrika*, 1938, **30**, 134–148.

Pearson, E. S. *The selected papers of E. S. Pearson*. Berkeley: University of California Press, 1966.

Pearson, K. On the criterion that a given system of deviations from the probable in the case of a correlated system of variables is such that it can be reasonably supposed to have arisen from random sampling. *Philosophical Magazine and Journal of Science*, 1900, **50**, 157–175.

"Student." The probable error of a mean. *Biometrika*, 1908, **6**, 1.

Wald, A. *Theory of statistical decision functions*, New York: Wiley, 1950.

Index

Abelson, R. P., 246
Allen, F. H., 154
Allport, G. W., 382
Alternative hypothesis, 111, 134, 195, 197, 239, 265, 268, 271, 274, 276–278
Anastasi, A., 55, 87–89
Anderson, E., 377
Anderson, N. H., 7–8, 23, 25, 39–41
Anderson, R. L., 191, 329
Anderson, S. L., 312
Anderson, T. W., 391
Anscombe, F. J., 145
Arnold, M. P., 55
Assumptions, 15, 49, 54, 65, 67, 105, 116, 121, 122, 195
  for covariance analysis, 191–192
  of homogeneity of variance, violation of, 312–313, 317–318, 327
  of independence, 53
  for t test, violations of, 198, 311–329
  for variance analysis, 331–332
  violations of, 198, 311–346
Asymmetry, 45, 47, 49, 322, 369
Atkinson, R. C., 256, 265
Averages, computation of, 3, 6
Axelrod, D. W., 204
Bakan, D., 198
Baker, B. O., 8, 39
Bald, J. G., 142
Bancroft, T. A., 191, 329
Barbacki, S., 394
Barnard, C. I., 369
Barron, L., 155
Bartlett, F. C., 379
Bartlett, M. S., 99, 158, 169–172, 174–176, 182, 185, 192, 312, 328, 332, 334, 341
Bartlett, S., 164–166
Bartley, S. H., 383
Beall, G., 174
Berkson, J., 137, 292, 303, 308
Beta-response, 89–92
Bettelheim, B. H., 383
Binder, A., 2, 197, 256, 257, 268, 269, 272–274, 306

Binomial distribution, 60, 94, 95, 111–112, 121, 123, 137, 172–173, 177
  goodness of fit test of, 63–64, 140–144
  negative, 134, 173
Bliss, C. I., 145, 176
Block, J., 39 n.
Blodgett, H. C., 218
Bolles, R., 255, 297
Boneau, C. A., 24, 26, 42, 44, 47, 198, 311
Borel, Émile, 390, 396
Boring, E. G., 303, 304, 379
Bose, R. C., 391, 395
Bower, G. H., 257
Bowley, A. L., 393
Box, G. E. P., 312, 320, 327
Braithwaite, T. B., 239
Brandt-Snedecor formula, 145–146
Breuer, J., 380
Brody, A. L., 257
Bross, I., 251
Bruner, J. S., 382
Brunswik, E., 378
Bryan, W. L., 379
Buckland, W. R., 259
Burke, C. J., 39–40, 53–55, 102–104, 107–117, 120–132, 196, 205, 216, 220–223
Burtt, H. E., 380–381, 384
Bush, R. R., 246, 257
Calendar time, scales of, 14–15
Campbell, D. T., 13, 283
Cannon, W. B., 380
Cantril, H., 378
Cattell, J. M., 304, 305
Cattell, R. B., 382
Chanmugan, J., 377
Chen, H. P., 55, 92
Chi-square test, 53–156, 292, 356
  additive property of, 64–70, 110–111
  applications of, 63–93
  "approximate accuracy" of, 115, 117, 118
  categorizing, incorrect or questionable, 56, 77, 87–89

Chi-square test (*continued*)
    contingency tables and, 79–80, 82–86, 113–119, 132, 135–136, 145–155
    correction for continuity, 86–87, 99, 116, 118, 135
    distribution function of, 58, 61–76, 102–103, 107, 109, 121, 123, 124, 134, 136
    imposed restrictions and, 74–76, 80, 82, 96–97, 104
    of independence, 53–54, 61, 77–89, 123–128
    intermediate theoretical frequencies, and, 56, 93–96, 127, 128
    limitations in application of, 60
    linear regression, 93, 125–126, 138, 149–151
    of non-frequency data, 56, 92–93
    non-occurrence frequencies, neglect of, 56, 57, 60, 91, 104–106, 121–123
    observed frequencies and, 56, 69–70, 73, 76, 81–82, 86, 92–95, 103–104, 109–110
    power function of, 134–135
    small expectations and, 133–136, 142, 143, 155
    small theoretical frequencies and, 55–57, 62, 68, 77, 80–81, 84–87, 98–99, 113–118
    sources of error in application of, 56
    strengthening of, methods for, 132–156
    two-category case, 58–60
    See also Degrees of freedom; Goodness of fit test
Cleckley H., 383
Clelland, R. C., 24
Cochran, W. G., 2, 23, 24, 54, 132, 134, 158, 172–173, 177, 179, 185, 192, 198, 262, 328, 330, 394
Cohen, J., 294
Cohen, L. D., 383
Conclusions, 67–68, 85, 199
    decisions and, 347–348, 364–377
Confidence-interval theory, 240–242, 254, 255, 371
Confidence level, 64, 65, 68, 71, 76, 82, 85, 86, 91, 184, 200–202, 209, 214, 225, 240, 277, 371
Contingency tables, 43, 77, 79–80, 82–86, 113–119, 132, 135–136, 145–155, 358–360
    multiple, 108
Coombs, C. H., 19, 27
Cornell, F. G., 253
Cotton, J. W., 24 *n.*
Coulson, A. H., 383
Covariance, analysis of, 158, 179–193, 342–343

assumptions required for, 191–192
    principal uses of, 180–185
    standard computations, 185–187
    technique of, theory of, 188–191
Cox, G. M., 262, 329, 394
Cramér, Harold, 391, 393
Cramér, M., 99, 114, 121–122
Cramming, problem of, 4–5
Cronbach, L. J., 33, 205, 303
Cross, K. P., 382
Culler, E., 380
Daniels, H. E., 312
David, F. N., 312, 328
Day, B., 183
Decision theory, conclusions and, 347–348, 364–377
    degrees of belief and, 233–236
    and hypothesis testing, 371, 376
    and significance tests, 306–307, 369–370
Degrees of belief, decisions and, 233–236
Degrees of freedom, 35, 44, 68, 73, 74, 103, 109, 111, 134–137, 349–363
    appropriate number of, determination of, 56, 107, 110, 125, 132, 357–363
    concept of, import of, 355–357
    motion of a point in space, 350–353
    sample by a point in N-dimensional, 353–355
    single, in goodness of fit test, 139–140, 144
    subdivision of, $2 \times N$ contingency table, 145–151
    in variance test, 137
Descriptive statistics, 33, 41, 49
Diamond, Z. K., 154
Dill, J. B., 55
Distribution free tests, 35 *n.*
Distribution functions, chi-square test, 58, 61–76, 102–103, 107, 109, 121, 123, 124, 134, 136
    goodness of fit test of, 63–76, 123, 135–145
Dixon, W. J., 25
Dukes, W. F., 348, 378
Duncan, D. B., 25, 35
Dyke, G. V., 153
Ebbinghaus, H., 379, 384
Eddington, A. S., 388
Edgeworth, F. Y., 387, 389, 392
Edwards, A. L., 54, 74 *n.*, 113, 116 *n.*, 120, 128–129, 205, 292, 305–308
Edwards, W., 197, 268–276, 278–281, 283–286, 289
Einstein's relativity theory, 265–266
Eisen, N. H., 383
Epling, M., 39 *n.*
Equinormality, 23–25, 27, 28

Estes, W. K., 246
Evans, J., 382
Experimental designs, 25, 125, 127–128
Exponential distribution, 42, 44, 47, 321–325
F test, 23–24, 26–29, 33, 40, 41, 182, 185, 191, 198, 202, 203, 215, 299, 312–314, 328, 336, 337
Factor anaylsis, 13
Factorial design analysis, 25–26, 35
Federer, W. T., 25, 182
Feldman, S. E., 257
Feldt, L. S., 25
Ferguson, L., 298
Finney, D. J., 188, 334
Fisher, R. A., 98, 107, 114, 115, 125–126, 128, 130, 133, 137, 144, 176, 179–181, 183, 208, 256–260, 289–291, 297, 299, 302–303, 306, 307, 334, 341, 348, 349, 353, 373, 387–399
5×4 contingency tables, 82–84
Fix's tables, 135
Foley, J. P., Jr., 55, 87–89
"Football numbers," 7, 11, 19–22
Franklin, M., 55, 84–85, 114, 115
Freeman, G. H., 135–136
Freud, S., 380
Friedman, G. B., 257
Fry, T. C., 62 n.
Gagné, R. M., 204–205
Gaito, J., 24–26, 255, 298, 303
Game theory, 265
Gayen, A. K., 312
Geisser, S., 24 n.
Goldberg, S., 246
Goodfellow, L. D., 95
Goodness of fit test, 58, 61, 122, 132, 137
    of distribution functions, 63–76, 123, 135–145
        multi-category case, 70–73
        two-category case, 63–64
    of functions with frequency as the dependent variable, 89–92
    of regression lines, 107–108
    single degrees of freedom in, 139–140, 144
Gosset, W. S., 158
Goulden, C. H., 116 n., 117, 118
Graham, C. H., 383
Grant, D. A., 2, 26, 55, 89, 104, 118, 197, 244, 245, 248, 256, 257, 259–265, 268, 269, 272–275, 280, 281
Greenberg, B. G., 183
Greenhood, E. R., Jr., 62 n., 96 n.
Greenhouse, S. W., 24 n.
Guilford, J. P., 56, 73 n., 93, 303–305
Gwinn, G. T., 219

Haldane, J. B. S., 136, 142, 397
Hale, R. W., 340
Halton, J. H., 135–136
Hardy, G. H., 397
Hardyck, D., 8, 39 n.
Harrow, M., 257
Harter, N., 379
Hayes, C., 384
Hayes, K. J., 384
Hays, W. L., 39–40
Heinemann, E. G., 383
Hey, G. B., 332, 341
Hick, W. E., 196, 211, 214–216, 220
Hickerson, G. X., Jr., 204
Hodges, J. L., 305
Hoel, P. G., 99, 114
Hogben, L., 258
Holland, M. A., 55
Holt, R. B., 55
Hornsell, G., 312
Hotelling, H., 391, 393
Hyman, R., 246
Hypothesis testing, 40, 41, 195–346
    Bayesian approach, 196–197, 239, 241, 269, 272–275, 277, 279, 280, 283–284
    conclusions and, 370–371, 376
    decision-theory approach, 371, 376
    statistical information and, 249
    See also Chi-square test; Null-hypothesis testing; Significance tests
Independence, chi-square test of, 53–54, 61, 77–89, 123–128
Interactions, interpretation of, 34–35
Interocular test, 283
Interval scales (see Measurement)
Invariance, 9–11, 28–33, 37, 40–41
Irion, A. L., 379
Irwin, O. C., 55, 92
Jacobson, E., 380
Jensen, G. D., 383–384
Johnson, N. L., 312, 328
Jones, F. N., 23 n., 35
Jones, H. L., 344 n.
Jones, L. V., 196, 207, 216, 217, 218 n., 219, 220, 225–227
Jones, M. C., 380
Kaiser, H. F., 297, 300–301
Karlin, S., 397
Kaswan, J., 35
Katz, D., 378
Kellogg, L., 379–380
Kellogg, W. N., 379–380
Kemeny, J. G., 246
Kempthorne, O., 26, 33, 397
Kendall, D. G., 170, 175
Kendall, M. G., 99, 114, 259, 395
Kendler, H. H., 218

Keynes, J. M., 289
Kimura, M., 397
King, H. E., 55
Kircheimer, B. A., 204
Kolmogorov, A. N., 392, 393
Krasnow, E., 39 n.
Krauskopf, J., 383
Kruskal-Wallis H test, 24
Kuenne, M. R., 55, 85–86, 116–117
Lacey, O. O., 297
Lanarkshire milk experiment, 157, 159–168
Lancaster, H. O., 138
Landis, C., 55
Lehmann, E. L., 305
Lenneberg, E. H., 383
Lewis, D., 53–55, 93 n., 102–104, 107–117, 120–132
Lewis, H. B., 55, 84–85, 87, 114, 115
Lewontin, R. G., 397
Lindley, D. V., 269–270, 277
Lindman, H., 269–272, 279, 289
Lindquist, E. F., 24, 26, 82, 298, 312, 313
Linear regression, chi-square test of, 93, 125–126, 138, 149–151
Lord, F. M., 7, 19, 28, 39–40
Lower, J. S., 197, 273
Lubin, A., 39, 41
MacCorquodale, K., 34, 218
Mainland, D., 135
Malécot, G., 397
Maltzman, I., 218
Marcus, M. J., 35
Markov model, 256–257, 265
Marks, M. R., 195, 196, 199, 211, 213, 214, 216, 217, 219, 220
Massey, F. J., Jr., 25
Mather, K., 260, 397
McCurdy, H. G., 382
McGeoch, J. A., 379
McGuigan, F. J., 378
McNemar, Q., 39–40, 262, 294, 384
Measurement(s), 6–52
   dimensionless factors of, 10–11
   perfect, 41, 42, 49
   scales of, 6–18
      interval, 7, 9, 11, 14–15, 27–31, 33, 35, 37, 40, 41, 47–48
      nominal, 6, 9, 11–13, 31, 40
      ordinal, 7, 9, 13–14, 27–29, 31, 40, 41, 47–48
      ratio, 7, 9–11, 15–18, 40
         derived, 16
         fundamental, 16–17
      statistical tests and, 27–37, 40–41
      transformed (see Transformations)

theoretical considerations in, 27–37
   weak, strong statistics and, 39–52
Median-type tests, 24, 25
   Mood-Brown, 35 n.
Meehl, P. E., 219
Melton, A. W., 293, 295, 297
Messick, S., 255
Mettler, F. A., 380
Miller, D. R., 380
Miller, H. L., 197, 269, 273, 274, 280, 281, 285
Miller, R., 395
Mood-Brown generalized median test, 35 n.
Moran, P. A. P., 397
Mosteller, F., 14, 36, 257
Murphy, D. P., 146
Murphy, G., 380
N = 1, 348, 378–387
   frequency and range of topics, 381–382
   rationale for, 382–384
   selective historical review, 379–381
Nair, K. R., 185
Newtonian theory, 265–266
Neyman, J., 134, 202, 212, 258–261, 264, 265, 292, 298, 306, 348, 376, 387–389
Nominal scales (see Measurement)
Nonparametric tests, 23–38
   See also types of tests
Normal distribution, 15, 42–45, 102–103, 107, 121
   goodness of fit test of, 63, 73–75, 135, 144–145
   sampling from, 312, 315–320
Norris, E. B., 55, 89, 104, 257
Norton, D. W., 24, 26, 42, 312–313, 325
Null-hypothesis testing, 115, 132–135, 187, 195, 196, 207–209, 215, 228–287
   classical bias against, 197, 268–287
   decision-theory approach, 306
   multiple models and, 285–286
   theoretical models and, 244–268
   Type I error and, 201, 214, 215, 217, 231, 258, 259, 293–296
   Type II error and, 201–202, 214, 215, 217, 231, 258, 306
   See also Hypothesis testing; Significance tests
Nunnally, J., 292, 293
One-stage Markov model, 256–257, 265
Ordinal scales (see Measurement)
Parametric techniques, 157–178
   Lanarkshire milk experiment, 157, 159–168
   transformations, use of, 158, 169–178
Parametric tests, 23–38
   power function of, 25
   significance level and, 24–25

Parametric tests (*continued*)
  measurement scales and, 35–37
  versatility of, 25–27
  See also types of tests
Pastore, N., 54, 109, 120, 121, 126–128
Patterson, H. D., 153
Pearson, E. S., 134, 212, 258–260, 264, 265, 292, 298, 306, 312, 332, 334, 376, 388, 395
Pearson, K., 74 *n.*, 102, 104, 105, 107, 116 *n.*, 120, 122, 125, 130, 257, 260, 308, 359, 387–390, 392, 397, 398
Peatman, J. G., 74 *n.*
Peters, C. C., 54, 56, 57 *n.*, 74 *n.*, 93, 102, 109, 120–126, 129–131
Petrinovich, L. F., 8, 39 *n.*.
Poisson distribution, 60, 123, 148, 171–173, 176, 340, 355
  goodness of fit test of, 63, 75–76, 135–140
Prince, M., 380
Product-moment correlation coefficient, 10–11, 14, 93
Pronko, N. H., 56 *n.*
Psychological research, significance test in, 288–310
Quensel, C. E., 312
Raiffa, H., 272
Rank-order correlation, 14, 15, 40
Rank-order tests, 24, 25, 28–29, 33, 34
Rao, C. R., 26
Ratio scales (*see* Measurement)
Rayner, R., 380
Rectangular distribution, 42–44, 322–323, 325
Reynolds, B., 218
Rosen, E., 382
Rosenthal, R., 298, 303, 378
Roy, S. N., 391
Rozeboom, W. W., 2, 196, 197, 228, 259 *n.*, 306
Satterthwaite, F. E., 328
Savage, I. R., 24, 40, 255
Savage, L. J., 269–272, 279, 289
Sawrey, W. L., 24
Scales, measurement (*see* Measurement)
Scheffé, H., 391, 395
Schlaifer, R., 272, 289, 307, 308
Schlottefeldt, C. S., 182
Scientific hypotheses, statistical hypotheses and, 258–261, 268–272
Seibel, J. L., 383
Senders, V. L., 8, 27–28, 39, 51
Seward, J. P., 55
Shapiro, M. B., 384
Shaw, F. J., 55
Shipley, T., 380

Sidowski, J. B., 23 *n.*
Siegel, S., 8, 27–28, 39, 51
Significance tests, 23–26, 28, 40, 41, 113, 134, 143, 191, 198, 200–201
  common misinterpretation of, 296–299
  conclusions and, 369–370
  decision-theory approach, 306–307, 369–370
  inductive procedures, 299–306
  inference model, 297–298
  null-hypothesis, 228–243
  one-tailed, 24, 27, 46, 47, 195, 196, 202–227, 293, 300–301, 333
  in psychological research, 288–310
  two-tailed, 49, 195, 196, 202, 207–223, 285, 293, 300–301, 332–333
Smith, F., 192
Smith, S., 55
Snedecor, G. W., 25, 35, 99, 260, 342, 394
Snell, J. L., 246
Sperling, H. G., 383
Split-plot designs, 24 *n.*
Spooner, A., 55
Stake, R. E., 51
Statistical hypotheses, 200
  scientific hypotheses and, 258–261, 268–272
Statistical inference, 8, 28, 33, 35–37, 41, 195
  automaticity of, 198, 299–301
  classical procedures of, 268–286
  logic of, 195, 196, 257–259, 300–301
  See also Hypothesis testing
Statistical information, hypothesis testing and, 249
Statistical tests, 23–38
  measurement scales and, 27–37, 40–41
  See also types of tests
Stevens, S. S., 7–8, 9 *n.*, 13, 14, 27, 28, 31, 33, 39–43, 49, 51
Stratton, G. M., 379
Strong statistics, weak measurements and, 39–52
"Student," 157–158, 390
  See also t test
Subscripts, 4
Sukhatme, P. V., 134
Suppes, P., 256, 265
Symbols, 3–4
t test, 24, 25, 40–51, 107, 157, 158, 185, 188, 202–204, 207–210, 257, 299, 302, 332–333
  violations of assumptions underlying, 198, 311–329
Tate, M. W., 24
Tchebycheff's inequality, 21

Teska, P. T., 383
Tetrachoric correlation, 108
Theoretical models, 244–268
Thigpen, C. H., 383
Thistlethwaite, D., 218
Thompson, G. L., 246
Threshold effect, 36–37
Tinker, M. A., 204, 205
Tippett, L. H. C., 363
Tokarska, B., 202
Transformations, 9–11, 42–51, 158, 169–178
  angular, 175–177
  empirical, 40
  inverse sine, 175–177
  isotonic group, 10, 13
  linear, 10, 11, 27, 40, 51
  logarithmic, 173–175
  monotonic, 13, 27, 29, 32
  non-linear, of cardinal numbers, 42–43
  normalizing, 13
  order-preserving, 13
  permissible, 7, 10–11, 27–29, 31–34, 40
  permutation group, 11
  square root, 171–173
  symmetric group, 11
Transposition behavior, occurrence of, 85–86
Tukey, J. W., 308, 347–348, 364, 376, 395
$2 \times N$ contingency table, 145–151
2×2 contingency tables, 77, 79–80, 85–86, 113–119
  auxiliary, 147–148
  combined, 151–155
Underwood, B. J., 199, 218, 297
Van Voorhis, W. R., 54, 56, 57 n., 74 n., 93, 102, 109, 122–126
Variance, analysis of, 23–25, 145, 157, 312
  assumptions required for, 331–332
    violations of, 198, 311–346
  correlations among errors, effects of, 340–343

gross errors, effects of, 335–336
heterogeneity of errors, effects of, 336–340
multivariate, 26
non-additivity, effects of, 343–345
non-normality, effects of, 332–335
purposes of, 330–331
transformations and, 169–178
  See also Transformations
homogeneity of, 23, 99, 122, 158, 17: 172, 312
  violation of assumption of, 312–31: 317–318, 327
Variance test, 137, 141–143
Versatility of parametric statistics, 25–2:
Voelker, C. H., 76
Wald, A., 257, 306, 372, 376, 393, 395–39:
Walker, H. M., 2, 347, 349
Washburn, A. L., 380
Watson, J. B., 380
Weinstock, S., 257
Weitzenhoffer, A. M., 19
Welch, B. L., 312, 319–320, 328
Wilcoxon T test, 24
Wilk, M. B., 26, 33
Wilks, S. S., 391
Williams, E. J., 35
Wilson, K. V., 297
Wilson, W., 197, 269, 273, 274, 280, 281, 285
Winsor, C. P., 153
Wishart, J. W., 391
Witte, R. S., 257
Wright, S., 397
Yates, A. J., 382
Yates, F., 86–87, 99, 116, 117, 176, 185, 191, 208, 335, 340, 341, 394
Yerkes, R. M., 380
Yule, G. U., 99, 114, 257, 258, 260, 387, 393
Zeigarnik effect, 84
Zubin, J., 55